I0004444

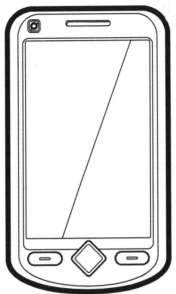

SMARTPHONE

MOBILE REVOLUTION AT THE CROSSROADS
OF COMMUNICATIONS, COMPUTING AND
CONSUMER ELECTRONICS

Majeed Ahmad

This publication is designed to provide accurate and authoritative information in regard to the subject matter covered. It is sold with the understanding that the publisher is not engaged in rendering professional services. The advice and strategies contained herein may not be suitable for your situation. You should consult with a professional where appropriate. Neither the publisher nor author shall be liable for any loss of profit or any other commercial damages, including but not limited to special, incidental, consequential, or other damages.

Copyright © 2011 Majeed Ahmad Kamran
All rights reserved.

ISBN-10: 1461033152
ISBN-13: 9781461033158
LCCN: 2011905006
CreateSpace, North Charleston, South Carolina

*Dedicated to my mother, Tahira, whose care and love
have been a guiding light in my life.*

CONTENTS

PROLOGUE

Often in the technology industry, the conversation revolves around the search for the next "killer app." The fear is that if you miss the window, profitability and market stature can be seriously damaged. Over the past four decades, we have seen the introduction and incredible growth of mainframe computing, the PC, the mobile phone (remember those "bricks" that we used to carry around?), the Internet followed by dotcom boom and bust one and boom two as well as a whole host of behind-the-scene revolutions in communications infrastructure from cables crisscrossing oceans to wireless base stations appearing throughout the world.

All of these disruptive technology eras are fuelled and enabled by the global electronics industry. This industry is covered by the product that I have worked on for the past twenty five years, *Electronic Engineering Times* (*EE Times*), and it is at this product that I met Majeed Ahmad, the author of this much-needed book on killer app that is the biggest game changer we have seen—the smartphone.

EE Times is a global website and trade publication that focuses on the people, products, and companies that drive the technology advances that many of us take for granted. Often the stories are steeped in deep technology terminology and acronyms that are meaningless for those of us who are not engineers or scientists. *EE Times* explains what is possible in the technology realm and for this reason it also attracts readers from outside of the engineering disciplines—CEOs, academics, and analysts—who are looking to identify the next big thing.

With nearly a decade of experience at *EE Times*, Majeed is perfectly positioned to explain the past, present and future of one of the most important technologies to emerge from the engineering labs of many of the world's most successful companies. This is not a book that merely chronicles the smartphone story—Majeed clearly sets out to produce the definitive history and up-to-date description of the product and goes further to help business leaders understand the ramifications that the smartphone will have on their companies, enabling them to create world class IT strategies for the future.

There is no doubt that Apple's introduction of the iPhone changed the game forever. At the time of its introduction, many seasoned mobile phone experts wrote of Apple's chances, stating that the margins were too thin and infrastructure too complex for Apple to make a difference. Fast forward just a few years and Apple is a winner in the smartphone space. In fact, it has redesigned the entire space with touchscreen capabilities, web surfing and, most importantly, a brand new software applications infrastructure that has been mimicked by Google and the Android-focused gadget makers as these companies realize that it's the ecosystem that wins, not just cool hardware.

Majeed covers all of the major players competing in the industry today, focusing not just on the winners but also those who pioneered along the way (Nokia, Palm, and others) but did not continue to craft a world-leading position. Today, giants such as Nokia and Microsoft are reeling from the ecosystem competition they are seeing from Apple and Google—Majeed covers this new turf war and the implications it has for businesses in the future.

The mainframe era led to the PC era and today we have the smartphone era and we are already seeing glimpses of where this is leading with the introduction of tablets and the dawn of the cloud computing age. For business executives, these are exciting times as these technologies are flattening the world. However, with the opportunity also comes the threat of new and agile competitors who are able to harness the new technologies and disrupt current business models, making this book a must-read for any executive who desires to understand the full implications of this new technological era.

In the late 1990s, the buzzword that was uttered the most in technology hubs such as Silicon Valley was "convergence"—the mashing together of consumer electronics with computing and communications. It took almost a decade for the first true convergence device to hit the mainstream and now the consumer, communications and computing convergence exhibited by the smartphone has encouraged other industries to get involved—from media to commerce to services business. Indeed we could be experiencing a transformation of the way we live our lives and conduct business as the smartphone becomes the home security device, digital wallet and, in many cases, virtual workplace. People who have made a FaceTime call to their family on the iPhone can see the potential for videoconferencing and beyond. As more businesses build cloud computing solutions into their IT strategies, this book suggests that the future of the smartphone is also in the cloud.

Free of the buzzwords that are off putting for many executives who did not achieve an engineering degree, Majeed has produced a must-read book that is deeply engaging and, most importantly, is full of ideas and suggestions to help leaders craft winning strategies for their companies. The smartphone is here, it's disruptive and it will change the world—this book will not guarantee your future success but it will make sure that you are well prepared for the inevitable changes.

Enjoy!

Paul Miller
CEO, UBM Electronics and UBM Canon (Publishing)

SMARTS IN THE PHONE

"I grew up with *The Jetsons* and *Star Trek* dreaming about stuff like this, and here it is."
—Steve Jobs at the launch of the iPhone 4 in June 2010

Star Trek, first aired in 1966, fascinated the world with its sophisticated collection of gadgetry that included communicators, phasers, and tricoders. The iconic television serial drew inspiration from a century-old strand of technological utopianism prevalent in American science fiction. *Star Trek* was nothing without its optimism about technology. Much of the technology displayed in it was abundantly imaginative; pundits called it twenty-fourth century inventions. At that time, hardly anybody could have imagined that the world would be seeing such gadgets in real-life working environment just in a few decades.

In the late 1990s, at the height of cellular boom, wireless industry masterminds began conceiving *Star Trek*-like gadgets alongside the constraints of contemporary technologies. They were convinced that a vintage *Star Trek* communicator was in fact an equivalent to a modern-day cellular handset. So, at the 1996 CeBIT fair, thirty years after the inception of *Star Trek*, mobile phone bellwether Nokia, which had won the handset wars by turning an expensive, gray business tool into a cheap, colorful consumer product, exemplified the nascent wireless data market by introducing the first multipurpose device on commercial scale. The Nokia 9000 Communicator was a large and pricey hybrid product that combined voice telephony with an organizer function and a modest Internet access.

Nokia's Communicator had identified the rise of a new segment in wireless data: the smartphone. The handset-cum-organizer made use of an operating system (OS) developed by a California-based company, Geoworks Corp. Shortly after this launch, Nokia Corp. released an enhanced version with smaller size and larger memory; the Nokia 9110 Communicator could download web pages during off-peak hours and could connect to a digital camera for transmitting pictures. The glitzy device weighing 397 grams got a much publicized product placement in Val Kilmer's 1997 movie The Saint. Although 9110 Communicator was an expensive product with a price tag of US$1,000, by enabling services like fax and e-mail on a hybrid platform, it accomplished an initial recognition for a new category in the marketplace that could gather momentum in the future. Earlier, to complement such services, the Finnish wireless titan had produced the first cellular data card and software-based data suite for GSM phones.

In a country where people aim to get things done before boasting about it, Nokia, true to its Finnish roots, didn't create a marketing buzz while launching this high-tech toy, calling it a mere new category that could evolve into a more meaningful product in the future. This was quite unusual in the telecom world, known for its hyper-marketing practices. Such an unconventional, modest, but practical approach had been credited for leading Nokia to an unprecedented stardom in the digital wireless realm. Its ability to think about the future had, so far, been a prominent factor in its startling transformation. On the flip side, however, when the notion of the mobile Internet with software portfolios began to crystallize, instead of pushing cell phones as next-generation Internet appliances and inventing innovative new products, Nokia remained focused on shaping the entire market.

Eventually, when the next sequel, Nokia Communicator 9290, the first smartphone with an open operating system, arrived in the market, reviewers saw it akin to a 1980s' science fiction villain. The astronomically expensive price bracket alienated the majority of public. In retrospect, though not very successful, the unfolding of these devices started a transition from a voice-only strategy; until this point, the mobile phone industry had been built on a single application: voice. The goal was to create a smart gizmo that would allow people to check their e-mails, surf the Internet, plan their schedules, and, of course, make phone calls: in other words, an electronic organizer, a personal computer, and a mobile phone—a device that nobody could afford to be without.

To get there, in the late 1990s, telecom companies, software houses, and handset makers started revving up strategies for the next tech revolution: wireless access to the Internet. They said portable and constant link to the information world would change lifestyles—and offer business opportunities—just as much as the web. So for Nokia, after bringing handsets the capability to support data services, such as Internet access, the next step was to develop a powerful operating system for these phones.

SYMBIAN PARABLE

In June 1998, Nokia and its Scandinavian cousin Ericsson embarked on a venture with British handheld computer maker Psion PLC to develop a common software platform enabling mobile phones access the Internet and other data services. Motorola Inc. joined the consortium—Symbian—at the last minute. Symbian announced a schedule to release the first OS-based mobile phone in a three-year time frame. The claim that its still unfinished operating system would be the new standard for smartphones and wireless Internet appliances created a stir in Europe and across the world. The alliance got a head-start with support from all three wireless industry titans. The mobile phone industry had reached one of the first crossroads in its brief history with the inception of Symbian.

When Apple's charismatic chieftain Steve Jobs launched Macintosh in 1984 with a mantra of "The Computer for the rest of us," a tiny British startup came with its own slogan: Computer in every pocket. Hardly anybody took notice of Psion, the company founded in 1980 by a Zimbabwean-born electrical engineer David Potter. A small but highly focused firm, Psion started with manufacturing hand-held computing products and, later in 1984, introduced an organ-izer operating system it called EPOC. The pioneering British company strove for the quality and conviction born out of Europe's expertise in Global System for Mobile Communications (GSM). The Symbian venture, fostered by Psion, was to create a standard built around Psion's EPOC operating system for wireless information devices such as smartphones and personal digital assistants (PDAs). The EPOC software was perceived as especially suitable for mobile devices as it claimed to make better use of battery power and memory, a crucial advantage in the wake of high usage of these resources in non-voice communications.

Psion quickly established itself as the best horse to back, as with its EPOC operating system, it was perceived as the most impressive technology for the mobile Internet age. Symbian was a darling of the technology press for the first two years of its life. Nokia, at this juncture, created a parallel track to deal with the smartphone business and got on with its life, which was quite good at that time. In all fairness to Nokia, it gave Symbian the strategic worth that the smartphone initiative deserved, but in the end, the overall execution stumbled. Few realized early on that Symbian was going to be a product designed by a committee of archrivals. Fed up with Symbian's bureaucratic structure, several key staff members quit the joint venture, and Symbian's work on new phone software started slipping behind schedule.

Product delays and lack of commercial focus had beset Europe's bright hope in portable data communications. A couple of years after its inception, Symbian began to see the U.S. rivals eating away its promising early lead. Symbian's plans had stalled by the summer of 2000. EPOC operating system drove fewer than 5 percent of the handheld computers and smartphones sold worldwide. Now the big three mobile phone makers—Ericsson, Motorola, and Nokia—were looking in other directions, and as a result, Symbian was no longer the front-runner in the operating system race for the soul of next-generation mobile handsets. Losing faith, Motorola and Nokia vowed to work on a Palm OS-based smartphone, while Ericsson allied with Microsoft Corp. in a joint venture. The consortium members were still committed to Symbian, or so they said; however, they had been keeping their options open.

SMARTPHONE: A MOVING TARGET

Stakes for the brain of next-generation mobile devices were getting high, and so were the efforts of software behemoth Microsoft and the PDA pioneer Palm. The popularity of the Palm operating system was a major factor; its proponents believed that the software for portable computers could be quickly adapted to bring wireless capabilities. And Palm, like Symbian, was based on an open standard. Moreover, Psion's keyboard-based handheld devices had been losing market share to Palm-like pen-based systems. Microsoft was also keen to ensure a key role for itself in this new market. Bill Gates had singled out the Symbian consortium as one of the greatest threats to his company. Not surprisingly, therefore, the personal computer software giant was doing everything in its powers to ensure that Windows CE—a slimmed-down version of the operating system that ran on PCs—would find its way into mobile phones.

Microsoft wanted to turn its Windows CE software into an operating system for a range of devices—from set-top boxes to cable television to handheld computers. The PC titan was steaming ahead, in typical fashion, throwing money and resources to try to catch up. Microsoft claimed it had advantage over Symbian—or any other technology— because it could tie mobile devices into back-end corporate data. The Redmond, Washington–based company also claimed it would offer technology and services that would span the entire range of wireless supply chain, including device software, server software, content and applications, and even the tools to create content for wireless devices. Clearly, Microsoft saw the wireless market as valuable as that of personal computers.

But skeptical analysts termed both Palm and Windows CE as computing-centric products, slow in handling voice transmission and representing more of a U.S.-centric computer industry viewpoint. Moreover, as the roaring 1990s passed, a few events significantly influenced the notion of the phone that wanted to be more. First, a spectacular failure of the European wireless Internet experience known to the world as Wireless Application Protocol (WAP) had a critical impact on the impetus for smartphones. WAP, an early attempt to reproduce the web for mobile phones, failed miserably, and it struck fear among enthusiasts of wireless Internet initiatives. Then there was this grand vision of third-generation (3G) wireless networks so advanced in their fabric that they portrayed phones downloading data-intensive graphics and music files and even transmitting video.

The story of 3G wireless proved a mystic tale of another technology bubble, leaving another dent on the smartphone's standing. The smartphone was a daunting prospect whose fate was intrinsically linked to two critical elements: mobile Internet and faster networks embodied by the 3G metaphor at that time. Suddenly, smartphone expectations began to scale back. Wireless operators were no longer talking about watching video clips on the train or videoconferencing in a taxi. Instead, they started focusing on more realistic goals, such as monetizing the proliferating short message service (SMS) traffic on mobile phones. The sudden, miraculous transformation promised by smartphones seemed like an illusion at that very moment, and it gave way to a gradual evolution of wireless systems. The dream of a smartphone was lying next to dreadful carnage.

The smartphone had been a slow mover since its inception, but WAP and 3G episodes brought all kinds of questions about its future. In the hindsight, the wireless juggernaut opened up these three fronts—smartphone, WAP and 3G—almost at the same time. They were humongous on scale, and they all were standing on their own, while intertwined at the same time. Trade media and the industry at large had also treated these three tech artifacts in isolation. These were the days of the smartphone's awkward adolescence, and people trusted Nokia, the king of the road, for making all the right moves. To become a fact from fiction, however, the notion of smartphone desperately needed a success story. And that wouldn't come from Nokia's doorstep.

2 SMARTPHONE'S THREE ACTS

"Work is no longer a place; it's a state of mind. It's become less about when I turn off the office lights and more about when I turn off the inbox."
—Christa Carone, chief marketing officer at Xerox Corp.

The smartphone marked a new era in mobile communications by meshing the magic of radio wonderland with computing might. According to research firm Informa, in 2008, there were almost 162 million smartphones sold, surpassing notebook sales for the first time. The search for the next big thing was suddenly over. The giant idea of a dream phone had finally reached its time. People were talking about smartphones again. In retrospect, the advent of the smartphone can be chronicled in three major acts, and the rest of the story more or less emanates from these three episodes. For that reason, this book starts the coming-of-age story of smartphones with these three juggernauts.

ACT ONE: BLACKBERRY

At a time when Nokia and its Symbian allies got themselves into a textbook formula for the riches of the smartphone, a small outfit in Waterloo,

Ontario was quietly charting its way into the smartphone bandwagon. After GSM took the world by storm, traditional paging went into doldrums across the world, except in North America, where a little-known wireless firm began to carve out future with mobile e-mail and two-way messaging on pagers. Wireless data became just another phase in the evolution of paging services when the BlackBerry, an awkward cousin of the PDA lineup, allowed people to wirelessly collect e-mails and reply on a tiny, but workable, built-in keyboard.

At a time when the wireless industry was plotting the dreamy convergence of the mobile phone and the handheld computer, BlackBerry emerged from the Canadian techno backwaters and made its way through corporate America. Research In Motion (RIM), an obscure supplier of two-way pagers, launched a small handheld that quickly became a cult item among employees at some big American companies for its always-on e-mail service, which synchronized with a user's office e-mail and didn't require dialing-in. The BlackBerry, which featured a full-alphabet thumb keyboard, provided users with continual wireless access to their e-mails and became a must-have for business executives and mobile workers in the United States.

With the BlackBerry RIM had championed a new product category: neither a handheld computer with a wireless connection nor an e-mail system shoe-horned into a mobile phone, but a device that combined an always-on wireless connection with a useable screen and keyboard, yet didn't ruin the cut of one's jacket. The BlackBerry's claim to fame was that it enabled users to send and receive e-mails using their own corporate e-mail accounts—not a separate address. E-mail messages and organizer entries that originated on a BlackBerry, or that were sent to one, were also stored and their status recorded on the user's personal computer.

What really made BlackBerry such a stellar? RIM was probably the first company to understand that it would have to put together a complete, end-to-end solution, though the underlying idea was rather simple. A special box sat next to the corporate e-mail server and piped messages out to BlackBerry devices. It also handled replies sent from those devices so that messages sent from a BlackBerry appeared in the "out-tray" on the sender's personal computer. RIM provided the box or computer server, arranged airtime with the wireless operator, and handled network configuration. RIM installed servers—the machines that dish out network data—inside its customer facilities.

That meant all e-mailing took place behind a corporate firewall and away from prying eyes. The BlackBerry functioned as a mobile version of Microsoft Outlook or Lotus Notes with a mirroring magic hooked into these widespread corporate e-mail systems. The early BlackBerry device communicated using a little-known, text-only network called Mobitex, the one that also provided refuge to the trivial Palm VII launch. After that, General Packet Radio Service (GPRS) data technology, a GSM overlay, which delivered an always-on connection, emerged with its ability to jack into the Internet with a faster connection and provided a better support for data-centric gadgets like BlackBerry.

The BlackBerry's initial success was based on its appeal to the so-called enterprise market—users with big companies and government agencies. A devoted army of corporate users relied on the BlackBerry to maintain access to their e-mail in-boxes while in a taxi, on a train, or in a meeting. Beyond financial services, doctors, nurses, and hospital administrators could have access to real-time lab results and patient status updates with a BlackBerry in hand. But still, the portable device could be a success in specific vertical markets, where the ability to respond quickly was worth paying for. For common consumers, the BlackBerry's main benefit—integration with corporate e-mail system—didn't apply as much.

Ironically, Microsoft had been toying with the similar idea in collaboration with the wireless bellwether Ericsson. But the initiative went nowhere despite hefty R&D resources of both companies. On the other hand, RIM refined an existing application—e-mail—that already had mass-market appeal by adding value through wireless communication. The small Canadian firm set forth a quiet revolution by taking the power of the Internet on its side and embedding it into an always-connected wearable device. Although a highly specialized tool for a highly special-ized task, this brilliantly simple wireless e-mail device brought a new twist in the trenches of handheld computing wars.

However, a mere 32,000 BlackBerry devices were sold in the first three months of 2002, compared with 3.25 million handheld computers and around 100 million mobile phones. That meant the BlackBerry and its like would remain niche products for executives and mobile workers. Even so, the BlackBerry's influence was being widely felt. It made the previous handheld computers look clunky and outdated. It also demon-strated that fitting a proper alphanumeric keyboard into a small device was possible. Most important of all, it showed that a useful, simple, and seamless service is powerfully addictive. But in retrospect, RIM's defin-ing moment came in April 2002 when it turned its pager-like PDAs into a real platform by launching handsets that acted as cell phones while retaining their regular BlackBerry e-mail feature.

In October 2001, when Handspring posted information about its upcom-ing phone–PDA combo for the making of Treo 180, from then on, it was probably a no-brainer for RIM. After RIM released the BlackBerry 5810, a phone optimized for wireless e-mail use, it achieved a customer base of 32 million subscribers by the end of 2009. Canadian historian and author Alastair Sweeny proclaimed in an interview with *Forbes* that the September 11 attacks and the subsequent anthrax scare made a case for BlackBerry's necessity as a secure communication device. The U.S. government had roughly half a million BlackBerry devices in operation, making it RIM's largest customer. The first truly successful smartphone was born. Why did it take so long for RIM? Probably, too much focus, an attribute that would become self-evident in the coming years anyway.

In April 2002, when RIM launched the BlackBerry 5810, the smartphone began reaching the wider market. Although its use was still mostly confined to business executives that wanted to take their office with them everywhere, it was this model that set the trend for the rest of the smartphone market. The RIM success story not only crystallized a new direction for the handheld gadgets industry at large, it was also a stark reminder of the fact that a portable gadget with right focus on specific applications could go that far. The BlackBerry episode also reinforced the long-held notion that wireless would eventually become the main vehicle to communicate the pervasiveness of a myriad of portable computers.

Mike Lazaridis, who, at the age of twenty-three, co-founded RIM with Jim Balsillie, was a *Star Trek* fanatic and was known to have taken inspiration from *Star Trek* communicator for all the machines he built. Because of strong security features and data capabilities, BlackBerry phones long remained the handset of choice for companies. But regardless of features like simplicity, economics, and power of communication, BlackBerry was an e-mail machine and was designed for that purpose. It made employees more responsive because they could reply to messages at any time. It made people more productive because they could catch up with their e-mails while on the move rather than back at the office, and being able to check e-mails from outside the office gave people more control over their workflow and eased their anxiety and stress.

RIM had become a great Canadian success story. But with success came the pains of a growing organization. The growing company became bureaucratic. There was probably a lack of realization that the technology that had made BlackBerry famous—wireless e-mail—wouldn't remain a game changer for long. It's also worthwhile to note that e-mail was the first killer application that popularized Internet among IT professionals. More than a decade later, e-mail became the first hit on smartphones as well. But here, RIM seems to have ignored a major lesson from Internet history: e-mail was the first killer app of the Internet age, but as the Internet chronicles tell us, it was more of an appetizer. The real crush came with the advent of the web. In retrospect, RIM didn't take the web experience as seriously as it should have. The company could have done more other than e-mail, surf the web, and watch video, for instance.

The wireless industry had been poor at explaining why any capability of the smartphone was worth doing, rather than just evangelizing an interesting technology. The BlackBerry succeeded because it was easy to explain what it did and why it was useful. But just when BlackBerry became one of the top choices of PC-loving professionals for its business-friendly composition and reliability, a far bigger story was to land on the smartphone playbook. BlackBerry wasn't going to keep the smartphone market to itself for long.

ACT TWO: iPHONE

Back in 2002, at the time when BlackBerry was starting to become a cult device among business types, and shortly after the launch of iPod, Steve Jobs, one of the most strategic thinkers in technology business, began contemplating about developing a mobile phone. He could see millions of Americans lugging separate cell phones, PDAs, BlackBerry devices, and now MP3 players. Consumers would naturally prefer one device, he thought. Next year, in 2003, he saw how consumers flocked to the Palm Treo 600, which merged a phone, PDA, and e-mail device into one slick package. According to the January 2008 *Wired* magazine article "The Untold Story: How the iPhone Blew Up the Wireless Industry," written by Fred Vogelstein, to Jobs, that was a clear testament to the demand for a so-called converged device. But it also raised the bar for Apple's engineers.

By that time, Apple's hardware engineers had spent about a year working on touchscreen technology for a tablet PC, and they had convinced

Jobs that they could build a similar interface for a phone. Plus, thanks to the release of ARM 11 chip, cell phone processors were finally fast enough and efficient enough to power a device that combined the functionality of a phone, a computer, and an iPod. So in fall 2005, Steve Jobs tasked a team of two hundred Apple engineers with creating the iPhone. Internally, the project was known as P2, short for Purple 2; the abandoned iPod phone was called Purple 1. Although Apple seems to have maintained the highest level of secrecy, by some media accounts, it had telegraphed its mobile phone intentions a couple of years before going public.

Zoom ahead to January 2007, when Jobs ended all speculation by making a wake-up call to the wireless industry from a phone that didn't exist. He touted Apple's prototype iPhone, due in June 2007, as a music and video player, a smartphone, and an Internet device rolled into one. The market witnessed a euphoric reaction from would-be users. The message: people craved a much better mobile computing experience. The sleek new device from Apple would generate innovation in the rapidly expanding smartphone market and new attention from consumers and businesses.

But there were also naysayers and prominent among them was the PDA luminary who had now set its sight on the promising new smartphone market.

Before the release of the iPhone, at the 3GSM Conference in Barcelona in February 2007, Palm CEO Ed Colligan told Swiss newspaper *Sonntagszeitung*: "I have great respect for Apple, but it won't be easy to create a good smartphone that will function on networks worldwide. Nokia, Motorola and Samsung have worked on this for 25 years and have only partially succeeded today. Our Palm Treo already has 90 percent of Apple's iPhone features at a much lower price..." He must have come to regret his words after June 2007, when Apple fired a shot across the bow by launching the iPhone, the device that would forever change the mobile phone business. It was a far superior product compared to any-thing that mobile phone users had seen. Someone had finally delivered a mobile phone with a compelling web experience.

Before the iPhone, a user could access the Internet after sorting through a dozen clicks to see mobile-optimized feeds on a tiny screen. The iPhone refused to rely on the "baby Internet," as Jobs put it during his Macworld 2007 keynote address, and instead featured a mobile ver-sion of its own Safari web browser with tap- and pinch-to-zoom for an elegant, unprecedented browsing experience, along with a powerful e-mail feature. Apple not only launched the most compelling mobile experience for access to the web, it also defined the smartphone battle by creating the biggest and most active ecosystem of software devel-opers. It rallied developers behind a solid operating system based on a subset of its well-known desktop software.

Web surfing via the device's large screen became the iPhone's meal ticket. Apple made this possible through a combination of the touch-screen interface and the app—a small piece of downloadable soft-ware that launches an application and can draw information from the Internet without the user having to open a browser. Widgets, or small applications that don't need to launch a browser, now began to pro-liferate on mobile handsets. But the mobile Internet experience, while fairly stunning, was only part of the story. The real genius behind iPhone was a smart, usable interface design. With its intuitive touchscreen, the iPhone revolutionized mobile web browsing; the phone was easy to use, enjoyable to navigate with, and looked far more attractive than did all its predecessors.

Apple had done wonders with the touchscreen by enabling a radically sharp screen and an impressive resolution. The touchscreen, which made so many apps come to life, was the heart and soul of the iPhone. Earlier, one of the biggest challenges the designers faced was how to provide a suitably sized, usable keyboard—something every PDA or smartphone maker had struggled with. Unlike anything on the market, with the iPhone, Apple bypassed the need for a button-based keyboard by providing a 3.5-inch touchscreen display with a soft keyboard and incorporating various tricks to enhance the typing experience. Intuitive user interface and consumer-focused design are things Apple does know quite well. After all, the user experience is the hallmark of Apple's consumer identity.

The navigation system of Apple's iPod was both radical and daringly simple—and it was with this same philosophy in mind that Apple mounted its charge on the cell phone industry.

By eliminating intermediary input devices such as a keyboard or a stylus, the iPhone made control tactile again. A magical equation of device, service, and application, the iPhone had an interface that was digital in every sense of the word. The iPhone's touchscreen interface fascinated gadget enthusiasts, and so did the Wi-Fi capability, the software, and the multitude of applications. Apple forced mobile carriers to seriously consider handsets with Wi-Fi applications. Mobile phone carriers, who didn't want to have anything to do with Wi-Fi, were now coming around to Wi-Fi as an optional feature recognizing the inevitability. Ironically, the U.S. mobile phone operators had also been resisting the introduction of touchscreen handsets, saying they won't work in the United States.

The iPhone had descended and wireless industry would never be the same again. Not even a $599 price tag at its launch could dent enthusiasm for the iPhone. The device relied on AT&T's EDGE network—3G would arrive a year later—and touched off frenzy from the moment it landed. In July 2008, Apple introduced its second-generation iPhone, which had a lower upfront price and offered support for 3G network. It also created the App Store with both free and paid applications. The App Store could deliver smartphone applications developed by third parties directly to the iPhone over Wi-Fi or cellular network without users having to download to a PC. The next incarnation—the iPhone 4—was launched with multimedia galore in June 2010.

Apple took the route of adding mobility and connectivity to an established and thoroughly considered operating system to offer a phone, web browser, and media player that worked well for each function. In a very similar episode to the personal computer saga, Apple took ideas pioneered elsewhere and brought them to life in the iPhone in a way no one had to date, using unassuming off-the-shelf hardware and a stunningly simple and fun user interface. Apple took disparate ideas from technology wasteland, got its arms around them, and developed a watchful synergy among these ideas.

The Silicon Valley star had this well-known ability to discover, not invent, products. The products already existed; it's just that no one had ever seen them, and Apple was the one that discovered them. The Macintosh already existed in the form of Alto personal computer at the Xerox PARC during the 1970s; it was only a matter of discovery. However, this time around, going a step further from playing out the old Macintosh strategy, Jobs was not only insisting on owning the software and the system; he also wanted to own the services the iPhone used and the retail shops where it was sold.

And, like before, Apple created a marketing firestorm that captured the imagination of the public as well as astute technology watchers. Apple's iPhone woke everybody up to the smartphone idea, and now the market had nowhere to go but up. Equipped with GPS, music, Internet, games, and a number of other features—just in a few short years—the iPhone became the greatest force in the world of smartphones. The iPhone era was akin to a brand-new day, the phone 2.0 journey from niche to mainstream in which there was no looking back. The device was so well-received that *Time* magazine named it the "Invention of the Year" in 2007.

In 2008, at the launch of the iPhone software development kit, John Doerr, a high-profile Silicon Valley venture capitalist, announced the

US$100 million development fund for the iPhone, calling it bigger than the PC. It made many people in the industry see the iPhone as the embodiment of a significant technology shift. The iPhone sported a PC-class operating system as well as a PC-class browser, and the emphasis from Apple really was on extending Mac OS X to a mobile platform. That was probably the moment when industry observers stopped seeing the iPhone as just being another mobile phone; they started to think of it as the reincarnation of the long-held pocket PC dream. Apple had, in fact, brought to the market what, in essence, was a full-fledged Internet experience and a fairly good representation of the PC experience to a device that fit in a consumer's pocket.

Apple's iPhone story is riveting. The iPhone initiated new market dynamics by building Internet browser, music player, and e-mail devices on top of mobile phone. When Apple proclaimed that it had reinvented the phone with the iPhone, the Silicon Valley technology titan had merits to its claim. Before 2007, Apple wasn't even in the mobile phone business; now the game-changing iPhone had become the most influential smartphone in the world. The iPhone had turned the mobile handset industry on its head. Amid an elaborate great leap forward with a dramatic design and consumer-marketing lore, Apple became the undisputed market leader.

ACT THREE: ANDROID

If the BlackBerry's release was huge, that of the iPhone was explosive. RIM had the first-to-market advantage and had carved out the indispensable and lucrative enterprise turf. It long dominated the high-end handset market via one, albeit essential, application: secure mobile e-mail. Then, Apple came along and introduced other mobile services, redefining smartphone to multipurpose multimedia devices. When the iPhone's web experience opened up a whole new world via apps, BlackBerry's e-mail focus essentially turned it into a one-trick pony, regardless of its success and utility. RIM and others were almost clueless when the iPhone sparked the apps revolution; all they could do was race to catch up. While Apple captured the leadership of the consumer smartphone, RIM continued doing a decent job hanging on to its lead in the corporate world.

But soon after Apple rocked the industry with the first successful web-ready smartphone, complemented by a brilliant marketing machine, another wireless surprise was about to unfold.

For a little history, let's first scroll back to 2002, when Silicon Valley–based Danger Inc. launched an all-in-one wireless device that improved on pagers by supplementing phone and Internet services. Danger's Hiptop smartphone—launched by T-Mobile as Sidekick—allowed users to send and receive e-mails and surf the web while an attachable camera was available as a separate purchase. The Sidekick II came with a built-in camera two years later. The Sidekick was best known as a smartphone for hip teens; the gadget was featured in some memorably awkward product placement on *The O.C.* teen drama series in 2005. Among the celebrities sporting Sidekick phones were Google co-founders Sergey Brin and Larry Page.

Danger co-founder and CEO Andy Rubin had chosen Google as the Sidekick's default search engine, despite it was trailing behind AOL and Lycos in traffic at that time. The up-and-coming Google wouldn't forget this unexpected endorsement. Rubin, a former Apple engineer who was also part of the pioneering team at WebTV, had founded Danger Inc. with Matt Hershenson and Joe Britt in January 2000. After his ouster as CEO of Danger in 2003, he co-founded Android Inc. with Richard Miner, Nick Sears, and Chris White. Miner was the co-founder of Wildfire Communications Inc., Nick Sears had once been vice president at T-Mobile, and Chris White had headed design and interface development at WebTV. Rubin had been hanging out on a beach in the Cayman Islands when he came up with the idea of creating an open-source operating system for mobile phones: Android.

Interestingly, later in February 2008, Microsoft acquired Danger and merged it into its mobile phone platform known as Project Pink. That acquisition was incarnated two and a half years later named Kin as an ambitious effort to marry smartphones with social networks. Microsoft killed the Kin project a mere forty-eight days after its launch. The Kin phones had a mediocre platform and were not priced competitively in the United Sates. So, the whole Kin smartphone strategy was crushed in less than two months. The larger irony, however, was that the ideas behind the innovative software store that Danger pioneered were not put to use by either Microsoft or Google; instead, it was Apple who reinvigorated the app store initiative. But that's another story.

At Google, back in 2005, Larry Page felt increasingly nervous watching Microsoft—one of Google's competitors—as the only computing firm cruising the mobile phone world albeit with a modest success. *Wired* magazine's Daniel Roth writes in his article "Google's Open Source Android OS Will Free the Wireless Web" that Page could see how Microsoft executives were keen on tying the mobile platform into its PC dominance and that they cared less about opening up the Internet to mobile phone users. Microsoft had less than 10 percent of the U.S. smartphone market share, but it could grow, he concurred. So when Andy Rubin of Android Inc. came to Page hoping for Google's backing, he was in for a surprise. Google wanted to buy Android: a free, open-source mobile platform that any programmer could write for and that any handset maker could install on its product. Rubin would take on the role of Google's head of the mobile unit.

Back then, Google seemed to have so much money that it didn't know what to do with it, so one of the richest companies on the planet quietly went about buying up a lot of startup companies.

Google dominated the web in the mid-2000s, but tomorrow could be a different story. Google could clearly see how smartphones were heavy on hardware and light on software; they had less memory and offered only a stripped down version of the mobile web. A second-class web could derail Google's grand strategy. The company's premise was that if browsing by phone could be made easy and fun, people would use it just like a desktop browser, with Google's search as the main port of entry. Under Google's wings, Android would have the spirit of Linux and the reach of Windows. It would be the global, open operating system for wireless future, envisioned the Google leadership.

Ironically, the first testimony to Google's mobile strategy came from the launch of its future nemesis, the iPhone; users in large numbers pointed to Google and drove more traffic to the search engine than any other mobile device. If Apple could generate that much for Google, surely Google could do better for itself. Then-Google CEO Eric Schmidt, a BlackBerry man at heart, was initially skeptical about the Android acquisition. But once things got going, he put his weight behind the new strategy. "That is the recreation of the Internet. That is the recreation of the PC story," he told business leaders at the World Economic Forum in Davos In January 2008. Just a couple of months earlier, in November 2007, thirty-four mobile and technology leaders had come together to form the Open Handset Alliance, calling Android the world's first truly open and complete mobile platform.

Android, a cross-platform operating system for smartphones released in 2008, was an open-source platform backed by Google along with major hardware and software developers. Before Android, Linux had lacked a user-friendly interface and a high-profile effort for attracting mainstream application developers. Interest had been high in Linux as an open-source environment for netbooks and other mobile devices. But Linux had fragmented into several camps and lacked a consumer-friendly user interface and a unified application framework. Android attracted much attention as relatively sophisticated software that could fill that gap. Its easy-to-use methodology and richness of applications could fuel more aggressive work on part of software developers.

The first phone to use Android was the HTC Dream, launched as the G1 in September 2008, manufactured by Taiwan's High Tech Computer (HTC) and branded for distribution by T-Mobile. The software suite included on the phone featured integration with Google's proprietary applications such as Maps, Calendar, and Gmail, as well as Google's Chrome Lite full HTML web browser. Third-party apps were later made available via the Android Market, which included both free and paid apps. A compelling mobile web environment from an industry darling was now available as free, open-source code to anyone interested who had basic credentials. When Google first announced Android, people in the industry didn't take it seriously. But then they came around. The search maestro had seized the moment with an impeccable timing.

In summer 2008, Motorola threw its weight behind Android operating system. At that time, its mobile phone business was losing more than US$500 million a quarter, and Motorola was struggling to update more than half a dozen software platforms, each customized for specific carriers around the world. Motorola, and its clamshell StarTac, introduced in 1996, dominated the nascent cell phone market in the early going, but it was overtaken in the next decade by Nokia's superior logistics and mass production as well as by South Korean rivals. Since then, Motorola had bounced between flops and hits. So the former king of wireless, which had seen its fortunes tumble when it wasn't able to come up with a successor to its smash-hit Razr phone, scrapped its own OS development efforts in favor of Google's Android in 2009.

Droid, the phone sequel that launched Motorola's comeback effort, ran the Android operating system. Droid was the first smartphone to really challenge the iPhone, though it did little to dent Apple's hegemony.

Nobody ever imagined how quickly the Android mobile phone platform would take off—not even Andy Rubin's engineering comrades in Silicon Valley who produced it. The software kept getting better, and top handset makers like HTC, Motorola, and Samsung jumped on board, rolling out dozens of Android-based phones. What began as a trickle now turned into a tidal wave. In August 2010, Google announced that it was activating 200,000 Android phones each day. Google's success in breaking into the smartphone world with its Android operating system had been nothing short of remarkable. Barely eighteen months after the first handset carrying its software was introduced, the Internet search company had become the clear challenger in the new category of web-enabled touchscreen devices pioneered by Apple.

There are many reasons for Android's success, but the combination of its distinctive relationships with wireless carriers and with handset makers was probably the most significant. Google was coming at it from a unique strategy to align with mobile carriers and handset manufacturers, at a cost that Apple would not be willing to accept. If Google achieved a land grab on handsets, it had two distinct advantages: where there is Internet, there is mobile search, and Google still dominated in search. Second, where there is search, there are ads, which Google could monetize. So Google could become people's mobile billboard, which portends all sorts of location-based advertising and coupons.

Android had been the only platform capable of garnering enough support from apps developers to challenge the iPhone. Google was trying to make programming for mobile phones analogous to programming for a PC or the web. As a result, supporting Android seemed like the best bet for handset makers to avoid losing market share to Apple. Android continued to improve in terms of usability and the number of applications available. However, with so many different Android phones floating around and with so much openness to the web, the search giant also risked delivering a crummy, fragmented, even disastrous user experience with security leaks, viruses, and customer service that fails when needed most.

The face-off between Apple and Google over the mobile gold rush had pitted two companies with very different technology approaches and philosophies against each other. This clash of visions could define the technology industry in 2010s and could shape the way in which hundreds of millions of people experience the online world. The two rival notions of the wireless future are taking shape, which could bring somewhat different outcomes for mobile Internet users. Google is counting on devices that are open to the web. Apple's handsets, by contrast, are vertically integrated products designed above all to make technology easier and more pleasurable to use. So eventually it could turn into a rivalry that will shape the next phase of personal computing.

REVERSAL OF FORTUNES

In the early phase of the smartphone inception, the computer industry tried to cram PCs into pocket-sized devices, while the mobile phone industry had arrived at the same point by adding new features. When the once distinct worlds of computing and mobile telephony began colliding, the giants of each industry—Microsoft and Nokia, respectively—squared up for pre-eminence. Smartphone and mobile phones continued to converge and collide during the early 2000s, and when the second act in the "smartphone" drama got underway later in the decade, Nokia turned into a sad story because of its loss of leadership. The Finnish giant had garnered 40 percent of the global handset volume at one time during 2007, according to research from Strategy Analytics. Now the company that embodied wireless industry's quest to inherit the efficiency of the PC industry was turning into an also-ran.

In general, mobile handset makers like Nokia, coming from a telecom background, prioritized hardware muscle and superior specifications, while computing players such as Apple and Google emphasized user interface and software to enhance ease of use. Although Nokia's focus remained on mobility and the company had no intention to extend its resources to a traditional computer business, Nokia also understood that fundamental changes were happening in computer industry. To its credit, Nokia foresaw a day when wireless handsets would be as powerful as computers. But now, being surrounded by competitors of PC origin, it seemed to have lost the battle for smartphone dominance, even the race to be a runner-up to the iPhone.

Although Microsoft claimed to be an incumbent in the smartphone realm, having sold some 35 million licenses of Windows Mobile since its release in 2001 through 2008, the mobile phone circles had turned their back on what they perceived as a PC industry interloper. After a decade of trying, Microsoft failed to conquer the smartphone business it had hoped to define. The dominant technology company of the past two decades found itself far behind in the new markets with the greatest potential. The smartphone turf embodied all the places where Microsoft had been before Apple got there. While its focus early on was making an operating system for smartphones that could send e-mails and run specialized applications, Microsoft largely ignored the consumer markets for such devices, which Apple's iPhone appended in 2007.

Palm Inc., a one-time mobile device pioneer with its Palm Pilot handheld, had also been eclipsed by Apple's iPhone, RIM's BlackBerry, and devices running Google's Android software. Computer giant Hewlett-Packard Co. scooped up the financially struggling maker of smartphones in April 2010 in a billion-dollar acquisition to stake its claims in the hyper-competitive smartphone market. Meanwhile, Apple continued to define the smartphone software battle, forcing others to catch up with the ease of use, web browsing, and a broad set of applications for its iPhone. And while Apple was winning the smartphone battle, developers seemed to acknowledge that Google's Android might shape into their second best bet. Apple's desire to dominate the future of mobile was only matched by Google's bid to turn the smartphone into an Internet node, a computer that was much more personal than the existing mobile computer, the notebook.

The fact that Simon—launched by computer prodigy IBM in 1992—was the very first smartphone could be seen as a harbinger of computing

DNA in smartphones. Ironically, both Ericsson and Nokia got their hands clean at computing business back in the 1980s and left it saying non-core. A decade later, while dealing with the smartphone conundrum, they found themselves on a somewhat similar crossroads. Also, by the late 2000s, Dell, HP, Lenovo, Acer, and others were singing of the same hymn book: the future of the computer industry is mobile. The notion that the future of smartphones seems more like the future of computing and that smartphones are becoming full-fledged computers leads to a stunning observation: all five major smartphone undertakings have their roots in portable computing arena.

The term "personal digital assistant," or PDA, came into being after the historic 1993 launch of Apple's handheld computer: Newton, the precursor of the iPhone. Second, the DNA of the first mildly successful smartphone—Palm Treo—came from pen computing and PDA pioneer Jeff Hawkins. Third, Nokia-backed Symbian had its origin in the British handheld computing outfit Psion. RIM co-founders Mike Lazaridis and Jim Balsillie launched BlackBerry in 1999 as a PDA that also acted as a two-way pager. And finally, Danger's Sidekick handheld, the forerunner of Android, started as a PDA offering e-mail and instant messaging services. But the million-dollar question is why Apple and Google, which were not even on the smartphone radar screen before 2007? If the smartphone is where next-generation computing is headed, then why has Microsoft missed the next big opportunity in mobile phone cycle?

Part of answer to this question lies in the fact that both Apple and Google didn't only carry computing DNA; they also possessed two intrinsic elements of the smartphone recipe: Internet and media. The smartphone revolution is not merely about a new software paradigm; there is a massive reexamination underway of how technology, media, and communications intersect on this ever-evolving landscape. And this could be hugely disruptive. It could transform outfits like Apple, Google, Facebook, and Amazon into more than just dominant technology companies; they could well become the news, entertainment, and communications networks of the twenty-first century.

Take Apple and its feisty leader Steve Jobs as a brief case study on how the computing and media worlds are increasingly intertwined. Jobs, who saw smartphones as actual computers in the full sense, probably appreciated better than anyone that computing is in transition.

After Apple introduced the iPhone in 2007, it changed its name from Apple Computer to Apple Inc., and started calling itself a mobile company. Many computer makers have joined the chorus since then. But Jobs was also one of the few industry leaders who boasted expertise in computing and consumer electronics along with stints in the media business. Back in Apple's heyday, industry commentators used to call Apple "a vertically integrated advertising agency," Apple's former CEO John Scully told *Bloomberg Businessweek*. It wasn't a compliment, but it partly showed Jobs's inkling toward being media savvy. In 1986, after leaving Apple, Jobs paid filmmaker George Lucas US$10 million for a small firm specialized in computer animation.

Over the next six years, Jobs injected another US$40 million of his own money into the company, came to be known as Pixar, as it set out to make the first-ever computer-animated feature film. The result was *Toy Story*, which since its release in November 1995 grossed more than US$177 million at box office. The Pixar IPO, timed to take place just after *Toy Story* opened, was a huge hit too. The share price more than doubled in the first hour of trading, and Pixar chairman Steve Jobs, who owned 80 percent of the company, was a billion dollars richer. In 2006, Pixar was sold to Disney, making Jobs Disney's largest shareholder. In the hindsight, it was an astute move; the next year, Jobs made his foray into the mobile world, and the rest, as is said, is history.

Within a few years, smartphones became the center of the mobile maelstrom with devices like the iPhone moving us toward a mobile society where our phone became the remote control for our daily lives. The smartphone became the platform of choice for embedded computing and the PC was no longer the center of gravity. Smartphones with higher average selling prices were also upsetting the traditional business models, and like the music industry before, now the cell phone establishment found itself in the midst of a shake-up. The wireless industry, however, welcomed this disruption seeing the revenue potential from new applications and services.

This chapter chronicled the three main characters of the smartphone story: BlackBerry, iPhone, and Android. Then this section tried to make sense of the smartphone's inseparable relationship with computing and established that the underlying DNA indeed comes from the computing world. I further explore the computing heritage of smartphones in the next chapter and the issues related to the smartphone's intrinsic association with the Internet and media will come in the fourth and fifth chapters.

UBIQUITOUS COMPUTING

"The most successful computing device of all times is the cell phone."
—Jeff Hawkins, inventor of the first successful PDA, Palm Pilot

In 1988, Mark Weiser, a technologist at the Computer Science Lab at Xerox Palo Alto Research Center (PARC), put forward the notion of ubiquitous computing as information technology's next wave after mainframe and personal computers. In this new world, what he called "calm technology" will reside around us, interacting with users in natural ways to anticipate their needs. Weiser coined the term "ubiquitous computing" to describe a future in which personal computers would be replaced with invisible computers embedded in everyday objects. He believed that this would lead to an era of computing in which technology, rather than panicking people, will help them focus on what is really important.

Weiser's work—based on research on human-computer interaction and PARC's earlier work on computing—initially sparked efforts in areas such as mobile tablets and software agents. Subsequently, these efforts morphed into pursuing intelligent buildings packed with wireless sensor networks and displays, where information follows wherever people go. Weiser's vision was shared by many in the PC industry. The first practical manifestation of ubiquitous computing emerged in the early 1990s when John Doerr, the legendary venture capitalist at Kleiner Perkins Caufield & Byer, started the pen-computing frenzy by funding Go Corp. By 1991, the pen-based computing wave had become the "next big thing" in technology circles. Yet, despite this pen-based computing rush, only a single product became commercially available from GRiD Systems, a small computer outfit on the east of the San Francisco Bay.

But then Apple Computer's chief executive officer, John Scully, fanned the flames of pen-based computing in a speech about a handheld computer he called the personal digital assistant or PDA. "Palmtop computing devices will be as ubiquitous as calculators by the end of this decade," he told his audience. Scully echoed Weiser's vision touting that computing would eventually go a step farther in the journey that started from mainframe to minicomputer to personal computer. In May 1992, Apple CEO announced the Newton, an amazingly ambitious handheld computer. Scully set the computer world on fire with his prediction that PDAs such as Apple's Newton would soon contribute a trillion-dollar market. He professed that this gadget would launch the "mother of all markets."

The overwhelming electronic gizmo was built around Scully's vision of handheld computers. Representing a new class of portable devices, the Newton was stitched together from several emerging technologies to perform specific tasks like memos and personal diary. It was capable of sending and receiving messages, and was equipped with modest computing and handwriting recognition capabilities. In August 1993, when Apple introduced the Newton, Scully foresaw a world in which a simple device would connect mobile users to vast networks of information.

But eventually Apple became the Dr. Frankenstein of handheld computers. Newton was not only ahead of the market; it was also fatally flawed from technological and commercial standpoints. It had the feel of an electronic brick both in size and in weight. A user could barely hold it in his or her hand. Newton failed to connect with the rest of the computing world, and five years after its launch, Apple abandoned the product to focus on its core Macintosh lineup. When the newly arrived chief executive Steve Jobs pulled the plug on Newton, it seemed more like a mercy killing. Apple's ambitious project had turned into a PR nightmare, literally. Newton's spectacular failure made most people believe that PDAs won't work.

But the Newton debacle proved a kind of start-over that led to a new generation of PDAs that would focus on more practical features. A plethora of such products sprang up—offering some sort of interactive capability—but the turnabout period for PDAs was to take a bit longer. Among those early pioneers was Jeff Hawkins, a loose-limbed, perennially boyish electrical engineer who carved his first prototype from a block of wood and carried it around pretending to scribble on it. He had been struggling for years to turn his long-held dream of mobile computing into a reality. After stints at Intel Corp. and GRiD Systems, where he designed laptops, Hawkins took a leave and developed an algorithm that allowed computers to recognize patterns and, hence, handwriting. Returning to GRiD, he applied some of the ideas to the GridPad, a reasonably successful pen computer.

THE ZEN OF PALM

GRiD was a pioneer in mobile computing, and many of the technologies present in notebooks and tablet PCs wouldn't exist were it not for GRiD. Hawkins was considered the main architect of GRiD's pen-based computing program. The success of his product designs gave Hawkins a modest amount of fame as a pioneer of the fledgling pen-computer industry. A self-assured and brilliant product designer, he firmly believed that the future of personal computing was in portable electronic devices. In 1991, Hawkins, like so many technology entrepreneurs before him, was in the grip of an idea for a new type of computer. To chase that dream, Hawkins left GRiD with a license for software he needed and started Palm Computing on January 2, 1992.

He only set out to create a new kind of computer; he had no intention of launching an entire industry. The handheld computer market was doomed after Apple introduced its infamous Newton, but Hawkins kept going with the help of Donna Dubinsky, a Harvard MBA with an impressive Silicon Valley track record, and marketing whiz Ed Colligan. A succession of handheld products that appeared soon after the Newton launch, including Palm Computing's first product, the Zoomer—short for *consumer*—suffered the same fate as its progenitors. While talking with customers, Hawkins came to realize that his competition was paper, not computers. PDA makers had so far been trying to replace desktop computers; Hawkins found out that people wanted to replace their pocket diaries and desk calendars.

He insisted that the computer be small enough to slip into a pocket. Hawkins also found out that the handwriting-recognition software other products were using was too complex, and that it slowed down the device. So he developed new software called Graffiti that required users to learn a simplified alphabet the computer would quickly recognize. The Palm Pilot used a quasi-handwriting-recognition system and didn't have a keyboard. Users could synchronize the device with personal computer via a cord and swap notes with other Palm devices through infrared beaming port. After a two-year design effort, a simple, no-frills compact device hit the market in February 1996 under a larger corporate umbrella of U.S. Robotics.

Hawkins, Dubinsky, and Colligan first worked out the technical issues, and then they turned to the financial ones. After being repeatedly rejected by investors, they found a white knight in U.S. Robotics, which first agreed to fund the handheld computer but later offered to buy the whole company. Palm Computing had come perilously close to running out of cash during these years, forcing the troika to sell their company to modem king U.S. Robotics, which, in turn, was purchased by network hardware maker 3Com. Although the merger between 3Com and U.S. Robotics didn't go very well, Palm Pilot turned out to be a prize for the new company. Palm Pilot became one of the fastest-selling high-tech toys of the decade. The elegant little computer that Hawkins and his team of idealists brought to life against all odds became an American icon. One million Palm Pilots were sold in the first eighteen months.

Pilot's benefits could be summed up in three compelling words: smaller, faster, and cheaper. A Palm veteran Rob Haitani coined the term "Zen of Palm" to describe the design philosophy of simplicity and immediacy that ultimately made the product a runaway success. It was the fastest-selling product in computer history and a major force to reckon with among all kinds of digital appliances. It irrevocably changed business life and became a ubiquitous sight in America's conference rooms, airport departure lounges, and commuter trains. And once the inexpensive models came on the market, students and women joined the PDA-totting ranks that had previously been occupied by mostly high-income professional men. Handhelds had entered the mainstream.

THE WIRELESS ACT

The advent of Palm seemed like such a portentous leap into the future, but the lack of some kind of interactivity could dwarf its widescale implementation, or so thought many industry observers. The PDA, once a glorified address book, was now increasingly seen as a stepping stone in the convergence of mobile platforms. The conventional wisdom had it that wireless—providing the ability to surf the web or check e-mails from an untethered handheld or cell phone—would be the biggest thing since the dawn of Internet itself. So industry analysts were almost unanimous in their belief that a handheld computer wouldn't succeed until it included, at the very least, fax and e-mail functions. They questioned the Pilot's lack of wireless capability and PCMCIA expansion card slot.

At that time, during the mid-1990s, Motorola and Sony were developing communicators, while IBM and BellSouth were collaborating on a cellular-palmtop hybrid, which would eventually become the first smartphone launched at the commercial scale. So about the future of the PDA, it merely came down to a simple choice: connected or not! The disconnected models were mostly perceived as fancy organizers. But was this wireless bonanza for real? Or would it distract the industry from far more important goals: coming up with innovative portable computing products? Had old-timers learned the right lessons from the exuberant pen-computing days? Jeff Hawkins, who'd seen it all, was contemplating on the answers.

He had set his eyes on creating the ultimate organizer and didn't want to dilute its effectiveness with communications features. According to Hawkins, there was a huge market out there if only there was the right kind of portable electronics product. People wanted a great mobile phone, a great pager, and a great organizer, he argued. Hawkins said that designers of wireless devices would subsequently grapple with many unknown factors. The challenges included not only choosing the right wireless network, radio, and battery but also determining exactly what consumers would want in the way of features. Did they want e-mail or mobile web or access to corporate networks? All these were pieces in the puzzle, all were interconnected, and all were moving targets.

Back in the mid-1990s, Hawkins felt certain that all handheld computers would one day have wireless functions, but that day was a long way off. The basic problem, as he saw it, was the shape of wireless networks and that the available network technologies were too immature for satisfying wireless palmtop. Wireless circuitry would impose a dramatic battery drain, would make the device far bulkier, and would deliver the Internet at snail-like speeds, which users would find unacceptable. Hawkins wanted to wait until the technologies had improved. But U.S. Robotics executives started pushing him to develop a wireless version, one that would send and receive information without the need for a phone line and a modem. Hawkins figured out it would take two years to come up with a wireless Palm Pilot.

In the end, it wound up taking three years, finally seeing the light of the day in May 1999, bearing the name Palm VII. The idea about Palm VII was to build a two-way radio, which, riding the BellSouth data network via a flip-up antenna, would allow users to tap into the Internet and send and receive e-mails without a modem and phone line. The vehicle of Palm Pilot's launch on airwaves was the decade-old RAM Mobile Data network, which had now acquired the name BellSouth Mobile Data. Despite noble efforts of RAM Mobile Data, which built the wireless data network in the early 1990s, users didn't flock to the service. But once BellSouth scooped up the company and backed it with marketing and branding, the network quickly filled up.

BellSouth—one of the last standard-bearers of Graham Bell—had been evangelizing the embryonic mobile date industry for quite some time; it was also the first telephone company to set up a subsidiary for data services. Now, the Baby Bell was offering wireless data services to 90 percent of the urban business population in its coverage territory in the southeastern United States through a network based on Ericsson's Mobitex technology. BellSouth wireless data network provided crucial services to the big U.S. companies to enable them stay connected with their workforce in the field. Naturally, 3Com had turned to BellSouth for providing communications services on its palmtop organizers. BellSouth, on the other hand, needed the small form factor for devices operating on its data network. That made up the foundation of this venture.

Taking the Palm Pilot form factor and its basic organizer features, the two companies envisioned tapping into the power of the web to bring the world's first successful Internet appliance. Hawkins, then at 3Com's Palm Computing division, sat with the president of BellSouth wireless data unit to hash out details of the newborn Palm.net. In the summer of 1999, using web clipping format, Palm VII version enabled users to check stock quotes and weather information and to buy tickets online through specific websites. Instead of trying to display a complete, downloaded web page on the tiny screen, the Palm VII would be preloaded with mini web pages, complete with graphics, blanks, and pop-up menus. Data from the web would appear in the appropriate blanks: stock prices, news articles, movie schedules, sports scores, phone book information, and so on.

Because only tiny snippets of text would be transferred, each web lookup would take only a few seconds. Within a year of Palm VII launch, the customer base of BellSouth grew from 200,000 users to 570,000. The Baby Bell added three hundred base stations to the network and renamed the network as Cingular Interactive. But though the long-awaited wireless data capability brought forth a new era of portability, serious technical and market hurdles still persisted. Palm VII was radically different from the handheld maker's other products; not only in having wireless features but also in its higher price (US$599). So despite initially revitalizing the BellSouth's wireless data network, it never became a hit like Palm III or Palm V.

Computer scientists had been predicting the advent of such information appliances for decades. And now Palm, which had single-handedly generated the PDA craze and had succeeded where many had failed before, was leading a drive into the brave new wireless world. But in retrospect, Hawkins had been right in rejecting the notion of adding superfluous new features that only a selected few would use. Palm VII was eventually regarded as an early entrant in the nascent world of wireless data services. The machine was years ahead of its time. Wireless mobile computing, however, would remain a hot button in the industry for years to come.

PDA–PHONE RENDEZVOUS

Just when the PDA started meaning a Palm Pilot, there came a turning point in the short history of handheld computers. Although Palm had become a flagship product of 3Com, the parent company was known as a networking powerhouse, and its limited vision for handheld computers was frustrating for the Palm founders. In 1998, after failing to convince 3Com to spin off the Palm unit and to invest more in the development of the product, Hawkins and Dubinsky left 3Com to start their own handheld-computing company, Handspring Inc. Ed Colligan would soon join ranks with his old comrades.

Ironically, later in 1999, 3Com did just that. To keep focused on its core networking market, 3Com decided to spin off Palm into a separate public trading company. The timing was ironic in another sense; only a few days later, 3Com's offshoot Handspring announced a new product based on Palm operating system. The trio who had brought the world the Palm Pilot struck again, bringing a faster, cheaper, and far more versatile handheld computer within a period of one year after founding Handspring. The new gadget, named Visor, introduced the innovative plug-and-play capability in handheld computers to add paging, audio, and camera modules. Communications meets computing meets consumer electronics!

Visor's biggest asset—its easy-to-use expansion slot, called the Springboard—could virtually turn the device into any consumer or communications appliance. The savvy Palm clone with an expansion slot could transform the gadget into anything from a GPS receiver to an MP3 player. After sliding the module into its slot, the module automatically installed its software and began running no matter what the Visor was doing at that time. Once a user was done with the module, he or she could take it out; the program uninstalled itself instantly and disappeared. The advent of Visor had in fact marked the birth of the smartphone all over again, and once more, the prodigy came from the computing side. The expandable PDA gave handheld devices a new direction, making Palm handheld look like a pedestrian.

Meanwhile, Palm, whose role in the marketplace was fast changing from lead innovator to an also-ran, was having a tough time in correcting its missteps. Handhelds were outgrowing their roots as electronic organizers and were mutating into Internet appliances and wireless communications tools. The biggest technological surprise was in the offing: the emergence of voice as a key technology for handheld devices. Handspring saw writing on the wall and vowed to turn its PDA into a mobile phone. Visor already had a built-in microphone; all it needed was a phone module. Palm, on the other hand, was seen as lacking this thrust into the hybrid-device market.

Visor was absolutely stunning; it became a Wall Street star overnight. The upstart founded by Palm alumni had mapped out an effective media blitz. Now the Palm fans were anxiously waiting for this magic device, but what happened next could serve as a textbook example for any startup company. Visor, which suddenly emerged on the commercial scene, ended up in a severe inventory crisis. A brisk web sale followed the product launch, and then, Handspring crumbled under pressure from component shortage and hardware problems. Product glitches and delivery delays led to a logistic nightmare; people were looking for gadget all over but could not find it for many months to come. Visor became victim of its own success.

Once the dust settled, after Visor's pop icon standing and logistic debacle, however, it turned out that its biggest asset—easy-to-use expansion slot—could also become a liability. Two inches long and half an inch deep slot used so much space that it made users wonder why they have to carry a bag full of Springboard stuff. Consequently, Handspring abandoned the Visor line of organizers to focus on developing mobile phone combined with a data gadget. Though Visor was a modest business success, the notion behind the realization of Visor carried far more revolutionary implications. Soon a stream of elegant concepts began to appear in the marketplace, evangelizing what a PDA could do as a portable electronic device.

PDA: WHERE DO I GO FROM HERE?

PDA proponents argued that the portable computer is not synonymous with a just one killer application and that there is a room for a whole range of PDA devices in the marketplace. They started to convince business segments that mobile PDAs are going to deliver utility beyond personal information management (PIM) systems. Case in point: Doctors and hospital administration officials initially used PDAs for billing and keeping schedules. But eventually these devices were increasingly seen as lifesaving tools that could allow doctors keep track of the patients' medical records. Doctors and nurses could reduce medical errors by scanning the bar codes on medications. Likewise, hotel personnel could check in guests at the curb; police officials could verify a vehicle's warrant status from anywhere on their beat.

Companies had started looking into ways to wirelessly deliver real-time inventory lists and acquire histories of their sales force and distributors. On the retail side, for example, sales clerks could stop punching order numbers into handheld data-entry terminals that were about the size and weight of a brick. Here, PDAs could perform a much wider range of functions than the old data-entry tablets. With a palmtop, sales people didn't have to key in every number for every unit in an order. Instead, they could just pull up the customer's usual order and make necessary changes. Compared to its brick-like predecessor, a PDA could do a lot more than just take orders. It could update critical information such as product specials and promotions as well as make any changes to a customer's account information every time the PDA synchronizes to the company network.

Some U.S. companies began encouraging their employees to use PDAs because that could boost productivity away from office. These firms were seeking ways to take advantage of the personal organizers that many people thought of as simple consumer gadgets that merely stored phone numbers, addresses, and notes. With tons of data storage capacity, PDAs could move beyond helping execs get organized to fundamentally changing business operations. And as PDAs proliferated, employees began pressuring their companies to integrate these devices into the corporate networks.

The PDA concept was branching out into new niches in the manner exemplified by Garmin, a Kansas-based leader in the market for specialized location-finding devices that relied on the U.S. Department of Defense's network of GPS satellites. In November 2002, it introduced an integrated palmtop with a GPS device that could determine location anywhere on earth by communicating with satellites, displaying detailed color maps of the location, and showing a route to a destination. The sensor-laden GPS-ready handheld foreshadowed what was likely to infiltrate into personal electronics in the coming years. With longitude/latitude, detailed mapping, altimeter, and electronic compass, all in a waterproof package, it looked more of a technology smorgasbord.

In order to function for such applications, the palmtop must include hardware with modest memory and a limited web browser capability, a wireless data connection, a GPS receiver, and some sort of orientation device—in this case, a magnet compass. Beyond wireless realm, the promise of a new world of usefulness was further empowered as PDAs acquired more robust multimedia applications such as games, video streaming, and the ability to take pictures and e-mail them. PDAs could now incorporate digital cameras, allowing the device to behave like a handheld most of the time, but could conveniently expose its camera features as the user needed or wanted them. The appeal of the PDA, as it turned out, could be very broad.

There was a time in the early 2000s when the possibilities for the PDA seemed endless. In the gadget nirvana, how big one could dream? Pocket-size computers were now far closer to laptops in capability than electronic day-timers. They could surf the web in color; double as a phone, a camera, and an audio/video player; and even maintain web-sites. PDAs had come a long way from the days when all it could do was maintain an electronic calendar and address book. But the crucial question was how the special aura of connectivity would be realized as PDAs went through this radical change. PDAs could either merge with mobile phones or could find their identity in a breed of new gizmos, each catering to a specific killer application.

INTERACTIVITY GAMBIT

Once a curiosity, the PDA eventually became a near-necessity for untold thousands of busy people. With the realization of various wireless technologies, the PDA was seen as evolving beyond just a replacement for paper day-planners into a veritable communication hub. New technologies in wireless, computing and multimedia arenas just reaching maturity would make PDA-like devices even more powerful in the coming years, or so said the technology pundits. But the home-run product, the one that would recreate the wireless market the way the Palm Pilot had created the thriving handheld segment from the ashes of the Newton and the Zoomer had yet to appear. In fact, the only bona fide mega-hit remained the simple, elegant pocket organizer that Hawkins had invented.

It'd been more than a decade since the three-person firm called Palm Computing had opened for business, but the handheld industry was still in its infancy. Hawkins saw the handheld computer industry as embryonic; he would quote the famous analogy "It's like 1982 of the PC world." With all the attention on wireless-ready PDA, it was becoming hard to remember that most PDAs in use didn't have a wireless connection of any kind, and a good number of their owners probably didn't want one. A spate of consumer surveys indicated that wireless wasn't necessarily everything when it came to PDAs, despite what market analysts tended to say in unison.

Wireless Internet had been touted as the next big thing since the late 1990s. The technology gurus argued that once the wireless Internet began to take off, giving users access to the web through portable computers would be a no brainer. That could put the PDA at the heart of always-on wireless nirvana, changing the landscape of communications as well as of computing for many years to come. That's one reason why "always-on" was something of an object of great optimism to PDA zealots. But was the PDA the ideal platform for web surfing? The first problem was limited coverage. Although companies such as BellSouth provided Internet access to handheld devices like Palm, having a built-in wireless modem, network coverage was spotty at best.

While Palms worked well in major metropolitan areas of the United States, they had difficulty connecting beyond those boundaries. Another problem with wireless data networks was the slow rate at which they delivered the web pages. These networks were mostly working at a speed of 9.6 Kbit/s, which was fine for plain text but was too slow for ordinary web pages with any kind of graphics. There was a possibility of a wireless setup offering higher data rates of up to 14.4 Kbit/s, but that required the use of a modem and cell phone and, hence, wires to establish connections. Some industry circles closer to the wireless sector hoped that GPRS would also complement the PDA. But the new data-enabled wireless networks like GPRS were taking shape at a slower pace than makers of personal computing devices would have expected.

Besides this wireless gambit, another camp in the handheld tribe figured that the PDA made more sense as a multimedia device. They conceived devices that were flashy and filled with multimedia goodness. The cheerleader happened to be one of the world's best-known consumer electronics company: Sony. The handheld computer manufacturers desperately wanted to expand the device beyond the "just PC peripherals" category, and there was a crucial factor influencing this add-on campaign. The handheld industry was rapidly becoming a commodity business, just as Palm veteran Hawkins had feared. "The organizer business is going to be like calculators," he said. "There is still calculators business, but who wants to be in it? They are cheap, and sort of the backwater of consumer electronics."

If handheld organizers became low-profit goods, companies who specialized in low-cost manufacturing, not product development and innovation, could make a living by producing them. That was one reason why Hawkins was now eager for his new venture Handspring to get into communication business as soon as possible. In 1996, the "Zen of Palm" was a good idea. The philosophy allowed Palm Computing to sidestep most of the problems that had plagued the too-ambitious Newton and other pen-based computers of the time. People were willing to learn Graffiti if it meant nearly perfect handwriting recognition. Fast-forward seven years: technology had improved by several orders of magnitude.

The PDA without wireless connectivity was doomed, and the PDA with wireless connectivity just couldn't contest with mobile handsets. While the PDA had the big-screen prerogative, camera-equipped phones were more popular, and software vendors wanted to be on the phone side because it was a far bigger market. We have seen time and again how economy of scale would completely transform the business landscape. PDAs with multimedia goodies were able to grab only short-term interest as cell phone residing on a much higher volume plateau gradually shaped into a convergence device. While the idea of having a digital companion was alluring, in the end, the PDA proved to be a valiant attempt at lifestyle organization that found fans only among a special class of consumers. Its other big promise—connectivity—paled to more ubiquitous cell phones.

CELL PHONE: FRIEND OR FOE?

By the mid-2000s, the PDA had tumbled from high-tech stardom, and its promise of a connected life in the wireless age was seen as elusive. The quest to take computing off the desktop and onto the road via the PDA had turned into a maddening transition. But despite the failure of wireless-enabled Palm, RIM's BlackBerry PDA was rewarded for its right vision. In the hindsight, RIM probably looked at the PDA market and said, "Yes, this is what it should be." In the contrast between the failure of wireless-equipped Palm PDA and the triumph of cellular-hooked BlackBerry PDA was hidden the ultimate predicament: the mobile phone was the future of PDAs.

While PDAs were still in the early stage of their evolution curve, striving to bring wireless and multimedia capabilities, telecom manufacturers mostly remained wary of the PDA market, terming it a turf of computer players. Many of them predicted the same fate for the handheld devices as of pagers in the wake of emerging smartphones that could eventually push PDAs into obsolescence. Why someone would buy a separate device if all the organizer features were available on a mobile phone? From data to music to video, everything was going to become available on smart wireless phones. What else do we need!

This section of industry, mostly comprised of wireless handset manufacturers, was convinced that the handheld computer and the mobile phone were fitted to fuse into a device that could both display and fetch data on demand. The stakeholders in the opposite camp, however, argued that converged products had been tried and failed and that a great phone doesn't necessarily make a great organizer. Merging a mobile phone with a PDA might actually turn out to be more of a liability than a benefit. For any PDA to run for only a few hours on a single charge is unacceptable, phone or no phone.

Certainly PDAs had improved since the Newton was introduced, but so had mobile phones. So far, the average mobile phone was essentially a dumb device: good for allowing people to chat, but hopeless when it came to managing the information. Now, the wireless telecommunications industry had taken the cues and was in a bid to redefine itself by facilitating a marriage between data and voice-centric cellular systems. Data gradually becoming an integral part of cellular services was also reinforced by the fact that cellular handsets had a natural advantage over handheld organizers when it came to wireless Internet access. Rather than snap on a modem, a user could simply flip open his or her phone and browse the web.

PDA proponents had long advocated that people would need two devices—a phone to talk on and a wireless device to stay connected to important people and information. But because the so-called wireless PDAs didn't come up to expectations, more and more industry observers were heeding the prognostication "cell phones will eat PDAs like pagers." A testament to this shift was Palm Inc.'s decision to make its operating system compatible with all sorts of mobile phones instead of an all-in-one solution. The company finally gave up to the market pressures in 1999 when Palm handheld was integrated into Qualcomm's cellular handset.

In cellular phone versus PDA debate, the phone came up as a truly personal digital device. Sales of handheld computers, or PDAs, at around 10 million a year, were dwarfed by 600 million shipments of mobile phones in 2004. Friend or foe, the future of the PDA seemed deeply intertwined with that of cellular phone, which itself was about to embark on an ambitious new journey. After an initial boom in the late 1990s and early 2000s, PDA sales began to level off as more and more consumers decided to buy cell phones with built-in PIM features. On the upside, however, the PDA turned out to be the first vital building block in the formation of the smartphone roadmap drawn on the portable computing playbook.

THE DAYS OF COMBO

Soon after the inception of Symbian in 1998, and consequently the "smartphone" splash, handheld designers began their DNA splicing experiments will cellular phones. The most exciting competition in personal technology was the race to design the best communicator, a new kind of handheld device that combined a wireless phone and modem, a digital organizer, and an e-mail and web terminal. These smartphones were primarily light voice terminals that got their smarts from the ability to display calendars, send and receive e-mails, and play music files. Conversely, a slight alteration came in the form of PDA phones: handheld computers that were mainly intended for wireless data communication, but they also doubled as phones.

At this stage, Palm realized that as cell phones incorporated more PDA functions, a huge new market lay before it, if only it could just get its operating system in the new handsets. However, such hybrids, except hyper-focused BlackBerry, were neither fish nor fowl, so initially they seemed an unattractive compromise to most users. Cellular stalwart Qualcomm's big, ungainly Palm-based phone flopped, and so did Kyocera's cell phone integrated with a Palm PDA, which it had been shipping since 1999. The original Palm handheld had flourished on its elegant simplicity and the original concept of "connected organizer" that shared data with a personal computer. But satisfying the feature demands of licensees as diverse as Nokia, Qualcomm, and Sony proved far more challenging, which subsequently slowed the Palm operating system uptake.

Although the marriage of computers and wireless was a truly exciting development, this was a transitional period in both wireless and computing domains. In those early days, the clunky hybrids, blending different functions, often proved too complicated and costly. Although Palm had the bragging rights over the PDA side of the convergence field, its operating system was fast becoming an afterthought in the phone market. There was an exception, however. In late 2001, in the midst of this historic transition, Handspring introduced the Treo Communicator, a hybrid phone and palm device that initially offered consumers a choice of a Graffiti version or one with a small keyboard.

The Graffiti version found few takers so Handspring decided that all future products would use miniature keyboards. Treo, a dramatic shift in the Handspring product line, would send and receive e-mails and instant messages, wirelessly browse the web, and of course, make phone calls. "We feel the game is changing and we can combine a cell phone, a pager, and a PDA without compromising any of them," said its chief scientist Jeff Hawkins.

Adding paging and e-mail functions to a cell phone–palmtop hybrid wasn't exactly a new idea. In the early going, when customers beseeched the boyish handheld pioneer to combine the Palm with a cell phone or a pager, he'd answered, "If I built it, you wouldn't like it. It would have to be too big. It would have a short battery life." Designing a cell phone–organizer combination device—a communicator, as Handspring would eventually call it—was an undertaking of staggering complexity. Smart technology and goofy gizmos didn't go well together. Hawkins wanted to first find a way to create a traditional handheld computer that could also be turned into a cell phone later. The plan was to have an organizer business and then to transition to the communication business.

The bits and pieces of Hawkins's master plan for a piece of the pie in the vast communications market gradually fell into place after the inception of module-centric Visor. Now when the technology marched on, components miniaturized and batteries improved, the challenge was to design a single device that was a good organizer, a good pager and a good cell phone. A communicator that was a great organizer but a mediocre phone would never find mass market. Treo, to the joy of gadget junkies, had a good first stab at an integrated cell phone and

PDA. It was smaller than most men's wallets: the first communicator that fit into the pocket!

Handspring had made an unhedged bet, shutting its old organizer business to focus solely on communicators, which offered higher margins. The company had the brains, the soul, and the creativity that were at the roots of Palm's success. But critics predicted that the cell phone companies, with their massive clout and enormous research budgets, would eventually win the smartphone market. Handspring's dynamic trio obviously didn't agree. They argued that cellular companies had little experience in handheld computer arena, and they had never cultivated a community of software developers or designed an attractive, efficient user interface for handheld screens.

This time around, however, skeptical analysts proved right about Handspring's ability to shape a new generation of smart but economical gizmos on its own. In June 2003, in a Palm Pilot déjà vu, Palm Inc. announced to buy Handspring in a deal that would unite the pioneers of handheld computing industry. The market Handspring bet on failed to materialize as quickly as it had hoped. "I thought the Treo would take off faster," conceded Hawkins, based on the pattern of Palm Pilot. Handspring had entered an arena where the rules of business would be entirely new. Palm Inc., on the other hand, was interested in Handspring primarily because the PDA market was seen as dying, but the Treo smartphone was becoming popular as a phone with extended PDA features.

So Palm and Handspring swallowed their pride and merged. It was like prodigal sons coming back to the flock. At this very time, Hawkins said that the most important application in the mobile phone space is still voice: "The next one to evolve will be broader data communications, for things like transactions and browsing. Messaging is also a killer application and it's bigger than e-mail." From his vantage point, Hawkins outlined four killer applications: voice, wireless data, messaging and PIM functions for calendar and address book. By 2003, the market for smartphones was still small, but it was growing fast as new features were added to handsets making them even smarter.

Smartphone shipments surged 400 percent in the first quarter of 2003. Slowly and steadily, the pieces of the smartphone puzzle were falling in place. Then in 2007 came the defining moment in the PDA's metamorphosis into the smartphone. Apple took its largely unfinished business that it had started with Newton and effectively turned it into a PDA which could make calls and most importantly had excellent web-browsing ability. The form factor of the iPhone did strike some resemblance with the all-screen look of the Newton MessagePad. Fourteen years after Apple had released the Newton—a handheld device with handwriting recognition, desktop syncing, and an embrace of third-party applications—the Cupertino, California–based company was proudly showing its descendant hailed by the industry as a revolutionary and magical product.

THE PDA IS DEAD; LONG LIVE THE PDA

The PDA had quietly emerged into a powerful and rich tool for personal digital management. Then the rude awakening came when industry witnessed portable computer pioneers like Handspring and consumer electronics bellwethers as such Sony giving up PDA product lineups one by one. For some, it was about the time to declare the demise of the handheld computer or the PDA, which many people in IT industry thought would become ubiquitous someday. "PDA is dead," proclaimed Symbian chief David Levin in fall 2003. The PDA—the portable monument of the late twentieth century and early twenty-first century vision of ubiquitous computing—was indeed folding into the annals of technology history.

In less than two decades, portable computers had gone through a remarkable evolution before they morphed into smartphones. They first outgrew their roots as organizers and then mutated into Internet appliances and wireless communication tools. The next logical step was packing computing smarts into all sorts of portable devices. But while the PDA was heralded as the electronic potion to organize one's life and bring the PC to pocket, the PDA could neither bring volume nor margins. And it was still painfully difficult to use for common consumers. In the midst of this identity crisis, it became evident that though the overall concept behind PDAs as a mainstream productivity tool was sound, PDAs needed to expand into serious vertical markets like medical, field force automation, and education to stay relevant.

That didn't happen, either. So what about those utopian ideas that called for legions of machines communicating with each other, producing streams of information that would follow people nearly every step of their lives. Given the pervasiveness of cellular technology and the rise of Wi-Fi hot spots, and the fact that lines between the PDA and the mobile phone were continuously blurring, the dream of ubiquitous computing suddenly seemed within reach by the mid-2000s. The PDA—a synonym for handheld computers—symbolized a great experimentation in the mobile computing realm. A part of it folded into the smartphone recipe, and the other part was morphing into its next self: a host of gadgets as new manifestations of portable computing would emerge in the coming years serving a variety of markets. It was about computing, not computers, after all.

The story of portable computing would take many shapes and forms in the following years, sparking innovative attempts on multiple fronts, ranging from ultra-mobile PC (UMPC) to mobile Internet device (MID) and from netbook to tablet PC. The death of the PDA, as it turned out, was greatly exaggerated. Portable computing would be embedded in many more devices, and most of them would communicate over wireless networks.

In the final analysis, however, it was now beyond any doubt that the world's most popular electronic gadget was the mobile phone. The old and new mobile phone establishments were now betting that enough of the things that people wanted from the Internet could work without a personal computer's big screen. Despite the smaller display of cellular handsets, users were still able to send and receive e-mails, perform e-commerce transactions, and get real-time updates on such things as travel, weather, news, and sporting events. They claimed that with the wireless Internet, computing might achieve its apotheosis: simple, reliable, ubiquitous, and pervasive.

THE MAKING OF THE MOBILE INTERNET

"To be truly wired, you must be wireless."
—Simon Garfinkel, *Wired* magazine, October 1997

Nokia's Anssi Vanjoki first heard about the Internet in 1993 at a time when his company was heavily betting on digital cellular phones. One day Vanjoki saw a new hire hunched over a strange-looking database on his PC. It turned out the newcomer was online, using the Gopher menu system to browse through a library at the University of Texas. Vanjoki thought that if he could do this on the PC, he should be able to do the same on a digital handset. Across Scandinavia, many engineers were having similar thoughts. It might have seemed implausible at that time that people would surf the Internet and exchange e-mail messages from their cell phones given their tiny screens and awkward keypads. About six years later, mobile phone users were unmistakably doing so, and in reasonably large numbers.

The notion of the mobile Internet first emerged on the technology scene during the mid-1990s when the mobile phone was on its way to ubiquity. When dreamers started talking about accessing the Internet from cellular handsets, the mere idea of coalescing two of the most transformational technologies of twentieth century sounded fascinating. Science fiction writers had long been dreaming for a communication upheaval as a means to perfecting the society—wearable devices like the *Star Trek* communicators and the *Dick Tracy* watch phones—and it was actually accelerating toward reality by the end of twentieth century. Cellular phones and PDAs—already smaller than the *Star Trek* communicators and Captain's electronic logs—would bring the world on the verge of a highly mobile society.

The wireless euphoria of the 1990s ended with a sense of enchantment. The mobile phone's triumph over the fixed-line telephony as the primary means of voice communications seemed assured by now. It was the sudden arrival of data, however, that was rewriting the rules of the mobile communications game. After cellular handsets became nearly as commonplace as a wallet, the telecommunications industry found itself at a new crossroads where a unique combination of mobility and connectivity promised a whole new world of possibilities. In fact, the natural progression of wireless data technology was only a few years behind its wired-line counterpart: the Internet.

Although an exhilarating phase of growth was just around the corner, the fabric of the mobile data market was far more complex than the traditional cellular settings that made up the voice-centric business model. Such an ordeal, however, applied to almost every technological endeavor, so hardly anyone was anticipating that the realization of data communications in wireless space would happen overnight. Historically, mobile communications had focused on handling simple telephony services. The first two generations of cellular systems were designed purely for voice communications.

Wireless data applications had been in place for quite some time in specific industry areas and vertical markets. Vehicle dispatch services were a prime example. But for consumers and high-end users, a new plateau was required to close the gap between the wireless and the data networking worlds. Short message service (SMS), through its mass-market appeal, had provided an initial impetus to wireless data delivery. Now the industry was gradually coming to terms with a new paradigm that would transform wireless data into a new market dynamo: they called it the "mobile Internet." And reminiscent of SMS's role in popularizing wireless data transport, the crystal ball for the mobile Internet came from a place the industry gurus were looking the least.

But before we look into that, here is a little recap of history. The evolution of the mobile Internet could be classified into three classical episodes, and they span across three continents acquiring distinct spaces as well as time frames. The fact that Apple and Google subsequently became the key figures in the mobile Internet coming-of-age story speaks well about tech-happy America's bragging rights on smartphones. After all, Americans not only invented the Internet but also much of the mobile telephony. In retrospect, the U.S. companies figured out how to seize the moment in mobile Internet trajectory, partly because smartphones incarnated from computers and partly because the mobile Internet was the inflection point of two great technology stories that America help build in the first place: the Internet and mobile telephony.

But in the heydays of digital wireless boom during the 1990s, Europe had gotten ahead of America in mobile telephony precisely because of its successful policies. While the United States left its cellular systems to the forces of the market and wound up with a fractured system, Europe settled on a single technology standard, Global System for Mobile Communications (GSM); charged little for licenses; and raced ahead. When Europe became the capital of the wireless universe in the 1990s, the notion of the mobile Internet started to be seen as Europe's best chance to extend its wireless lead into cyberspace.

The Europeans thought that while they had lost out on the first round of the Internet, they could catch up and even leapfrog America with the mobile web. In the 1980s, Europe missed on the PC revolution as Americans creamed the European protected national champions such as Bull, Nixdorf, and Olivetti. In the next decade, the continent lagged in the Internet buildup. So for Europe, the mobile Internet was yet another rich opportunity: it offered the continent its best chance to launch its own tech-driven New Economy. Mobile Internet, Europeans said, would create a new wireless world and perhaps even give birth to another economic miracle after GSM.

But at the dawn of the new century, the action was gradually shifting to Asia, which emerged as the new ground zero in the wireless wars. Much of the credit went to NTT DoCoMo who beat its European rivals out of the starting gate with i-mode. The next wireless course now seemed to belong to Asia where countries such as China, Japan, the Philippines, and South Korea were increasingly becoming mobile phone zealots. The firing shot on mobile Internet also came from the Land of the Rising Sun when DoCoMo created an island of success in the form of i-mode service by adapting the wired Internet-like business model.

THE JAPANESE TAP DANCE

In early 1997, NTT DoCoMo president Koyji Ohboshi called the managing director of business marketing, Keiichi Enoki, in his office and asked him to launch Internet services over mobile phones. Personal computers in Japan were expensive and homes were small, which left the Japanese far behind the United States in the PC arena. Japan was slow in its uptake of PCs for several reasons, but one large factor was that the early desktop machines were difficult to use for those not fluent in English. Moreover, people had to commute in trains for a number of hours, so they had relatively less time to use the desktop at home as a portal to the Internet. DoCoMo saw a window of opportunity here. The company set a new mission: to lead Japan with the world's first instantly accessible mobile Internet service, i-mode.

Enoki instantly set up a new department that would aim to develop a range of free and pay-per-use services: send 500-byte e-mails, do banking, reserve tickets, and buy and sell goods. Mobile phone users could also check restaurant guides, phone directories, dictionaries, weather, stock prices, news, horoscopes, music charts, as well play games and download cartoon characters to their handsets. In its carefully crafted marketing campaign, DoCoMo emphasized convenience and avoided phrases like "advanced functions" and even the word "Internet." Mr. i-mode, as colleagues would call Enoki, also decided not to alter the shape of the handset, realizing that mobile phone users in Japan were used to having small and cute gadgets. i-mode would be fashionable yet functional.

DoCoMo won a million i-mode customers in less than six months. The Japanese people could trade stocks, make reservations, and check headline news by paying for the data they transmitted or downloaded through i-mode phones. Consumers loved the simple-to-use service and quickly adapted it for displaying Hello Kitty on their color screens and for diversions such as online gaming. For business people, i-mode brought the convenience of being able to use the phone like an office. Especially for people who didn't want to sit at their desks, it was a blessing. For others, i-mode simply meant fun. Teenagers, for instance, found their own uses of i-mode. DoCoMo had collaborated with multimedia behemoth Sony to bring games and cartoons onto mobile phones through i-mode handsets.

Within a few months after its launch on February 22, 1999, about 4,700 web pages had been designed for small handset screens and were made available to i-mode users. The screen-mode packet data service operating at a speed of 9.6 Kbit/s provided access to mini-websites through a specialized NTT portal. Each i-mode phone had a small button that granted easy access to thousands of DoCoMo-endorsed sites. Some were free; others charged up to US$2.50 a month. Now people stopped talking and started thumbing. By the turn of the century, thanks to i-mode magic, DoCoMo was winning a million new subscribers a month.

i-mode, through smartly tailored applications and limited web access, had established the world's first cradle of mobile web culture. In just two years after its launch, DoCoMo had attracted more than 22 million users and 30,000 content providers. The service became the epitome of what could be achieved in the mobile Internet space, the end-all, be-all of wireless data. DoCoMo's ordeal with i-mode was nothing short of inspiring; no wonder everyone in the industry wanted to know its secret. To understand i-mode's acceptance, it was necessary to look at the service from both the technological and the business-model perspectives. On the technology front, the Japanese wireless starlet decided to go for packet switching instead of conventional circuit switching technology being used in cellular networks. i-mode was an always-on service: a user clicked a button and he or she was there. Packet switching saved users the hassles of calling up a website every time they needed to interact with one.

Then, the Japanese mobile operator came up with its own proprietary platform based on the Internet's standard script: HTML. That made it easier for developers to write applications without having to learn an entirely new computer language. i-mode used common HTML tags so that pages created for display on handsets could also be seen on the PC. Logica PLC of London developed a compact form of HTML that they termed as cHTML; Access Co. Ltd, a Tokyo-based supplier of embedded browsers for non-PC products, developed the micro-browser for i-mode handsets. Access also conducted the overall technology integration in collaboration with Fujitsu, Matsushita, Mitsubishi, NEC, and Sony.

Nevertheless, the real genius behind i-mode success was the way DoCoMo's business model optimized a relatively simple technology. What i-mode offered to end users was rich content; DoCoMo had established partnerships with hundreds of companies. The mobile phone operator offered an attractive pricing model to its partner websites. After collecting the fee and taking a 9 percent service charge, DoCoMo gave the remainder to the website. By giving content producers a means to charge users, DoCoMo ensured that there was plenty of content available. Tekeshi Natsuno was credited for designing a profitable, info-entertainment platform that morphed into Japan's biggest market success in years. The i-mode content guru lured high-quality content providers with a micro-payment scheme that enabled them to charge fees for their services—ranging from weather updates to horoscope.

JAPAN'S WIRELESS MIRACLE

Tokyo was now the capital of the mobile universe. The success of i-mode service had made DoCoMo the poster child of the wireless revolution. While other mobile phone operators were still trying to figure out the math, DoCoMo had shown the world how to make money from the wireless Internet. At the height of i-mode euphoria, the mobile phone operator buoyed the Japanese electronics industry—then swimming in the red—when it raked in more than US$18 billion on its initial listing of shares on the Tokyo Stock Exchange. In a matter of months, DoCoMo became Japan's largest capitalized firm with more than US$100 billion in stock value, which placed it ahead of Toyota and even its parent company, Nippon Telegraph and Telephone Corp. The pioneer of the mobile web also became the second most valuable wireless company in the world after Vodafone.

DoCoMo had won 32 million subscribers in just three years after i-mode launch. i-mode turned out to be a social phenomenon that transformed Japan forever. But the story of nation's ascent to wireless glory has a couple of intriguing twists. The country had been in a habit of protecting its markets from foreign influence through somewhat tricky regulatory arrangements. For instance, mobile phone users couldn't even own a handset; they had to rent it from the service provider. Eventually, under a strong pressure from the United States, Japan opened up its mobile phone market in 1994. As it turned out, it was a blessing in disguise.

Among the most notable achievements that Motorola's Christopher B. Galvin—who later became CEO of the company his grandfather founded in 1928—counted was having helped to open the Japanese telecommunications market to foreign mobile phone companies. Charlene Barshefsky, then the deputy U.S. Trade Representative, was reluctant to start a trade battle until the younger Galvin assured her that Motorola would share the heat if Japan resisted. As a result of their efforts, Japan permitted more foreign competition and began letting consumers own their cellular phones. Consequently, the country became the fastest-growing cellular market in the world, and by the end of the decade, the number of mobile phone users in Japan had crossed 50 million. Motorola, however, could not make a breakthrough in selling handsets in Japan when Japan's love affair with mobile phones began.

Japan's manic sweep to the mobile Internet is also attributed to one of its great wireless failures: Personal Handyphone Service or PHS. Pushed by the government in the early 1990s, Japan's electronics and telecommunications industries tried to create a global standard for personal communications gadgetry. PHS—launched in Japan on July 1, 1995—functioned as a cordless extension at homes and offices and as a portable phone in the street for high-density pedestrian traffic. Initially a huge success in Japan, it boasted a data rate of 64 Kbit/s that was many times faster than standard cellular networks. But the success proved short-lived as PHS couldn't stand against the rising tide of mobile phone services. The technology that provided a cheaper alternative to cellular services flopped abroad and gradually shrank in Japan, but it provided the ground on which Japan could build robust new wireless platforms.

Also, earlier in 1991, as part of the Japanese government's cautious, or rather calculated, drive to diversify the national telecom monopoly, NTT spun off its mobile phone unit. In the coming years, NTT Mobile Communications Network Inc. (NTT DoCoMo) turned itself into a futuristic technology company by adapting the speed and brutality needed to compete in the global marketplace. And by embracing new technologies, it quickly developed the future vision of a service-oriented company. DoCoMo was the abbreviation of its corporate slogan "Do Communications over the Mobile Network." It also happened to be a Japanese colloquialism for the word *everywhere*.

The Japanese wireless operator embedded cell phones deeply into local culture when it launched its phenomenally successful i-mode service. DoCoMo sorted out how to bring data services on handset screens when the mobile Internet was still a virgin territory. Its relentless focus and willingness to go to painstaking lengths to make life easier for consumers was what turned DoCoMo into an incredibly successful company. The world's most valuable success story in wireless web adaption was also credited to its senior executives and their un-Japanese willingness to challenge convention and to pursue aggressive strategies in meeting objectives. With i-mode, a service that had been viewed as an inflection point for the mass mobile Internet, the company was suddenly far ahead of its rivals.

Now DoCoMo was like a huge sumo wrestler overpowering the marketplace, and there was nowhere left for it to go but overseas. So it started taking its expertise around the globe through relationships with mobile operators in America, Asia, and Europe. When its share price peaked in spring 2001, DoCoMo was the eleventh-largest company in the world, with enough cash to buy a 16 percent chunk of AT&T Wireless, the third-largest mobile phone company in the United States. Japan, a nation of compulsive dialers, had reached the pinnacle of the great wireless transition. So the company that made cell phones a must-have item in the now-frugal Japan aimed to become a formidable design arbiter of the next pervasive computing platform.

THE WAP SAGA

The story of the mobile Internet wasn't all a win–win phrase. Almost parallel to the making of i-mode, the wireless industry at large was at the helm of a much broader initiative to provide an early taste of the mobile Internet. At the time when Japan was thinking over how to bring cyberspace onto mobile phones, Ericsson and Nokia were also chalking out their own Internet strategy. The European cellular titans, however, were merely following a lead from a small American outfit.

The affair began in July 1996 when AT&T Wireless rolled out the PacketNet data service on mobile phones using a technology from Unwired Planet. The California upstart, led by an Apple Computer veteran Alain Rossman, had developed the micro-browser technology that enabled Internet access over wireless handsets. A year later, Unwired Planet, rather than go it alone, approached cellular industry heavyweights Ericsson and Nokia to persuade them that its mobile phone browser should become the standard for wireless Internet communications. Ericsson and Nokia agreed to join the company on a forum for specifying WAP, short for Wireless Application Protocol. Motorola's last-minute joining to complete the wireless troika was reminiscent of its jumping onto the Symbian ship.

There were other parallels with Symbian as well. Just like Psion, Unwired Planet had gained a technological lead over large mobile phone firms and software rivals, so WAP was going to be an open standard rather than a proprietary one like Microsoft's Windows. It was quite unusual to see a little startup creating a de facto standard by turning its idea into a global initiative with support from major industry players. The company that was to give WAP to the world boasted to be the first ever wireless dotcom company when it changed its name to Phone.com at the height of the Internet bubble.

As for the handset business, by the late 1990s, it had become a monopoly of three wireless manufacturers: Ericsson, Motorola, and Nokia. Only Korean makers were able to break into the CDMA handsets market because, except for Motorola, these wireless magnates were not much focused in that particular segment. The question for the so-called wireless troika was how to share the market and maintain dominance in this flourishing but rapidly changing marketplace. A new industry segment was in the pipeline that renewed the hopes of the stagnant wireless data market. So the trio decided to acquire the role of wireless industry's standard-bearers in a very subtle way.

After its inception in June 1997, WAP gathered a lot of support around the world as a technology-independent underlying link with transport delivery mechanism. It was designed for a wide range of applications on smaller screens of mobile phones with the focal point on economical uses of Internet resources and air-link capacity. The effective use of wireless bandwidth—a precious commodity in this case—was crucial in the realization of the mobile Internet. Another important goal was to rework the web into simple text that would be readable on small display spaces. For this purpose, wireless markup language (WML) was developed for wireless handsets; it was meant to be very similar to HTML.

The protocol was to cater wireless carrier-centric applications with WAP gateways performing encoding and script compiler functions for delivery of data onto handsets. The gateway servers were required to do the protocol conversion from HTTP and TCP/IP to WAP, and content conversion from WML text to binary data. The task was to be handled by mobile phone operators equipped with WAP servers; wireless manufacturers were only to produce WAP-enabled handsets. A small piece of software installed in a cell phone—a micro-browser—would help display data sent from a WAP server.

To stake full claim to the riches of the mobile Internet, the first WAP service was rolled out in the fall of 1999. By then the press was full of rosy predictions: it was going to take over the web, kill off the PC, and change everything. But Europe's phone companies flunked their first Internet exam. First, there were not enough handsets, and when the handsets finally arrived somewhere in 2000, users found themselves struggling with slow data speeds, primitive applications, and sky-high connection charges. Consumers didn't like it, wireless carriers didn't like it, and content companies didn't like it. The over-hyped WAP service failed miserably in Europe as it did elsewhere in the world.

The user expectation had been raised so high that major disappointment was likely when people got a look at the technology, especially those who had wired web to compare it with. WAP phones were too difficult to use, they were too slow, and they did not provide anywhere near the "Internet on a phone" experience touted in the ads. A crop of GSM phones outfitted with WAP required users to place a call to a web portal each time to send and receive data. Wireless carriers' data-connection plans were extremely expensive and average consumers found it next to impossible to find WAP websites that worked on their mobile phones.

Only a modest number of wireless portals became available to support WAP applications, and there were some genuine constraints in redoing the web pages for WML. Sluggish performance, dropping connections, and paltry content turned users off in droves and they logged off for good. The much vaunted but now derided WAP technology could only attract 7 percent of mobile phone users a year after its launch. The WAP fiasco also brought new questions about Europe's credentials to do the Internet and the future roadmap to 3G wireless. Industry analysts pointed to the fact that wireless companies, while copying the wireline model, were not used to data services mastered by the Internet service providers.

i-MODE VERSUS WAP

The Internet-everywhere buzz in Europe came down to the reality of 1-inch screen. The critics of WAP said that tiny screens couldn't hold enough information to be useful in most applications and the numeric keypad was an abomination as an input device for controlling any amount of complexity. Moreover, the bandwidth was so minuscule that even small amounts of data took forever to transfer. But the i-mode experience had made one thing clear: a great deal could be achieved with tiny displays, small keypads, and low data rates. After a year of its launch, there were less than 2 million WAP users in Europe, a far cry from 13 million subscribers DoCoMo's i-mode had won in the same period.

While many observers reasoned that DoCoMo's mobile Internet success was a distinctly Japanese phenomenon, due in part to Japan's relatively low penetration of home PCs, i-mode's hitting of sweet spot was not that shallow. The success of i-mode had proved that fun new services could attract a big market. DoCoMo had managed to create a vibrant mobile Internet experience largely by staying clear of the content business: it merely provided the platform and opened it to the web entrepreneurs, who enticed people with applications in thousands. Another important lesson: getting hung up on a technology and not thinking about the service a company was delivering could be disastrous. DoCoMo aimed at fulfilling user expectations; the company didn't even promote i-mode as a wireless Internet service.

The "i" stood for information, not the Internet. "WAP failed in Europe because operators concentrated too much on the technology rather than the content," said Keiichi Enoki, who led i-mode project team. "It's like worrying about the quality of television sets before you have any programs." i-mode did have technical advantages, too. Unlike WAP phones, which required a slow dial-up connection to a server, i-mode offered an always-on link to the Internet. Another reason why WAP failed while i-mode succeeded was the fact that DoCoMo prefixed hardware platform. That facilitated better display characteristics due to standard hardware modules in handsets. Furthermore, it was a much smoother shift to cHTML, the compact version of the web lingua franca, than making a wholehearted transition to WML.

Now the signs of what might be called "i-mode envy" were seen everywhere in the wireless world. There was a possibility that Western excitement about mobile Internet services could eventually find refuge in a number of i-mode myths. The i-mode showdown had brought a Japanese player to the offense on the global communications front for the first time. Now DoCoMo began aggressively promoting the service as a standard for mobile Internet appliances through multibillion acquisition deals in Asia, Europe, and America. For businesses outside Japan, however, achieving i-mode experience was not simply a matter of carbon copying a successful concept, but of attaining a greater understanding of the driving forces behind i-mode. Despite all its exuberance, i-mode was still a proprietary system. And unlike the Japanese market, where one operator, NTT DoCoMo, defined standards that breathed life into i-mode, wireless companies elsewhere seemed to be unable to agree on a single standard.

WAP, an early attempt to reproduce the web for mobile phones, was a chic project of its time. It failed because its backers inflated the user expectations to a level almost equivalent to the Netscape-browsing Internet. But as the stories of an impoverished user experience came out of Europe, the technology handily became subject to headlines like "WAP is crap." The growing epidemic of "WAPlash" struck fear among enthusiasts as the wireless web initiative was blamed for the root of all mobile ills. The whole idea of wireless web was exhilarating, but it also brought enormous engineering and business challenges.

Did WAP stand a second chance to engineer its resurrection? The WAP Forum thought so. Despite its failure in the first go, many experts believed all WAP needed was bigger transmission pipes and better consumer education. The market would grow rapidly, they argued, once 2.5G services such as GPRS and subsequently 3G became widely available, helping to make WAP connections many times faster. The WAP proponents were firm in their belief that this web-on-the-go initiative won't join the ranks of failed technologies no matter how good a story trade press would make. The truth of the matter, as always, was probably somewhere in the middle: WAP was at an initial stage of the value-adding ladder.

The idea of wireless web was just another stage in the evolution of consumer behavior that was taking shape since cellular phones kicked off in the early 1980s. The Internet we all knew was a smash hit on the PC but could easily flop on the handset. It was because the wired web was built for a different class of users. Using the same path to track down information via the phone proved to be a torture, so the phone makers needed to construct a different Internet experience altogether, one that would focus on mobility and would send services directly to the phone with a minimum of clicks. That was the moral of the WAP versus i-mode story.

REALITY CHECK

The mobile Internet had promised to shape into a full-fledged industry. The idea of Internet in one's pocket spread like wildfire, but once the dream turned sour, marketers were no more talking about the mobile Internet. The mobile Internet catch phrase sounded tedious by the end of 2000. The construction of wireless Internet infrastructure was way behind schedule and there was no clear business strategy in place for mobile data services. In the wireless industry at large, no one seemed to have the faintest idea what form the new data services would take, which services potential customers would buy, or even how to market the new products. Seldom had so many seemingly shrewd executives gotten their signals so crossed. The wireless revolution literally went on hold.

So what went wrong? For starters, wireless networks the world over were already struggling to accommodate the surging volume of voice traffic, so they could deal with data only at a snail's pace. Moreover, the problems of dropped calls and limited coverage, which were already frustrating for the voice users, could prove devastating for data services. It is one thing to drop a voice call, but it is another thing when a user was in the middle of a stock trade or a banking transaction. Combining the Internet with mobile phones would pose technical, business, and cultural challenges.

The Internet users expected things to be free, and were prepared to accept a certain degree of technological imperfection. Mobile phone users, on the other hand, were accustomed to paying but expected a far higher level of service and reliability in return. Another big question mark was why people would move to a wireless data platform when fixed-line Internet access was so cheap, and in some cases almost free. Other doubts surfaced as the wireless industry entered the uncharted waters of mobile data. How would data contribute to the bottom line— profits? How would the user want to pay for the data services? Then there were technology issues.

The notion of mobile Internet as the offspring of two such spectacularly unpredictable technologies—computing and telecommunications— cut both ways because the two worlds represented starkly different visions of the wireless Internet. Those on the Internet side of the fence complained that wireless companies didn't really understand data networking; those on the wireless side grumbled that the Internet technology was flaky. To lay claim on the mobile Internet, wireless telecom industry was stretching far from its core market. But unlike the telecom guys, who pursued wireless to bring connectivity and left portability to come later, the computing companies were dreaming of a portable machine from the start.

The whole PDA story is a testament of the fact that the comrades in the computing world thought of connectivity only after the Internet was in full bloom. To get there, the computing companies concentrated largely on a mobile extension of the Internet as we knew it. But the wireless Internet as a miniature version of the web didn't credibly demonstrate the potential of mobile data services; it simply made consumers skeptical about what they were being sold. Nevertheless, these were still early days for the mobile net: people tended to think of it as the Internet without wires rather than as something entirely new. Wireless access to the Internet, however, would almost certainly be different from the wired Internet experience.

Marketplace diversity aside, the path to a wider take-up of the mobile web was never going to be an easy one. An obvious weakness, as mentioned earlier, was that the mobile net campaigners put too much emphasis on technology rather than making it beneficial to users. Moreover, they failed to recognize the importance of person-to-person interaction, instead putting money on access to centralized content like WAP. The wireless industry should have been quicker to pick up clues from the way earlier telephone technology had become popular, and more recently from the text-messaging conquest. They could contemplate, for instance, that success of non-voice applications involved setting up the right environment to allow such services to succeed.

In the hindsight, the phenomenal success of i-mode and SMS had built a foundation for the future success of smartphones. Text messaging had achieved an astonishing growth at a time when the mobile phone industry was trying to dictate the deployment of WAP. A perfect irony, the success of SMS left some industry experts with a new puzzle: whether it was pointing to a huge marketing demand for mobile net services or what most people wanted from mobile data was services that were simple, cheap and more like e-mail than the web. After all, text messaging made the best of the tiny handset screen while WAP failed to compensate the same small screen.

Furthermore, though WAP had failed to grab consumer imagination, a segment of the wireless industry hoped it was the fault of implementation rather than inherent resistance to mobile data. Here, they were taking some comfort from the popularity of text messaging and from the roaring success of i-mode service in Japan. The Europeans claimed to have learned valuable lessons from the WAP failure. Their mantra now was selling applications, not technology, and the most triumphant application by far was text messaging. They were keen to point out that WAP hadn't gone away; it just branched into more practical applications, such as multimedia messaging service (MMS), the descendant of popular SMS.

The MMS technology with picture messaging at the heart of it was essentially WAP: MMS made use of WAP components for implementing transport and push mechanisms. After the WAP fiasco, wireless industry in Europe borrowed a page from the WAP book and rallied behind MMS, which made it possible to zap photos and graphics to and from handsets, just like the way text messages were sent in droves. MMS was not a giant technological leap forward, but a feasible feature made available on the handsets equipped with built-in cameras. The ability to take pictures with a handset and send them to other people was something that people wanted and were willing to pay for. Phones with color screens had also provided a strong impetus for snap-happy users.

MMS—which allowed the sending of still and moving images, audio, and text—was expected to become a springboard for mobile entertainments services of the future. It accelerated the adaption of digital photography and became a stepping stone to more advanced multimedia services that could become the staple of the coming 3G networks. The battered mobile phone industry hoped there would be more message-based entertainment to come, especially when advanced messaging services featuring color, moving images, and audio were rolled out. MMS had evolved from text messaging, a runaway success, and because MMS was based on WAP standards, it was possible to add a plethora of new features to it.

But as it turned out, MMS was no miracle. The flipside was that MMS handsets and infrastructure from different vendors didn't work together smoothly, so users found it more convenient to stick with plain, old text messaging. MMS enhanced the consumer's experience of messaging, but in itself, it was merely an enabling technology. So the expectation of MMS to replace text messaging eventually faded; MMS was more likely to be used on special occasions like birthday greetings or photos from holidays instead of postcards. Again, the industry's great hopes for advanced mobile messaging were not fulfilled. The telecom industry is notorious for hyping new technologies, only to find difficulties and delays in their development later.

The transition from SMS to MMS was critical in creating new revenue margins per message for mobile operators. At the same time, wireless telecommunications companies needed to focus on a new Internet, the one with services dedicated toward people on the move. Fortunately, DoCoMo had provided a role model in the form of its celebrated i-mode service. Japan had made the most impressive contribution so far in the field of mobile net by tactfully blending applications and content with advanced wireless technology. The first-ever network service linking mobile phone users to the Internet, i-mode was a ringing example of how to hook customers onto a new technology and turn a fat profit as well.

GREAT IDEA, SHAKY START

The unambiguous message the industry could draw from the failure of WAP as well as success of text messaging and i-mode was that people would go for wireless data services if they were useful and inexpensive. The success stories also showed how important it was for telecommunications companies to innovate. With the ever-increasing potential of wireless data services, mobile phone operators needed to come up with creative new strategies to attract users. People wanted their mobile phones small because they looked cool and screens large so they could easily access the Internet. And the quality of wireless data would need to be as good as voice to come up to the user expectations.

Nokia's chief executive Jorma Ollila acknowledged the challenges ahead, calling it "a big paradigm shift." Data were new, and that changed everything. Mobile phone companies were first taken off the guard by the surge in voice services; they were surprised even more by the success of prepaid phones and text messaging. While struggling to match demand, wireless operators could do no wrong, as very little marketing was needed for a voice-centric environment. Next-generation mobile services, particularly data ones, were far less likely to be a success by similar fluke.

Beyond this mobile maze was a remarkable unanimity on the belief that the combination of wireless and the Internet would lead to a new revolution that could dwarf the one initiated by personal computing. Although we didn't see an explosion in wireless data, a fundamental shift was still quite evident. The sheer volume of SMS proved there was an appetite for mobile data services, and Japan had shown the world what could be termed as a crystal ball for the future of the mobile Internet. The mobile data market, which looked like a poor stepchild of wireless voice in its early years, began making significant strides once the notion of personal communications came into a holding pattern.

Clearly, there were all these pieces of action happening in a relative isolation. Some industry analysts owed i-mode miracle to Japan's growing band of data junkies who spent most of their time, outside homes and offices, on commuter trains. And according to them, text messaging was only popular because it was remarkably cheap. But then what could be said about South Korea, where wireless web services began to thrive as data-capable handsets became affordable by 2002. Over there, it didn't happen just because the content was attractive right from the start, but also because of the availability of slick new handsets for whom downloading games, cartoons, and richer-sounding musical ringtones were made so easy.

The real issue was not i-mode versus WAP, but the realization of a new type of data services. A lack of applications, for instance, was a big complaint from consumers during the launch of WAP. Wireless handset makers, new-comers in the software territory, rushed to sell steroid-injected phones while ignoring the potential trouble spots. Consequently, an embarrass-ing episode of software glitches raised questions about the software acu-men of mobile phone companies. Wireless industry probably believed in a more simplistic notion of the mobile Internet: build a network platform and let all sorts of capabilities or applications grow on it. Just like when Bob Metcalfe built the Ethernet, all he was looking for was to link a few computers to a printer at the end of the hall at Xerox PARC.

But such simplistic notions were in a stark contrast to the futuristic visions that WAP backers instilled in the minds of the consumers. In fact, WAP became a case study on the dangers of overhyping the technol-ogy. Despite all the debacles, however, the mobile Internet offered some tangible benefits by making mobile devices connected anywhere, any-time. And WAP was still a milestone in the coming together of mobile phones and the Internet. So the promise of wireless web was not lost despite enormous engineering and commercial challenges.

In many ways, the mobile Internet during the mid-2000s was at the same stage of development as the wired Internet was in 1995. Nobody really knew which technologies or business models would win or what consum-ers or corporate users wanted. The only difference was about expecta-tions. Some people wrongly expected the mobile Internet to be the same as the wired version, only mobile. Here, it'd be worthwhile to remember that the Internet had been around for more than fifteen years before it became usable on a wide scale. Scroll back to the early days of the Internet when web 2.0 wasn't there and most people used CompuServe, AOL, or Prodigy—all on dial-up connections—and endured limited content and network capabilities, browsers with primarily text-based services, no graphics, and a few bits of information posted by even fewer people.

The wireless industry had to start somewhere. What mobile Internet went through over the period of a decade after its inception is somewhat reminiscent to wired Internet's journey into the commercial realm. The notion of wireless Internet would subsequently turn into a made-for-mobile information highway where one could get on from nearly anywhere, and it'd continue moving in faster speeds. The next chapter encompasses the coming-of-age story of the mobile Internet sparked by Apple's celebrated entry into the mobile arena in 2007. As popular as i-mode was, it was nothing compared to the next i-product that would come along. The mobile Internet—after its humble beginnings in the late 1990s, its growing pains, and its misses and hits—was finally ready for the prime time.

5 MOBILE INTERNET 2.0

"Why shouldn't you be able to use a data communications device that you can take anywhere you want to go?"
—Martin Cooper, who evangelized cellular phones back in 1973, saying that people could carry them around in their hands, as opposed to AT&T's idea of using them as portable car phones

When Steve Jobs revealed the iPhone at the Macworld Conference and Expo in San Francisco in 2007, hardly anyone anticipated that this sleek new gadget would set off the mass adaption of handheld Internet access. The iPhone turned out to be the breakthrough Internet communications device that the wireless industry had been longing for since a decade. Until now, mobile phones ran on disparate pieces of software, had less memory, and operated under the constraints of pay-per-byte wireless networks, and consequently, the mobile Internet we had was a mere stripped-down version of the real thing. Apple understood better than anyone that instead of leisurely browsing and searching like on a PC, the mobile Internet would take off by focusing on sending and receiving timely, relevant nuggets of information.

The way i-mode created the first mobile Internet success story by balancing technology excellence with content innovation in fact provided a classical case study for future projects. The phenomenal success of i-mode boiled down to NTT DoCoMo's ability to understand the constraints so well and then make the best of what was available at that time. And the fashion in which DoCoMo conceived a functioning ecosystem for i-mode to succeed by letting small outfits run the content show was also quite remarkable. In retrospect, it would be safe to assume that Jobs and company learned all the right lessons from the success of i-mode and the failure of WAP. Because that's exactly what we witness in the mobile Internet's renaissance set off by the Cupertino, California–based company a decade later.

Probably better than any other player in the market, Apple understood that one of the biggest draws for smartphones is their ability to access the Internet without having to find a PC. Because handsets with small screens unsuited to Internet browsing, it took Apple's fanatical attention to usability to bring Internet-connected devices to the masses. Apple borrowed the PDA's large-screen prerogative and the use of icons from personal computers, and inspired the next-generation form factor for mobile phones. Moreover, Jobs and his comrades knew very well that GSM pipes were too narrow and accompanying wireless modems were too cumbersome, so they waited for the fat pipes that could transmit web pages, pictures, and other data at higher speeds.

But probably the most critical lesson that Apple seemed to have fathomed was that the software obligation was growing in the smartphone realm. So, first, Apple carefully calibrated its mobile Internet offering with the introduction of Safari, a graphical web browser, which was also the native browser for the iPhone iOS software. It was the first mobile browser to display mobile web pages identical to those displayed on a desktop computer without sacrificing usability. Previous mobile web browsers had been unable to capture web pages in the same format, often resorting to limited web viewing or incorrect formatting. Apple had developed the original Safari browser as part of the Mac OS X operating system. Remarkably, what worked so seamlessly for Apple had previously turned into fiascos for its old adversary Microsoft. That's probably because Apple chose to adapt a well-thought-out, holistic approach instead of going for disparate pies-in-the-sky.

Another crucial lesson that Apple must have learned from the WAP debacle was that browser was not end-all, be-all in the mobile Internet game. WAP was, in fact, a crude form of web browsing on mobile phones. So to make the Internet truly useful, the iPhone would rely on specialized applications called widgets, which combined device data on information such as the current location with online data and services. These small applications ran on mobile devices and accessed information on the back-end. That way mobility brought new aspects to the Internet, enabling cellular handsets to offer a truly comparable Internet experience to the broadband access that people had at their personal computers. Widgets became the key enabler for the Internet-on-the-go.

In the quest for the riches of smartphone, mobile Internet was the starting point. The notion of mobile Internet had long been an article of faith among the optimists in the wireless world. To lay claim on mobile Internet, however, wireless industry had to stretch far from its core telecom-centric markets. It was a fascinating idea on paper, but as we saw in the preceding chapter, mobility wouldn't come without a fight. Now after the ascending of the iPhone, in the smartphone business, the lines were becoming increasingly clear. The coalescence of mobility and the power of the Internet that visionaries had been dreaming for a decade suddenly became a reality. The humble radio was finally poised for the Internet touch on the mass market. Again, Apple's marvel of buzz marketing in the making of the iPhone archetype was somewhat reminiscent of DoCoMo's i-mode marketing triumph.

The technology press in the late 1990s was filled with predictions that most Internet activity will occur on mobile devices. And many people became agitated when the bonanza didn't occur within a year or two. The prediction had finally come to pass, thanks to the iPhone. Now, in the post-iPhone wireless world, the mobile Internet market was taking off even faster than predicted. These were the years in the late 2000s when mobile Internet adaption actually happened eight times faster than web adaption on the PC during the mid-1990s. No wonder Apple quickly began grabbing the market share. At this very junction when Apple was redefining the mobile handset through a knock-out web experience, industry watchers began to wonder if anybody could stop the Silicon Valley computing icon. Enter Android!

GOOGLE'S SECOND ACT

Android enabled handset makers like Motorola and Samsung to develop credible rivals to the iPhone, and when these companies started gaining traction, Apple's momentum began to slow down. But why would Google create its own mobile phone platform in the first place? The simple answer is that the Internet dynamo—which gave away a ton of services like Gmail, Google Docs, Google Reader, etc., and in turn, raked in tons of advertising revenue because of these and other services—wanted to take these services even further. There is strong evidence to suggest that in search and advertising domains users are likely do fifty times more data queries with a smartphone than on a basic wired connection. So the future of Google would be influenced through mobile devices one way or the other. And Google wanted to ensure that it positions itself for the mobile future with the development of its Android mobile operating system.

Initially, Android looked less about money and more about iPhone disruption. But the driving force behind Android, according to Google founders, was the vision to produce phones that are Internet-enabled and have good browsers because that didn't exist in the marketplace at the time of Android's entry into the smartphone fray, which more or less coincided with the launch of the iPhone. In a nutshell, it was imperative for Google to translate its web dominance into the mobile arena if it wanted to remain relevant. Google, lacking Apple's consumer market know-how, had its own struggles in adapting to the new world of smart devices. But while Apple's overwhelming influence on the media world was probably its biggest strength in the mobile Internet realm, Google's edge was better technology. Android's secret weapon was really the network effect.

The Apple-Google rivalry wasn't a mere squabble; it came down to the future of the web and subsequently Google's place in it. The industry was on the verge of a mobile revolution—riding on the back of smart-phones—as the biggest wave ever was to hit the world of computing. Just as mainframes gave way to minicomputers, which in turn gave way to personal computers, the PC was now being seen as displaced by smartphones and tablets. The smartphone users were download-ing services and apps in large numbers and every device sold was to generate an ongoing revenue stream. Former Morgan Stanley analyst Mary Meeker forecasted in 2010 that, by 2013, the mobile Internet ecosystem—capital spent on access fees, mobile-commerce, paid services, and advertising—will be worth more than half a trillion dollars per year.

Mobile phones were reaching into the furthest corners of the world, and by 2011, according to a study from Yankee Group, 5 billion mobile phones would be in service out of a total world population of about 7 billion. Though most of these handsets would be "feature phones" with limited capabilities, during this decade, the technologies would become so cheap that virtually every phone sold will be what today we call a smartphone. And every one of these smartphones would be constantly connected to the Internet. So what happens when most of the inhabitants of this planet carry a gadget that gives them instant access to pretty much all of the world's information? The implications for almost every aspect of life are dizzying.

Not surprisingly, therefore, this is a battle for literally every person on the planet and that's why companies like Apple and Google see this market as worth fighting for. Both Apple and Google are headed by recognizable visionaries. For Google, the desktop metaphor is fading. If phones are going to replace PCs as the main gateway to the Internet, why would consumers tether themselves to a PC, especially when phones are growing more and more powerful and becoming cheaper? That obviously made the online advertising giant very keen for dominating the mobile web experience the same way it ruled the desktop Internet domain. Google discovered the crucial reality just in time that it needed to find real success in the new world of the mobile Internet and become part of the next major evolution of the web.

There was another intriguing dimension of the iPhone versus Android juggernaut. Both handset manufacturers and wireless network operators were increasingly becoming commoditized, and they were desperate to find new sources of revenue. Between them, the most valuable thing they had was control over what goes on the phone before it reaches the consumer: content and applications. This is exactly what Apple and Google wanted to control as the future shifted toward the mobile Internet. Google, especially, had a lot at stake. Google had to pay Apple to keep Google as the default search engine on the iPhone. So if other major handset makers would auction off the default search services on the phones they shipped, Google might have no choice but to buy their support and it surely wouldn't come cheap. So in the endgame, Google needed handset makers as allies as much as they needed its free Android software.

Inevitably, Apple and Google represented two somewhat colliding visions of the mobile Internet. Apple, through its iconic offerings—the iPhone and the iPad—was generally perceived as creating a "walled garden" where content and games needed to be approved by Apple and accessed through its App Store. Putting such preconditions on which technologies Apple would support on its devices was increasingly seen as akin to trying to control how the mobile Internet developed. For instance, Apple saying yes to HTML5 but no to Adobe's Flash. Google's business model, on the other hand, is built on the openness and the anarchy of the Internet. The Internet search kingpin makes money on the web that is completely open, so any kind of control on the content of the Internet could threaten its revenue stream.

The smartphone turf now turned into a fight for the control of the mobile Internet market, where all digital convergence is headed, where the Internet is headed, where media is headed, where computers are headed, and where digital money is headed. Although we think of next-generation gadgets sporting the mobile Internet as a phone, as established in the previous chapters, in fact they are a computer. And we learned that for computers in the new ubiquitous era it's imperative to have a complete and functional connectivity. Ironically, in the next stage of the mobile Internet, when cellular communication was being invented all over again, Nokia and Symbian were nowhere to be seen. And Microsoft, which has a habit of being late to the web party, would again be playing catch-up in the mobile Internet space.

MOBILE INTERNET REVISTED

Mobile telephony allowed us to talk on the move, while the Internet turned raw data into exciting services that people found easy to use. To make the best of both worlds, smartphones kicked off the transition from voice-only strategy, and that was followed by mobile phone industry racing for data applications. As a result, these two technologies began converging to create a new communications nirvana they called mobile Internet. The convergence of the two fastest-growing communications technologies of all time—mobile phones and the Internet— would make possible all kinds of new services and create a vast new market as consumers around the world started logging on from their Internet-capable phones.

For a start, mobile phone operators tailored a variety of online services according to perceived users' demand, and accomplished a modest success. Users could check bank balances, weather updates, traffic reports, and news headlines either through SMS or by using a clunky Internet connection. Other notable initiatives in those early days included web media giant Yahoo! collaborating with cellular operators and Internet pioneer America Online jumping the bandwagon to bring instant messaging service to mobile phones. The notion of the mobile web kept crawling on the back of rather primitive smartphones, which had begun providing a reasonably good net connection by 2002. Meanwhile, the size of liquid crystal display (LCD) on cell phones kept on getting bigger in order to accommodate more data services.

Here, it'd be worthwhile to note that PDA makers, the descendants from computer industry—who had turned into phone wannabes—were the ones coming up with mildly successful mobile Internet calls. Inevitably, these devices looked more like PDAs than phones, and therefore, these smartphones in the early going were mostly targeted at high-end business users. RIM took an early lead with its BlackBerry devices, which added phone and other features to the wireless adaptation of Internet's first killer application: e-mail.

Not long after the BlackBerry phone landed, Handspring gave birth to yet another smartphone when it enhanced its flip-style Treo 90 PDA by adding dual-band GSM capabilities. The Handspring Treo, which later became Palm Treo, drew many customers with its winningly innovative design and got an early traction in the upper end of the phone market. With organizer, phone, and Internet features all on hand in a single, compact device, the Treo smartphone was a marvel of thoughtful design. The Treo featured options for connecting to the Internet on a descent-sized color screen along with a tiny but full keyboard. It was a terrific product for both phone-centric and data-centric people who wanted it all.

Another notable product that symbolized mobile Internet's awkward adolescence was Danger Inc.'s Hiptop phone, which T-Mobile marketed in the United States as Sidekick. Hiptop, which could be used to surf the web and to send and receive e-mails, was the first phone to offer both a full QWERTY keyboard and a built-in version of America Online's popular instant-messaging service. It quickly became a status symbol among the rich and famous in the United States in those early days of the mobile Internet.

Then came the mobile Internet's defining moment in summer 2007 when Apple launched the iPhone with a clever design and some nifty capabilities, and blew up the wireless industry. The vision of multimedia wireless Internet had finally come true through the iPhone's beautiful user experience; the mobile Internet had embarked on the next phase of its metamorphosis. Until this tipping point, the notions of wireless Internet and smartphones went hand in hand, galvanized by high-profile initiatives such as WAP, Symbian, and Bluetooth. Now the wireless industry had finally found the silver bullet it'd been longing for many years and the ultimate contest for the biggest prize had begun. This was the most dynamic time in the brief history of smartphones. As Internet moved from desktop to pocket, the nature of mobile phone usage was going to go through a profound change.

The mobile Internet technology had opened an entire new frontier for newcomers like Apple and Google. For the entrenched players, however, it was all becoming too much. Palm Inc.—a smartphone pioneer and the company that launching the PDA craze—eventually went up for sale amid financial woes and was acquired by Hewlett-Packard in spring 2010. But a far bigger shake-up was coming to the door of RIM, the maker of the first-ever successful smartphone. BlackBerry was growing wildly at the time Apple rewrote the rules of the mobile Internet. The famously addictive device popularized e-mail on-the-go and turned its Canadian maker into a US$34-billion company and the business world's leading supplier of smartphones. BlackBerry was everywhere: boardrooms, restaurants, and kitchen tables. The device that earned the enduring nickname of CrackBerry represented more than half of all corporate smartphone users in the United States.

The dark little devices would vibrate every time a new e-mail arrived, delivering a tiny thrill that millions of employees came to both loathe and desire. But RIM didn't take the web experience as seriously as it should have, and eventually it began to look like a tired remnant of yesterday's technology. BlackBerry's weak browser capability became a huge issue leading to significant erosion in its customer base. Later in 2010, three years after the iPhone launch, RIM completely revamped its browser; it was much faster, toggled more easily between windows, and was better at handling RSS feeds. By this time, however, BlackBerry had lost a significant market share to snazzier rivals such as the iPhone. In September 2010, three years after Apple delivered the first compelling mobile Internet experience, the iPhone surpassed the BlackBerry in quarterly sales with 14.1 million devices sold, compared with 12.1 million for RIM.

The failure to see the rise of the mobile Internet in the commercial arena and to grasp it in a timely fashion had put RIM into a transitional moment. Industry analysts were overwhelmed with concerns about the future health of its platform. But the disruption caused by the second coming of the mobile Internet didn't only mark the end of the "BlackBerry elite"; it also became a predicament for Nokia, which had once dreamed of reshaping the Internet. Nokia had been vocal and ambitious about its role in the Internet world. In fact, Nokia firmly believed that no other company was better placed to be the standard-bearer of what its marketing people endlessly called the "mobile information society."

No one had embraced the idea of wireless web more enthusiastically than Nokia. "In 10 years time, I would like Nokia to be dubbed as the company that brought mobility and Internet together," said its CEO

Jorma Ollila back in 2000. "It's not going to be easy, but this organization loves discontinuity; we can jump on it and adapt. Finns live in a cold climate: we have to be adaptable to survive." Now the fact that Nokia, though still the world's number one smartphone manufacturer in 2011, had stumbled several times in its efforts to catch up on the mobile Internet was a harbinger of the coming shake-up. The Finnish mobile phone maker was failing to fight back the onslaught of Internet-enabled phones like the iPhone. It was ironic that the company that got the whole smartphone thing going now found itself at a peculiar crossroads. What it really needed to do was deliver the promised land of the mobile Internet, and only then, this Finnish wonder could regain its star status.

MEDIA AT YOUR HAND

In 2010, Morgan Stanley projected that in five years the number of users accessing the Internet from mobile devices will surpass the number of users who access it from PCs. Mobile phones were now the focal point of all forms of digital convergence. The Internet kingpins like Google and Yahoo! were saying that the future of the Internet is on the mobile phone. The PC behemoths such as Dell and HP asserted that the future of computing is on mobile phones. The media giants from TV to music to video games were promising all future media content to be available on mobile phones. With the advent of the mobile Internet, and by that extension, of smartphones, it became evident that once-dreamy notion of convergence between telecom and computer worlds was for real, and it also brought forward another intrinsic element of the smartphone anatomy: media.

The wireless industry had been hung up on the soap opera-like marriage of mobile phones and PDAs for so long that it took its eyes off the real story: digital content. If the mobile Internet was going to be the starting point of the smartphone revolution, then its architects should have better known that the Internet was all about the open and free flow of information. Eventually, when the real mobile Internet took off under the wings of the iPhone, then the wireless industry woke up to the battle over the control of digital content. Likewise, media industries across the spectrum answered the mobile call only after Steve Jobs reinvented smartphone at the crosshairs of the mobile, Internet, and media industries and gave smartphones a business model that actually worked. After that, all these pieces from both sides of the fence began falling in place, leading to a whirlpool that would turn both wireless and media worlds upside down much faster than anyone ever could have imagined.

In retrospect, however, the media ascendance in the smartphone fray was not that sudden. In 1998, for instance, one of the first success stories of selling media content via the mobile phone came through the sale of ringtones by the Finnish mobile operator Radiolinja. Soon afterward, other media content appeared on then-tiny mobile phone screens, including news, video games, jokes, horoscopes, TV content, and advertising. Most of the early content for mobile phones tended to be copies of legacy media, such as the banner advertisement or the TV news-highlight video clips. That was followed by unique content for mobile phones, ranging from ringtones and ring-back tones in music to video content, exclusively produced for mobile phones.

Here, given the benefit of hindsight, it could be safely said that the European WAP effort for the riches of the mobile Internet partly failed because they concentrated too much on the technology and too little on the content. Mobile phone operators were reluctant to share revenue with content providers. They failed to see that data wouldn't catch on until every party to the transaction made money. This is something NTT DoCoMo recognized early on. The Japanese firm established a model that provided revenue for every player in the information value chain.

Eventually, the mobile phone got itself acknowledged as the third screen after TV and PC; it was also called the "seventh of the mass media" with print, recordings, cinema, radio, TV, and the Internet being the first six. In 2006, the total value of mobile phone-paid media content exceeded Internet-paid media content and was worth US$31 billion. Next year, the value of music-on-phones revenues equaled US$9.3 billion and gaming revenues were worth over US$5 billion. The evolving processing power and robust software capabilities of smartphones radically enhanced content richness and helped create fantastic audiovisual effects. Available apps ran the gamut of interests that included games, news and weather, maps and navigation, social networking, and music.

For instance, if something was happening in the news, people went right to their *New York Times* or *Wall Street Journal* app. There was a clear evidence that consumers were quickly switching away from wired web pages and toward apps as their preferred way to access the Internet. Smartphones were now able to access all of a user's content and media, including work and personal documents, music, video, and so on. Some of this would be stored on the device itself; some would be stored in the network cloud. Smartphones, therefore, created a revolutionary new distribution channel, which represents the next big opportunity for the media.

With Internet on-board, the smartphone could be a person's content-consuming device—with this screen, users could watch a movie. This could also be a user's reading device—the screen was big enough for people to have a reasonably good reading experience. And all of user's reading content could go with him or her anywhere. Having access to content and being able to display and transfer large amounts of information would let people do more with their devices. That includes watching popular TV shows, listening to music, shooting and storing high-quality pictures and video, and sharing it via tools such as Flickr, Facebook, and YouTube.

Besides gaming, the next most profitable commercial apps are the ones that connect users with valued content for which they are willing to pay extra, for instance, apps for showing live game videos. For example, after the initial, low-cost downloads, annual user fees can exceed US$100 for apps like the Netflix and Hulu mobile video services. The thousands of apps built around platforms like Android and iPhone, mostly by third-party developers, encompass almost every topic. Meanwhile, the mobile Internet experience continues to optimize and innovate. Case in point: a year and a half after the launch of the iPhone, 3G version was added to improve the browsing speeds. The actual look and feel of the handset varied only slightly; most of the changes in this release were inside the iPhone.

TECHNOLOGY MEETS MEDIA

The above history of media's foray into the mobile world only entails to cell phones in general. The smartphone, on the other hand, isn't just another type of handset, but is a full-fledged computer, which comes loaded with software and doubles as digital camera and portable entertainment center. That's why the companies like Apple and Google don't see themselves as selling computers, but instead connected Internet devices—truly a sea change from the traditional PC business. Apple built one of the most successful media businesses of the Internet age: iTunes, a content distributor. Google, apart from being known as the web gatekeeper, is undoubtedly the largest media empire capitalism has ever created. And finally—both companies wearing the hallmark of Silicon Valley innovation—understand the true value of an entire ecosystem.

There is another critical dimension hidden in the contrasting worlds of technology and media. On the technology side of the Internet, there has been a lack of profound understanding of media formats as technologists are often wary of any involvement with content. Conversely, on the media side, few people know about technology. This led to disconnect between two fundamental building blocks of the mobile Internet fray. Steve Jobs, the Macintosh Man, had built two of the most successful media businesses of recent times: iTunes, a music content distributor, and Pixar, a movie studio. So he perfectly filled that void. But probably not even Jobs and Apple's senior executives would have foreseen either the speed or the depth of the transformation that would eventually take place at Apple and the industry at large.

Apple has a history in media business that goes as far back as 1984 when it defined the fundamental elements of desktop publishing on the Macintosh platform. About two decades later, Apple's industry-changing music player iPod effectively killed the promising portable media player (PMP) market. The iTunes store, which arrived in 2003, allowed users to download music from the Internet for their iPods. However, by adding the Digital Rights Management (DRM) to its offerings, Apple forced the user to only use it for iPods. The iTunes is one of the most successful software packages in history, installed on more than 125 million computers worldwide and used for about 70 percent of all digital-music purchases as of end 2010.

Apple now sold TV shows, movies, games, and even university classes through iTunes. The company also used iTunes for a medium that could be its trump card: mobile apps. With iTunes being so well integrated and music players that seem more like cultural icons, the iPhone came along as one of the best phones for music and general media capabilities. The iTunes having created a frictionless distribution system made getting an app on a consumer's phone as easy as getting a song on iPod. The fact that apps became synonymous with music on iTunes was a harbinger of new opportunities for brands and media companies. First, the onset of the iPhone transformed the mobile into a unique channel for media, and then in 2010, Apple's iPad sensation was aiming to save the magazine and newspaper industry.

Publishers have put a lot of faith in devices like the iPhone and the iPad. As traditional media struggles to adapt in an evolving Internet-centric landscape and its audience is dwindling, there is a strong drive on how this extraordinary mobile medium of smartphones can connect to audiences and make money. Still, the business model pioneered by the likes of the iPhone and the iPad is not without initial stumbling blocks. While publishers want their mobile content to be monetized with advertising, it's the issue of control over the medium as both publishers and companies like Apple would want to drive the overall business process. Nevertheless, the advent of media on the smartphone platform is producing all kinds of new possibilities in content access and playback domains through browsers as well as native apps.

Smartphones with the power to access digital content could have a profound impact on people's lives. While mobile initially doesn't have the volume of the wired Internet, it does have more engagement and targetability. The mobile Internet is also a lot less cluttered than its wired counterpart; there is one ad per page on mobile versus up to five competing for attention on the wired web. Moreover, the personal nature of the mobile Internet makes the smartphone a lot more intimate as compared to the online and the TV experiences. For an Internet-enabled smartphone, it's the users' screen, they see it every day, and that makes a lot of difference.

Both wireless and media worlds were gradually coming to terms with swaying each other's territories in the new order facilitated by the mobile Internet being launched on the smartphone platform. But then the media conundrum on the mobile screen would follow yet another showdown— a clash between two conflicting visions of the Internet in your hand.

THE INTERNET VERSUS THE WEB

In 2007, when Steve Jobs demonstrated the iPhone to a stunned audience, Apple endorsed only a single way for having mobile Internet functionality onto the iPhone: via web browser. Almost immediately, a jailbreaking community came into existence as developers—frustrated by some of the limitations of HTML and Safari mobile browser—created and distributed their own native applications. Apple didn't approve or support this in any way and that led to perceptions about Jobs's assertion that websites are more than enough. The term "native app" is used for applications designed specifically for a platform, in this case a mobile phone operating system like iOS.

Some industry circles have this opinion that, in the start, even Apple didn't anticipate that native apps from third-party developers would become the single most-requested feature from consumers in the months following the iPhone's launch. They opine that the company had this skeptical view that consumers would only embrace web apps, making native apps little more than a curiosity. Still, Apple was prepared to roll it out to have third-party developers on its side. But if you read between the lines, the Apple press release dated June 11, 2007 has an apps blueprint written all over it. Moreover, according to some media reports, soon after i-Phone's launch, Apple was already playing with a software development kit (SDK) for developers to create native apps for the iPhone.

Fundamentally, there are two things apps do well that the web doesn't: process information and secure transactions. In a 2010 *EE Times* story titled "Mobile Internet apps moving e-commerce off the Web," R. Colin Johnson discussed in detail how apps offer a faster connection to the information that people most want and need, and apps providers do that by cutting through the clutter of data. For users, apps offer speed to information without having to mess with browsers, search engines, and URLs. As for security, with web pages, we are subject to spoofing attacks plus all the vulnerabilities of the browser we are using. For example, data could be copied and pasted between windows. Apple's iTunes led the way in demonstrating that apps could be locked down. Apple claims a flawless security record despite having more than 150 million credit cards on file, processing 230,000 new iTunes activations per day, and serving a total of 120 million devices in the field as of 2010.

Of course, apps are only as good as their authors, but when done well, they bypass all the messy sifting required to mine the worthy information from the billions of web pages cluttered with banners, flashing animations and unwanted pop-up windows. Apart from the fact that native apps provide improvements in performance and storage, they could also work offline. Moreover, placement of apps on the phone's home screen made them more likely to drive repeated usage. In a nutshell, apps presented content providers a simpler experience that took better advantage of the phone's hardware features by exploiting the platform's built-in technology. Apps are programmed separately for each mobile platform, much as game code is tailored for specific gaming consoles, and as a result, apps fully utilize the special media-handling capabilities of each handset. However, as we'll see later in this section, this trait cuts both ways.

By 2010, essentially most of the phone-based Internet usage was coming from Apple and Android mobile devices. Apple had launched a new wave of the mobile Internet growth on a platform that now largely bypassed the browser and many saw this apps revolution as spelling an end to Google's web domination. Because the handset screens were smaller, mobile traffic suited well for specialty software, mostly apps designed for a single purpose. For the sake of an optimized experience on mobile devices, users were willing to forgo the general-purpose browser. So they used the Internet, not the web. And that led to a new conflict for the riches of mobile Internet gold. The new mobile Internet now found itself riding on two parallel tracks: legacy browsers for web-based environment and new-age apps pioneered by the iPhone.

Although Microsoft's Mobile Internet Explorer, Apple's mobile Safari, and Android's Chrome-lite were being refined over time, there were legacy issues that they inherited from the first-generation, freewheeling

Internet. For instance, while Apple's iPhone allowed users to browse any website that supported Safari, Microsoft's devices required that sites be built for its Mobile Internet Explorer. And the browser on Windows Mobile devices wasn't a standard browser, so it didn't work with a lot of websites. Not surprisingly, therefore, four out of five developers said that their users prefer native applications to mobile websites because of user experience expectations.

The trade media and public at large tend to use "the Internet" and "the web" interchangeably for the same network; the web, of course, is an application that runs on the Internet. And the web has been the Internet's killer app, transforming the network of networks created by the U.S. Department of Defense for collaborative research into a global phenomenon having reached more than 2 billion users. On the mobile platform, however, native apps grew much faster, up from virtually nothing in 2008 to more than 7 billion downloads in just two years. Here, technology history provides an interesting insight: even in the early days of wired Internet, users preferred apps over browser tools. Again, the reason was speed because an app is much faster.

APP STORE ENVY

Apps culture was transforming web's commercial cravings to more trustworthy, easier-to-use services, thus reserving the openness of the web for its original purpose: the casual consumption of information, free from the security and privacy protocols and facilitated by a browser interface otherwise known to be chock-full of vulnerabilities. It did that by excising from the web and creating complex and security-conscious commercial applications. One of Apple's motivations with making iTunes an app vehicle was to have a consistent user experience, which would not be possible with all the different browsers on the web, but the second reason was the much cleaner security that could be provided with a proprietary apps platform.

Apple's adversary RIM was initially critical of the ecosystem of applications for the iPad, the iPhone, and the iPod Touch, saying that users didn't necessarily need an app while they have the web. The maker of BlackBerry acknowledged that there was a role for native apps, but was adamant that the web browser remained the best way of getting information on mobile devices. To bring the mobile to the Internet, it countered, users didn't need to go through some kind of software development kit; instead, people could use the existing web browsing environment. But RIM, like other smartphone makers, was forced to go the apps route in the quest to offer a viable mobile Internet experience to its customers. Even though Google's preference was for services accessed directly over the web, rather than through pieces of specialized software, it was forced to follow the suit as well.

Smartphones upended the software world, and overnight, everyone wanted to build games and apps just because the iPhone had offered exciting new ways of using this technology. Now, apps were seen as essential in order to build and expand a thriving mobile ecosystem. The apps culture quickly spread to app stores for the Android, the BlackBerry and Symbian operating systems, creating a parallel universe of iTunes-like walled gardens that could grow to be just as expansive as, but much more secure than, the web.

Especially, Google needed to find real success in this new world where the Internet was evolving to suit mobile apps. Google boasted one of the best performing browsers on a mobile phone. According to Andy Rubin, the tech whiz who oversaw the Android empire, it was the fastest and it was the smallest. And he vowed to add more functionality to the browser to give it an updated user experience. Evidently, Google rushed to get beyond the world of apps ushered in by Apple, toward a category of devices more suited to its own technology preference; ones that are built exclusively to channel web services. The result was Google's second operating system, Chrome, which it developed around its lightning-fast web browser. The search giant, attempting to lead the web movement, was now pushing a web-only regime through its Chrome OS, its browser-based operating system that would run only web applications.

If smartphones continued moving toward Apple-pioneered app-store environment that require unique development for each platform or mobile OS, web service models wouldn't remain viable anymore. Especially, when Apple had demonstrated a proven revenue model in which users were comfortable making app purchases. On the other hand, however, software developers couldn't afford to support every operating system. So as the mobile OS space becomes increasingly crowded, mobile app developers with small staffs would become far more interested in designing one-size-fits-all in-browser web-based apps that provide mobile users experiences comparable to native applications. Take the case of Microsoft's Windows OS for desktop computers: that it became a de facto standard really made it easy and relatively cheap for software developers, and that has been a tremendous benefit to society since the mid-1990s.

So when Google engineers began advocating a new architecture that would blend the web and native apps, it came as a welcome relief almost across the board. Such a mash-up would enable developers to create an application for the mobile phone and then have the freedom to emulate it as a web application. In other words, if a standard web browser can act like an app, offering users a similarly clean interface and seamless interactivity, users perhaps will resist the trend to the paid, proprietary apps. For this very premise, the wireless web has its hopes set on an upcoming technology known as HTML5, designed to handle audio and video internally without the need for browser plug-ins like Adobe's Flash and Microsoft's Silverlight. It has gained a significant attention as a web-building code that has both Apple and Google behind it.

HTML5: THE EVOLUTION OF BROWSER

For Adobe, creator of a widely-used technology called Flash, which manages video and animations on many websites, fame came through a theatrical feud with Apple when Steve Jobs called Flash software a buggy, battery-sucking relic of the past. Flash is Adobe's highly popular platform for displaying interactive graphics, animations, and multimedia within a web browser. According to Adobe, 98 percent of desktop computers support Flash, which subsequently has led to its widespread use among software application developers. Jobs, on the other hand, called Flash the number one reason for Macs crashing; he didn't want to reduce reliability on iOS products like the iPhone and the iPad. No Flash, however, meant that the iPhone browser would be incapable of displaying a large portion of the web.

As free Flash was not supported, videos couldn't be streamed from the vastly popular television and movie sites like Hulu, because websites that used Flash to render content or navigation won't work on the iPhone. Adobe's answer: "when you're displaying content, any technology will use more power to display, versus not displaying content. If you used HTML5, for example, to display advertisements, that would use as much or more processing power than what Flash uses." But it wasn't just Apple; Google and even Microsoft wholeheartedly embraced HTML5 and dubbed it the future of the web. Technology companies and application developers alike proclaimed HTML5 is the cornerstone of web's second act.

Authored by Google employee Ian Hickson, HTML5 promises to be the "genes" from which much of the next-generation web will spring to life: websites, content, and web-based apps could be partly or wholly coded with it. One of HTML5's biggest sells is its flexibility. Regardless of what device a website or app shows up on—iPhone, Android tablet, desktop, laptop, all of which have decidedly different form factors—developers can use almost the exact same software code for each device platform, which cuts down on the time, effort, and manpower required by many mobile app developers to program native apps for the crowded mobile OS market. And users won't have to launch dedicated iPad or iPhone apps to watch their favorite shows—they'll watch content right inside their browsers instead.

At the height of the debate over the longevity of native software apps against the power of the web, Mozilla, creator of Firefox browser, claimed that its new browser for smartphones will contribute to the death of mobile app stores. "In the interim period, apps will be very successful," said Jay Sullivan, vice president of Mozilla's mobile division. "Over time, the web will win because it always does." Web proponents such as Mozilla and Google vied for Internet standards that would enable any app to run on any device, just as Java proponents touted a "write once, run anywhere" mantra in the 1990s. Ironically, here, they quoted Adobe's Flash as the software that had emerged as a cross-platform environment for creating animations, games, and apps for the web.

But many complained that software programs like Java and Flash exhibit bugs, performance problems, and security vulnerabilities, among other issues. Java's promises of universality didn't quite work out because different implementations of the Java Virtual Machine meant that Java coders needed to rework their apps for each target device. Web proponents, however, maintain that the wider acceptance of next-generation Internet standards, particularly HTML5, will win out where Java failed. After all, there are tons of applications that could be delivered through the browser at stunningly low costs. During the web's supremacy spanning many years, the browser became a platform of choice mostly due to economy of scale reasons.

As we just saw in the preceding section, both Apple and Google, with very different operating ethos, got squarely behind the same developer language: HTML5. Google's affinity for this free-for-all open standard is a no-brainer. The HTML5 could shift the focus from proprietary software offerings like apps to web-based tools and web environment is really the core of Google's business model: Gmail, Docs, Maps, and so on. In other words, anything that is web-based gives Google the opportunity to do what it does so well. Moreover, HTML5 as an open web standard would enable every browser implement Google product features nearly the exact same way and the development will be quicker and more cost-efficient.

But it was Apple's decision to not include Flash for the iPhone and the iPad tablet that really set the stage for the rise of HTML5. On the surface, Apple's stance might seem counterintuitive, considering iTunes, which includes the native Apps Store, continues to thrive with some 160 million registered users as of 2010. HTML5 could pave the way for the dominance of web-based solutions over natively developed apps. Research firm Gartner estimated Apple to garner US$4.5 billion in revenue from apps in 2010. So why push for an open web specification that could one day affect the company's bottom line? Apple must have found the answer walking in the corridors of technology history where companies going against the flow of the open standards often ended up as losers.

Moreover, for Apple, preserving user experience is of utmost importance, and HTML5 is poised to become the viable platform to solve many of the problems miring the web. Apple seems so confident in its ability to offer a superior user experience with the iOS platform that having a new web platform that provides a level playing field and is not run or monopolized by a single company is still a safe bet for it. On the contrary, if Apple can't technically lay claim to it, someone else will, and it could well be Google or Microsoft. Also, while the hype behind the quickly-emerging web specification had reached a fevered pitch by 2010, HTML5 and its related technologies still have some growing to do. So until HTML5 catches up with processing and bandwidth challenges, the mobile Internet playing field will remain more suited for widgets.

In terms of web video, HTML5 technology had already started to take off by the close of 2010; 54 percent of web video became available for playback in HTML5, a significant rise from a meager 10 percent in January 2010. But though it's getting better in terms of overall performance, the embryonic technology is still miles behind native applications on mobile devices. An iPhone app could outperform a web app doing the same thing by up to 100 times. There hasn't been much of a debate on HTML5's merits in shaping the web's future as it'd enable browser to rival apps with simpler tasks, but it might take some time before it could handle large data sets and low-latency type situations to be able to do the types of local storage that we see native apps doing. The sky is the limit for HTML5, but in all likelihood, it won't be ready for the prime time before we are well into 2012.

THE UBIQUITOUS INTERNET

Nineteen years after Mark Weiser envisioned the concept of ubiquitous computing and fourteen years after Apple's ill-fated attempt to turn this dream into a reality in the form of the Newton, its descendant, the iPhone completed the unfinished business, translating the vision of ubiquitous computing into a commercial proposition. In its very essence, ubiquitous computing needed a ubiquitous communications system, and the iPhone just filled that vacuum in 2007. The mobile Internet had reached an inflection point from where it'd start its quest to become ubiquitous. It's the notion of true Internet ubiquity that promises a world based on the certainty that wherever someone is, he or she is constantly connected to the Internet and its services. And once Internet access becomes truly ubiquitous—as taken for granted as electric lights or running water—it would become as much a fundamental part of infrastructure as bridges, roads, and tunnels.

The early stage of the mobile Internet was mired by mobile phones with screens that were too small and had low resolution; their keypads were diminutive and they couldn't display all formats on the web. Apple meticulously worked out these challenges and that's how the iPhone became an inflection point for smartphones. Fast-forward to 2011 and we see super-smartphones in the works with much larger on-board memory and faster multicore processors. If trend lines are extrapolated, smartphones would most likely be the conduits for majority of us to access the Internet in the coming years. It's that threshold point that is shaping the smartphone into the ultimate gadget of twenty-first century. From education to healthcare, energy to battlefield, there is hardly a discipline that doesn't have smartphones on its product roadmaps.

The smartphone embodies a fusion for all sorts of portable devices—from camera to music player to gaming gizmos—and now that it has been seamlessly connected to the Internet ecosystem of apps, services, and content, having a device that is always connected and is always with a user could radically shift the wireless paradigm. These new-age devices will provide us the ability to store and retrieve mountains of information and to perform tasks like navigating unfamiliar terrains. Smartphones equipped with cameras and Internet connections would be akin to mobile surveillance points and what would the world like with billions of such surveillance points.

The mobile Internet had finally unleashed the huge transformative potential that smartphone designers had envisaged, and from here, there was no looking back. Now that Apple had crystallized the concept of web-ready phone, the question is where the mobile phone will go from here. Probably, a simple answer would be that evolution will go on for now.

Despite all its exuberance, the truth is that the mobile Internet is still in an early stage of its development. After all, the most used service on mobile phones is still the voice call and the second most used service is text messaging. In 2010, these two offerings accounted for nearly 90 percent of consumed services and overall revenue generated by mobile phone industry. Moreover, a smartphone, while serving as an Internet node, would be a computer that's much more personal than mobile phone of 2000s. So the computing game is very turbulent right now for the stakes in the soul of the new digital frontier. Furthermore, it's worthwhile to remember that it was the critical mass that defined the success of the wired Internet, and so would be the case for the heydays of the mobile Internet.

Now, however we put it, things are changing in the mobile realm. The smartphone has marked a new era in portable electronics with connectivity and that connectivity came via the Internet. With the power of the mobile Internet on its side, smartphones are now establishing themselves as personal supercomputers. The coming together of mobile phones and the Internet is going to unleash the next tidal wave in consumer electronics that pundits have been talking about since almost a decade. Truly ubiquitous wireless broadband will be much more than a mere combination of media and communication platforms. It could become the nervous system of civilization, enabling a thin mesh of always-on smart devices that coordinate and facilitate almost every aspect of human life.

According to a 2011 study from Morgan Stanley, the growth curve for the mobile Internet could be around twelve times as steep as for the desktop Internet, which we remember how transformative that was during the 1990s. The report called the speed of the mobile Internet take-up a revolution the likes of which people haven't seen before. There are three basic components of the Internet-enabled smartphone make-up: the mobile hardware, the mobile operating systems, and the web-based ecosystem. Chapter 4 and 5 of this book encompassed web-based ecosystem under the banner of mobile Internet phenomenon. The next couple of chapters will delve into the other key tenet of smartphone software: operating systems. And in the following chapters, I detail insight on the hardware side of the smartphone and how it evolved over the years.

SOUL OF THE SMARTPHONE

"Software is like making a movie and building a sky-scraper. You're not quite sure how it's going to stand until it comes out in the end."
—Joe Belfiore, Microsoft's man in charge of Windows Phone 7 project and former leader of its Zune division.

At the height of the Internet boom, in July 1999, luminaries of technology and media businesses gathered at Sun Valley, Idaho for a high-tech retreat. Over there, according to a story published in the July 9, 2001 issue of *Bloomberg Businessweek*, Microsoft boss, Bill Gates, approached Nokia chief, Jorma Ollila, to discuss the mobile Internet and how their respective companies—one a titan in software, the other in mobile phones—were preparing for it. Gates was hoping that the Finnish handset maker could incorporate Microsoft software in its phones. "How come we don't merge our efforts with Nokia?" he later wrote to his colleagues. But Ollila and his comrades had seen long ago how Microsoft virtually took over the personal computer industry, so even when Ollila and Gates chatted, the Finns were moving to keep the PC behemoth from Redmond, Washington, at bay.

Phones were to become mobile computing platforms with a collection of operating systems, application suites, browsers, and user interfaces. So more and more importance now resided in software. The notion of the smartphone had erupted on the wireless scene in 1998 when Symbian promised to transform the whole idea of mobile gadgetry by launching phones with PC-like software programs and e-mail functions. But Symbian by no means marked the beginning of "smartphone," a catchphrase for handheld computers that in various combinations could send and receive e-mail, browse the web and make phone calls. In spring 1995, a year before the launch of Palm Pilot, Jeff Hawkins had visited Ericsson's headquarters outside Stockholm to present his views on a mobile phone that could evolve into a smartphone. The Ericsson managers were intrigued.

At around the same time, Palm managers had been wooing Motorola who was looking for an operating system for a simple, handy two-way communicator that it could brand as an "instant messaging in your pocket" device. Motorola had been successful in selling small electronic gadgets like pagers and cellular phones. Now it wanted to develop a smart pager, a gadget that would be able to both send and receive messages using a tiny keyboard. The Schaumburg, Illinois–based firm had become a dominant supplier of mobile phones worldwide just a few years ago. But Motorola, now among the last few vertically-integrated electronics giants, had seen its market share shrivel as it misread consumer demand and tripped over its own feet in developing new products. So on January 1, 1997, when Christopher B. Galvin became the third Galvin to lead the company his grandfather Paul Galvin had founded back in 1928, he immediately began focusing on the new opportunities in the wireless Internet realm.

After the advent of digital wireless telephony, Nokia caught the tsunami in the 1990s and gained a formidable market position. The next wave was the mobile Internet and Nokia's future seemed to depend on making the right call for the next generation of data phones. The drive to the mobile Internet was pushing Nokia from the simple, radio-based handsets it knew so well into the dizzying world of computers and consumer goodies all of which would communicate over the wireless Internet. Here, at this crossroads of digital convergence, Nokia was to smack into powerhouses like Microsoft and Sony who were making inroads toward similar goals. A report published by Nomura, an investment bank, questioned Nokia's ability to manage the transition to wireless data. "Is Nokia the IBM of our times," it asked back in the late 1990s.

The challenge that Nomura had pointed out was real, though the analogy looked flawed at that particular time. IBM became ossified after sitting atop its industry for decades. Nokia, by contrast, was no stranger to reinventing itself. Nevertheless, Nokia heeded the call by evading the "Nokia-against-the-rest" temptation that could translate into unavoidable technology risks. The market predicament led to the formation of the Symbian alliance, which in many ways was reminiscent to the inception of WAP initiative, owing its life to the industry clout of the wireless triumvirate made up of Ericsson, Motorola, and Nokia. And Motorola was a Johnny-come-late during the formation of Symbian just like the WAP affair.

Naturally, the top three wireless makers wanted to stay in control and keep all the money in the family. However, with the heyday of the 1990s behind them, the wireless troika was facing a strong competition from handset makers in Japan and South Korea. The smartphone was indeed a new class of device, and though it resembled ordinary cell phones, that didn't give Nokia and other leading handset makers a birthright to the market. The mobile Internet run was seen as about to sweep the wireless handset industry that was driven by incumbents like Ericsson, Motorola, and Nokia rather than outside Internet competitive forces. To lay claim on to the mobile Internet, these incumbents were stretching far from their radio roots and that quantum leap in technology was adding to the uncertainty.

The new data-centric gadgetry was to be based on Internet-style packet switching rather than circuit switching of conventional voice networks.

Even the least sophisticated handsets could be far more complex than high-end phones of the 1990s. Nokia's own reentry into the commercial smartphone market with the launch of US$600 Communicator 9290 was hailed as a weak start. The Nokia 9290 was a combination of mobile phone, organizer, e-mail, and web device with a color screen and a keyboard. The champion in designing cool wireless handsets had essentially rebuilt a Psion device as a mobile phone that made some initial grounds in Europe. The Nokia 9290 opened like a book and looked more of a little laptop that the user dug out of a briefcase or large purse, rather than a phone kept close at hand. Beyond its bulk, it was slow both in PC synchronization and in downloading e-mails.

There were other factors, too. What probably motivated Symbian's parents more than anything else was that as computing and wireless converged, they didn't wish to end up like PC makers—low-margin assemblers that were little more than a distribution channel for Microsoft's intellectual property. Microsoft's hypercompetitive market drive was apparently a non-starter for the wireless industry. Microsoft, however, was not easily thwarted. It started seeking handset partners and licensees, and joined hands with wireless operators like British Telecom and NTT DoCoMo. In the meantime, the wireless industry's disillusionment with Symbian continued.

THE MICROSOFT WAY

"A computer on every desk and in every home" had been the mission statement of Microsoft for many years. By the mid-1990s, however, much of the focus of personal computing shifted to the Internet, where Microsoft was not a clear winner. Furthermore, a new category of mobile, handheld devices was shifting computing further away from the desktop PC. Microsoft reckoned that computing would no longer be confined to the desktop, which the software firm controlled thanks to the dominance of Windows operating system. So it came up with

"Windows Everywhere" strategy to extend its reach into both high-end servers via Windows NT and small portable devices through Windows CE software. The titan of the computing industry had set its sights on entirely new markets, so it amended the statement to "Empower people through great software, anytime, anyplace, on any device."

Microsoft had scrapped two projects before it announced Windows CE operating system at the Comdex show in 1996. WinPad was a stripped version of Windows 3.1 and Pulsar was a paging operating system with little computing power. Windows CE, on the other hand, was a multipurpose operating system for portable consumer devices as compared to Palm, which had been specifically designed to serve handheld computers. The acronym CE stood for consumer electronics. What's most crucial about Microsoft's foray into the portable device world was the fact that handset makers had seen from the fringes how Microsoft virtually took over the PC industry helping itself a heaping share of profits. Handset makers worried that if Microsoft grabbed the software business, they would end up selling dumb terminals that were only valued for the software they ran.

Microsoft's strategy initially came out as disappointing as Windows CE received a lukewarm response among the consumers as well as the industry. But the PC software powerhouse continued its offensive efforts through joint ventures and industry alliances to encompass the operating system on a broad range of portable electronic devices. In its fourth attempt to create an operating system for PDAs that would finally catch up with the market, Microsoft introduced Windows CE 3.0 in April 2000. The company dubbed the new software as PocketPC to distance the new OS version from its failed predecessors. PocketPC was essentially miniature PC software that ran the full gamut of Microsoft's desktop applications: Word, Excel, Outlook, and Pocket Internet Explorer browser.

It goes without saying that Psion's EPOC operating system was conceived as being specifically designed for mobile devices because it made a better use of limited battery and memory resources. Since data, which would increasingly drive these next-generation devices, used a lot of power and memory, EPOC's composition was conceived as a crucial advantage. And while Symbian was tightly focused on the wireless market, Microsoft, on the other hand, wanted to turn its Windows CE software into an operating system for a whole range of devices. The result, according to some software experts, was an operating system that was too clumsy for mobile phones. A friend of all is friend of none! Rivals called Windows CE merely a small computer platform. Then there was the stance that a lot of communication software were involved in a phone and that Microsoft was not really a communications company.

Microsoft's early foray into the mobile world was full of hits and misses, mostly because Microsoft initially took a rather simplistic approach of cramming PC-like software into mobile phones. The software giant was forced to revamp its mobile computing strategy because devices based on the earlier Windows CE operating system were being killed by Palm. Microsoft had originally hoped to extend the PC experience to a mobile device by merely stuffing the desktop version of Windows into a tiny gadget. The company executives had thought everyone would flock to them just because who they were. But then Microsoft reevaluated what actually made sense to have in a mobile device. Although the launch of PocketPC with its improved interface and multimedia features received favorable attention from different corners of the industry, Compaq's iPaq handheld computer, introduced in summer 2000, in fact, rescued Microsoft's mobile strategy.

After launching a more flexible software platform for mobile devices, marketers at the software giant found it imperative to settle the nomenclature maze surrounding its PDA and phone products. The company silenced its critics by tailoring its Windows CE operating system into two forms: PocketPC, which ran on handheld organizers, and the Stinger operating system for smartphones. It subsequently evolved into the PocketPC Phone Edition having the pedigree of a PC-equipped PDA with phone functionality while the Stinger—later named Smartphone 2002—was a phone platform that also enabled PDA functions. The key difference between the two resided in the user interface. The PocketPC Phone Edition let users navigate through a bigger PDA screen with touch functions and handwriting recognition. Smartphone 2002, on the other hand, allowed operation via soft keys on a mobile phone.

Amid all the uncertainty of the new gadget realm, Microsoft first tried to master the handheld computers and then made a move into phone realm. The crowning glory of this effort was the Stinger, an adaption of Windows CE aimed at mobile telephony applications. In early 2001, Microsoft had announced plans to license its Stinger operating system to handset makers around the world. Although the Stinger handsets would foremost be phones, not computing devices, they would also offer features like remote access and e-mail via Microsoft's popular Outlook software. The days of "anything but Microsoft" rhetoric sparked by the company's browser war against Netscape now seemed to be folding into the history books. After early failures, the company swallowed its pride and now a more contrite Microsoft, willing to collaborate, was on its way to become an important player in mobile software infrastructure.

THE MOBILE SOFTWARE WARS

The battle was not for phones; it was for the platform, and the platform wars would continue in the coming years. Inevitably, a ferocious battle raged over who controls the operating system that would drive these new devices. Operating systems are the basic set of programs that control any device through hardware in the form of processor. Symbian and Microsoft's Windows Mobile—the name that Microsoft's mobile OS software ultimately settled on—had been marketed to both users and developers as the major stakeholders dating back to 1998. The smartphone OS tech wars ran almost for a decade. However, by the close of the decade, mobile software development playbook didn't belong to Symbian and Microsoft, but to Apple and Google. The history of high-profile failures and reversals of fortune in the mobile phone business makes a fascinating story divulged in the following sections.

While Nokia and Microsoft fiddled with their software strategies, Apple and Google were quietly making strides with iOS and Android operating systems, respectively. Both Nokia and Microsoft had been in this space since the beginning. It was theirs to lose, and they lost it. They had everything to create the right environment for powerful applications and a compelling user experience. Symbian and Microsoft could be credited with first mover status in the smartphone industry, yet despite valiant efforts to establish third-party development programs, neither achieved the kind of success that Apple and Google accomplished with their respective software platforms.

Apple, instead of desperately trying to defend comparisons of the iPhone platform to Windows Mobile in features or Symbian in reach, focused on the unique value it was adding and its ease of use. And rather than ceding control of the App Store to third-party developers, Apple maintained influence over its development environment to avoid allowing the software platform to develop a reputation of being crass, seedy, sloppy, or unfinished. "We weren't the first to this party but we're going to be the best," Jobs told his audience at the launch of iPhone OS 4.0 in April 2010. Apple stoked the smartphone market with its iOS software and rallied developers behind this solid operating system based on a subset of its well known desktop software.

Symbian and Windows Mobile didn't deliver the web experience or marketing push that Apple had already going for it. But Google was another matter. The twenty-first-century poster child of Silicon Valley innovation had managed to create a platform that genuinely rivaled iPhone for the best smartphone experience around. Industry observers were initially skeptical about Google's chances to find its footing in an already crowded market. By spring 2010, however, Android phones were outselling iPhones. Google estimated that 200,000 Android smartphones were being activated daily by wireless operators on behalf of its handset original equipment manufacturer (OEM) customers. Clearly, Google seemed to have a winner at hands in the smartphone realm.

Handset business had evolved in unexpected ways. In 2001, Palm veteran Jeff Hawkins had said, "The traditional handset doesn't have much of an operating system on it, so I don't think there is an opportunity to be the Microsoft of the cellular phones." By the time, however, technology trade media was mesmerized by the prospects of a Nokia-Microsoft standoff for the riches of operating system software that would serve as the engine of smartphone. But some people questioned if operating system was going to be be-all, end-all of smartphones. Among them were wireless bellwether Ericsson and top chipmaker Intel. Ericsson's high-end feature phones used its homegrown, real-time operating system (RTOS) software. The Swedish cellular firm at that time aimed to add features and applications incrementally and safely without necessarily adapting a full-blown operating system.

RTOS, an embedded operating system dedicated to a narrow set of applications, was often seen as the middle ground for easing the hardware challenges. The software was compact and efficient, and filled the critical need for a platform. RTOS was a rather simple software solution, but it worked. In early 2003, when Intel announced its long-awaited wireless-on-a-single-chip solution, Manitoba, it was also aiming to be run on an RTOS. As the cell phone industry started moving into consumer territory, where price was a sensitive issue, full-featured operating systems like Symbian carried additional costs due to associated royalties. Then there was a memory penalty as full OS-based handsets used a lot of memory, which cost money and space on an increasingly smaller footprint. So unless Symbian and Microsoft came up with more affordable licensing models for mobile phone vendors, at one stage, it appeared that the volume handset market would continue to be fragmented among independent RTOSes and related silicon platforms.

Beyond RTOS space, the browser captured the vast majority of wireless data applications at the front-end. And what didn't fit into the browser—where users needed richer environment to store something or stream media—that's where Java came into play. In 2001 Java Developer Conference, Ollila stood up to a packed audience imploring software developers to write wireless applications for Java because Nokia planned to ship millions of Java-centric web-ready cell phones in the coming years. But while Sun Microsystems earned significant industry clout from licensing Java's mobile version to handset makers, the widely deployed technology didn't actually create a functional platform, nor did it kick off any sort of popular, viable marketplace for developers and phone users with Java Micro Edition (ME) capabilities, largely because the Java ME platform was fractionalized between "almost compatible" variants.

In those early years of the smartphone's awkward adolescence, the most likely outcome was conceived as a fragmented market in which Symbian might have held on to smartphones and other phone-like communications devices hooked to the Internet. Palm could continue serving the mass market for handheld devices, while Microsoft dedicated to more demanding business users. That scenario was also seen as leading to a parting of ways between Europe and the United States. Smartphones had taken off in Europe and Asia, while America, where lack of a digital wireless standard hampered the growth, might have stuck to phoneless organizers by using them for data-only or pager-like connections. Fast forward to 2008, trade media was playing the Apple versus Google riff much in the same way they sang the siren songs of an imminent battle between Nokia and Microsoft a decade ago.

The fact that microprocessors became more powerful and that memory became abundant meant that mobile handsets would turn into handheld computers. This, in turn, shifted the handset's value toward software and data services. That's where America, especially Silicon Valley, was so comfortable innovating and operating because of its unparallel technology ecosystem. So companies like Apple and Google built overarching mobile platforms and stole the show from Nokia. The cell phone industry finally came around the notion of smartphones being defined as mobile devices running on an operating system for which third-party applications could be written. In the following sections, greater insight into mobile OS software business, along with one of the most stunning reversals of fortune in modern technology history, is explored.

APPLE'S iWONDER

Apple's Benjamin Button approach for its iPhone operating system turned out to be a stroke of genius. The iPhone OS got the single most important thing—user interface—right from the start. The user interface is the presentation layer of the OS software. In the hindsight, looking back at the agony of Symbian, which had been there since 2002, running on millions of phones, software experts were almost unanimous in their belief that if there was one fundamental problem with Symbian, it was the user interface. Otherwise, they said, Symbian was still the best OS from a technical standpoint as it was robust and consumed very little power. It became apparent when Google's first Android-based phone—the HTC G1—came in 2008 offering a user experience similar to the iPhone. Next up was RIM and its BlackBerry 6 operating system, torn between enterprise and consumer markets; its engineers burned the midnight oil to furnish touchscreen interface on its new smartphone models. Even Microsoft reinvented its Windows Phone 7 OS to have a user interface with new software chops.

Before the iPhone, smartphones mostly had resistive touchscreens and confusing menus that were difficult to use. The old, resistive touchscreen technology responded to a user physically pushing the screen, while the newer capacitive technology was manipulated by the electricity in the user's body. The iPhone effectively built usability into its wares as its excellent software design was coupled with a strong product sense. Apple's new phone raised the bar for user experience through a tight integration between software and hardware that was hard to replicate. It ushered in an interface that was clean and easy to navigate and was driven by icons rather than text-based menus. Now the software wasn't something that only geeks bought; it started being something that everyone bought, every day. The iPhone was indeed a key inflection point in software development.

The subsequent upgrades on the iPhone, including OS 4.0, have been mostly about adding stuff one would have assumed to be there from the start—such as multitasking, the ability to organize apps into folders, and file attachments that could be opened in third-party apps. But there was one difference: when Apple got around to implementing these features, it tended to get them right. "It's really easy to implement multitasking in a way that drains battery life," Steve Jobs told the crowd at the launch of iPhone OS 4.0. "If you don't do it just right your phone's going to feel sluggish and your battery life is going to go way down. We've figured out how to implement multitasking of third-party apps and avoid those things."

Apple's mobile software store provides a good case study on what's been so unique about its software strategy. For a start, it was built upon experience and previous successes. Well before it even launched the iPhone, Apple began selling iPod game software on a small scale to work out the kinks in packaging and delivering mobile software securely in a way that avoided casual theft. Apple also perfected micropayments for music and video on iTunes, paving the way for a high-volume, low-cost mobile software store. Apple also built on its decade of progress in developing Mac OS X desktop platform, so it was able to release development tools that were both mature and familiar to a large number of developers. Likewise, the iPhone's hardware clearly evolved under the influence of lessons learned during the development of the iPod.

The iOS is proprietary software tightly controlled by Apple and is the least compatible with the rest of the industry, to the degree of carrying on a public fight with Adobe Flash, which powered most videos on the web. Jobs slyly negotiated an arrangement with AT&T to carry the iPhone without even showing the carrier name on the phone. That was because Apple wanted to tightly control the design of the iPhone's OS and the hardware to deliver a mobile experience tailored for the phone users rather than the demands of a wireless carrier. The operating system subsequently helped to build Apple's reach beyond just the iPhone, supporting family of devices like the iPad and the iPod Touch, as did Apple's App Store, making it technically the best OS and related ecosystem out there. The fact that Apple offered multiple, non-phone platforms also hugely benefited the apps sale as it added the perceived value of software purchases for consumers.

If a company has learned hard lessons about the consequences of allowing third parties and rival vendors to hijack or destroy one's platform, it's probably Apple. It fended off companies trying to clone the Apple II and the Mac, and watched as Microsoft hijacked its Mac apps to create Windows. But Apple seems to have figured things out: don't give away your platform if you expect it to remain around as a viable contender. You don't allow third parties to put you out of business, and you don't funnel your success toward rivals, unless you're in the business of giving away technology for free and are closing up shop. But this very issue, which cut both ways, unleashed one of industry's highly contested debates. Holman W. Jenkins, Jr. summed this up as "Apple second date with history" in his *Wall Street Journal* article in May 2010.

THE MAC DÉJÀ VU

The story goes back to the heady days of Mac versus IBM compatible computers saga in the 1980s when the Macintosh ushered in a new era in computing. In 1984, when the original Apple Mac launched, it did change the whole PC industry forever by introducing the first user interface that was so simple that people didn't have to take a computer class to operate the computer. The screen had menus and icons that users could point to with a mouse. In that process, personal computer sales exploded, and the PC was embraced by almost all businesses. The early user interfaces that tried to mimic Apple's Mac were very poor; Windows 1.0 was truly horrible. So from 1986 to 1989, there was almost no contest because Macs were far and away better than early Windows PCs.

But Microsoft kept working at it, and by 1990, there was Windows 3.0 which took over the world and became the first mass market user interface on a PC that outsold the Macs by a big margin. It's important to remember that even Windows 3.0 was not as good as the Mac. But by 3.0, Windows was good enough. For the purist, like a professional in the graphic design arts, there was only the Mac. But for the critical mass, the Windows at 3.0 was good enough because the average consumer was no longer willing to pay the premium Apple wanted. When the PC mass market came off age, it went to the Windows version and very rapidly all PCs apps and services followed. This was regardless of Windows never being able to catch the Mac technically in any subsequent editions either. Apple, while refusing to open up its OS software to others, almost went out of business in late 1990s, and many would have blamed it on what seemed one of the seminal business blunders in technology history.

Apple resurrected itself as a maker of sleek computing devices for a segment of the public that valued tastiness. This story is resurfacing with Google's Android already outselling the iPhone; most consumers see the latest edition of the Android OS as "good enough" to be a viable and true rival to the iPhone. Because Apple insists on keeping its software and hardware under tight control, Google's platform is one that would benefit from competition among mobile handset makers. Moreover, because Apple insists on vetting all applications that run on its phones via the App Store, many pundits reckon that users would need an Android phone to capture the full benefit of openness to the web. That way, Android users could expect their available services and apps to outstrip those available to iPhone users through the App Store.

The jury is still out on whether Google will catch or pass Apple in making the best smartphone operating system, but the gap diminished to the point where by 2010 an Android-powered Samsung Galaxy phone could become a viable multitouch rival to the iPhone. The proponents of open-software platforms claim that Apple could not win by merely being the best in this race after the rivals become good enough. They argue that all Apple can do is hope to hold onto its customer base, which if seen from a broad mobile market perspective, is still tiny. Apple held a mere 2 percent of the global handset market in 2010, which was even smaller than what the Macintosh has traditionally held in the PC market. Macs have had about 4 percent of the global PC market during the same time frame. So commentators conclude that this is the wrong strategy for the iPhone and for Apple to be satisfied with this tiny sliver of 2 percent, half of what the Mac achieved.

The critics of Apple's software strategy are firm in their belief that developers will eventually abandon the iPhone OS for more open platforms such as Android. Conversely, the proponents are equally convinced that the absence of some developers on the platform won't really make the difference. As long as there are tens of millions of iPhone OS devices out there—and Apple keeps selling them at a fast pace—there will be developers who want to sell apps to run on them. Apple supporters assert that the more important factor is the ecosystem that drives people to the platform, and of course, the perceived value of the handset or operating system through the large amount of apps and development activity around it. What drives app-developing platforms is the embedded user base—how many shipments are being pumped into the system and how many people are sticking around.

Despite all the accolades for Google, Android and the rise of open source, Apple supporters point to the reports about some top smartphone manufacturers like Samsung wanting to build their own operating systems. The influence that Apple has is clearly making other guys a little bit envious, so they want to have control over their own smartphone experience that they bring to market. Google's move to acquire Motorola's handset business, as explained later in this chapter, apparently also underscores the allure of an integrated hardware-software business model epitomized by Apple.

Moreover, Apple's iOS is running on company's popular iPod Touch and iPad devices as well as the iPhone, and this larger, integrated OS footprint is also drawing attention from the gadget wannabes. If the iPhone was running on a closed ecosystem, so what! The mobile handsets are proprietary in their very essence, and they run on proprietary networks. In a nutshell, what company's proponents are saying is that all those things that haunted Apple's computers are non-issues in the phone market.

But then some critics credit Apple's robust development tools, its market buzz, and its media attention, or its status as the first mover in its field, as being the reason for the company's euphoric success. They say that these factors suggest that a challenger could simply swoop down and take Apple's business away by doing the same thing on a grander scale, an implication that frequently references how Microsoft Windows overshadowed the Mac in the mid-1990s. So to contend with this popular prognostication of the iPhone versus Android battle turning to be a replay of Windows versus Mac, Apple management is increasingly focused on growing an ecosystem rather than on creating devices. For starters, what Apple really needed was a good second act to the iPhone story and that came through the remarkable ascendance of the iPad tablet. Now Apple has its iOS seeded into not only one market, but three separate markets that it dominates: the iPhone, the iPad, and the iPod Touch.

Moreover, the next version of Mac OS would assume many key elements of the iOS software, which would translate into greater synergy within Apple's overall ecosystem. The Cupertino, California–based firm has defied conventional wisdom on many fronts. Take the case of iOS being stripped down and rewritten from Mac operating system to only a few hundred megabytes, roughly a tenth of the size of Mac OS X. By successfully revamping an elegant mobile OS from desktop software, Apple shattered a long-held myth that had been haunting Microsoft for about a decade. Industry watchers have been calling Microsoft's mobile strategy akin to "take desktop PC and put it in the wireless handset" philosophy. They said Microsoft was being hampered by its continued PC-centric view of the world. Apple and Jobs had rewritten the rules of the software game by succeeding where Microsoft failed.

Apple has so far managed to stay ahead of the curve through startling product innovation on the iOS platform. Furthermore, Apple advocates contend, mobile OS isn't end-all and be-all in the software battlefield for the riches of smartphones. As discussed in the preceding chapter, a conflict between browsers and apps is running in parallel to lay the claim for the next-generation mobile Internet. And here, a lot will depend on how HTML5 comes through in the years ahead. Apple, like its mobile OS adversary Google, is fully behind this open web standard. The fortunes of the smartphone and the mobile Internet are intrinsically linked to each other, so the future of HTML5 standard would inevitably influence Apple's standing in the mobile software realm despite its choice to carry a closed OS for the iPhone.

Apple has won the mobile game through a sustained level of differentiation, first through apps and then via cloud services. Apple's iOS 5 and iCloud offerings up the software ante and take Apple possibly miles ahead of its competition. These long-term platform enhancements could act as a deterrent against any disruptive threat to Apple in the foreseeable future. The launch of the iCloud, as explained in more detail in chapter 19, also marks Apple's transition from a Mac-centric company to a cloud-centric company. Predictions are hard, but the case for a Mac déjà vu might seem too good to be true. For any such likelihood to become a reality, we have to first see the smartphone affair as a mirror image of the PC episode. Steve Jobs had this strong conviction that Apple's woes during the 1990s were attributed not to the fact that Apple refused to open up its platform, but to the fact that Apple stopped building great computers.

GOOGLE'S SEARCH FOR THE NEXT BIG THING

Some pundits called Google's Android roadmap a Power Point presentation in its early days. Android's meteoric rise shows that being underestimated could be an advantage in technology business as much as in politics. While Android operating system contains 11 million lines of code, the whole program takes up only 200 Mbytes of space, about as much as forty MP3 songs. Yet despite its tiny size, Android is changing the mobile industry in profound ways, shifting the balance of power from Europe and Asia, the previous leaders, to Silicon Valley and reshaping the fortunes of the world's biggest wireless technology companies. Google offers its Android software to any maker, but it works exclusively with certain manufacturers to develop new phones that show off the latest features, a decision that is continually broadening its partnership gamut.

Although iPhone has set up a higher standard, Android's genesis Linux has the experience of defeating the market leader. In the late 1990s, Sun Microsytems' high-end server was the favorite of the IT industry, Internet venture companies and Wall Street investors, while Linux attempted and accomplished nothing back then. After ten years, Sun had gone, leaving Linux to dominate the data center market. Although the Linux-based Android initially needed improvement in the aspects of user interface experience such as the booting time and photograph performance, momentum seemed to be going its way.

Android enables developers to write managed codes in the Java language and controls the device with Google-developed Java libraries. In the start, Android wasn't quite as polished as the iPhone OS, but it quickly established itself as a worthy competitor to the iPhone. Instead of trying to modernize an older system originally created for voice-centric phones like Symbian, Android got on an equal footing with the iPhone as it started with a clean slate, developing a modern mobile OS for a new kind of device—a small computer that happens to make phone calls as well. Moreover, unlike older operating systems, Android matched the iPhone in being good at rendering web pages and running many applications at the same time.

In the mobile arena, Google seems to be playing the role of the new Microsoft with its cool handset software. Its Android operating system is a highly technical product that requires a lot of integration into the hardware platform users want it to run on. But in return, it provides a nice application framework for some pretty interesting features on its own. And because it's free, people who want to invest their time and resources can use it as a building block in the solution to whatever problem they're trying to solve. The adaption of Android helps ensure that Google's Internet search, maps, and other services like Gmail will be mainstay on mobile devices.

Android is widely seen as Google's second act because many in the industry perceive that the end of the Google search party is nearing, so the web icon is now running off to own the mobile search. There is also a growing feeling in the industry that wireless needs to embrace products like Android to facilitate the mobile web. At the same time, however, the industry watchers are keenly monitoring whether Google would try to leverage its dominance in web search and advertising to gain a foothold in the emerging mobile world. Some skeptics even doubt that Google wants to stop Apple and Microsoft more than to find a new revenue opportunity. Google licenses the software for free to hardware makers, and it is unclear when the adaption will start bringing in significant revenue streams. The Mountain View, California–based search giant makes money from advertizing and from collecting data it can sell to advertisers, and its customers reciprocate by wanting services for free, which means that Google's living is solely advertising-supported.

The cost of developing Android will likely be tiny compared to the opportunity for controlling the leading mobile platform, the perceived basis for the next big wave of Google's search and advertising business. When one already owns the portal that brings every PC to the web, the only way up is in owning the coming-of-age mobile portal. And Google won't cede that to Apple's iTunes without a fight. Android generates scant revenues from its mobile business so Google model is likely to build a killer application first and monetize it later. The faith behind Google's model is that if one owns the platform that is valuable, it can be monetized. A self-assured Eric Schmidt said, "You get a billion people doing something, there's lots of ways to make money. Absolutely, trust me. We'll get lots of money for it." Case in point: Apple is obliged to share with Google search revenue generated by iPhone users. On Android, Google gets to keep 100 percent.

In a nutshell, more web usage is what Google is after; more web usage means more Google advertising. But even if Android fails on its own, the search titan could still go on supporting its Google apps on the BlackBerry and wherever else it can without direct competition for ads and paid search. In contrast, if the iPhone were to fail as a platform, Apple would be completely out of the smartphone game. That's why Apple was fighting relentlessly to manage the future of its platform, while Google was initially keeping Android around as a viable software platform, running it mostly like an experiment. In the early going, this made Android a bit like Apple TV; if that product fails for Apple, it could simply give up and begin licensing iTunes playback on the Wii, PlayStation 3, Xbox, and third-party TVs. In retrospect, it wasn't until 2010 that Android hit its stride.

Android's appeal partly owes to its low cost and relative openness, but it also offers a rich set of tools and capabilities that permit low-price devices to compete against the iPhone. Android's next move could be down-market toward cheap smartphones under US$100 tags that could dramatically expand smartphone penetration. T-Mobile has been the first to hit this tipping point by launching an Android-powered phone priced in the neighborhood of US$100. The Comet handset featured 3.2-megapixel camera with camcorder, Google Maps navigation, digital media player, microSD card slot, USB interface, Bluetooth, tethering, and Wi-Fi hot sport capabilities. Historically, technology markets tend to standardize around a single platform that brings economy of scale. The network effect further complements this as developers devote more and more of their time and energy to that unifying platform. Android yields a story with an echo of such a platform.

ANDROID: OPEN IN A SPECIAL WAY

If Apple's second date with history seemed like a stigma for the Cupertino, California–based company, Google was facing an equally haunting challenge: fragmentation. Steve Jobs used Apple's quarterly earnings call in fall 2010 to question Google's claim about its mobile operating system, Android, as open. He said, "Google likes to characterize Android as open and Apple as closed. We find that a bit disingenuous…Android is fragmented…Compare this to [the] iPhone, where every app works the same…the multiple hardware and software iterations present developers with a daunting challenge…" He also questioned Google's claim that Android is hardware-agnostic, which creates a level playing field for handset makers and third-party software houses.

Jobs pointed out that some apps developed on Android will only work on certain Android-powered phones and not on others. The result, according to Jobs, is a nightmare for consumers and developers, whereas Apple offers a simpler and more consistent experience. "In reality, we think the open versus closed argument is just a smokescreen," Jobs concluded. The real issue, he said, is integrated versus fragmented. "We think integrated will trump fragmented every time." Jobs calling the Android development a daunting task went on to say, "Android is very, very fragmented and becoming more fragmented by the day," he said. "When selling to users who want the device to just work, we believe integrated beats fragmented every time."

Bruce Perens—who coined the term *open source*—has been working on behalf of this software community ever since open-source movement took hold. He says that "open" can be defined around three core traits: a license that insures the code can be modified, reused, and distributed; a community development approach; and, most importantly, assurance that the user has total freedom over the device and software. The Android OS is, in strictly legal terms, open source. Android is released under a software license, which allows anyone to use, modify, and redistribute the code. Android, however, falls short on the first two facets. It's the lack of community-based development that Android's critics say makes it no more "open" than Apple's locked-down, decidedly not-open iOS model.

Unlike major open-source projects like Firefox or the Linux kernel, one can't see what's happening behind the scenes with Android, nor can small developers contribute to the project in any meaningful way. Google typically releases major updates to Android at press conferences, not unlike those Apple uses to show off new iPhone features. Once the code is released, Android developers can download it and do what they want with it. Android basically gives people two options: accept what Google gives, or fork the entire codebase. Other than the ability to roll one's own version of Android, it's really no different than iOS, which works on a similar "take what Apple gives you" model.

The way Android works in this bewildering tangle of hardware makers, software developers, and wireless operators is that Google enables the wireless carriers to shape the Android platform from the user's perspective. In other words, it's wireless carriers first and Google second. So when a specific version of Android arrives at the hands of mobile phone makers, it's also tightly controlled like Apple's iOS. There is one difference though: Android is still less controlled by its parent Google in the sense that the Android Market is not tightly regulated like Apple's App Store counterpart. But then again, a closer look at the mobile phone world reveals that Android and iOS are not exceptions. The mobile phone carriers argue that open phones would threaten the network; Apple argues that an open phone would threaten the user experience.

The critics point to Google's Chrome OS project, which is run with a level of transparency and community involvement largely absent from Android. They call it a better representation of Google's open model values. But this whole debate of Android meeting the bare minimum definition of being open misses the point: Android is not about creating a software ivory tower to reinvigorate the open-source movement. Instead, it's about creating a sustainable business model that could succeed in the rapidly evolving mobile phone software landscape. Android embodied a winning strategy from the start. First, Google became an erstwhile challenger to Apple in a relatively short period without doing any of the fighting: its hardware manufacturing partners did it all for Google. Second, without having to build a single phone, Android took volume leadership in the smartphone space.

As to fragmentation, it has affected Linux and other open-source systems in the past. It occurs when a system has multiple versions with very different software components. Take the example of Linux distributions that have their own file system arrangements, different window managers, and different packaging systems. This makes it difficult for application developers to create applications for Linux because they have to presume that certain components will exist or be in the same location. Fragmentation used to be much worse in the early going when Java came with hundreds of variations in feature phones. Now operating system fragmentation is a major hurdle in apps and widgets take-off.

For Android, the real stunner was the rapidity with which it matured as an operating system. It was moving so fast that manufacturers couldn't roll out updates to the software at a speed to match. While Apple has cleanly deployed new operating system features at annual intervals, Android's development is managed more like Linux, where new platform features are incrementally advanced, starting with partial support and gradually getting to the point of being refined and usable. This factor works against a cohesive installed base, as different vendors introduce phones with different versions of Android, each with proprietary add-ons and modifications that prevent predictable upgrade cycles. The mainstream version of Android would always lag behind the latest version of Android, unlike the iPhone market where phones are rapidly pushed to the latest version under Apple's control.

The result: Google was struck with a number of slightly different variants of its Android portfolio. In sixteen months since the first Android phone hit the market, Google made four major upgrades to the operating system. But smartphone makers unable to keep up with that pace continued to introduce new mobile phones with older versions of the OS, leading to consumer confusion about what capabilities various Android phones have. Then there were different handset manufacturers dressing up Android with various add-ons and custom user interfaces. This variation in how Android looks and feels also led to fragmentation. One of unintended consequences of such fragmentation could lead to handset manufacturers fiercely competing for space in wireless carrier's lineup, but in turn, the carriers could play the manufactures off each other to get their prices down.

Moreover, both phone manufacturers and wireless operators aren't motivated to update handsets to the latest version of Android because they want to sell new models, not keep their customers on old ones. On part of Google, it was fundamentally about the math of the platform as the company's initial focus was on establishing volume. But all these factors put the Android landscape in a flux. Now if Google fails to unify the Android platform, a future looms where Android will be stuck in a myriad of software islands. That's not just an issue for hardware makers; it's a challenge for users and developers as well. It could lead to a splintered marketplace where one Android user can't necessarily access the same apps that another user can. Moreover, end users won't be able to pick up an Android handset and expect it to work in the same way.

THE EVOLUTION OF ANDROID

The fragmentation caused by the different versions of OS software was on the decline by the end of 2010. One of the obvious ways ahead was to try to force manufacturers to update the software on their devices. The majority of handset makers, including Motorola and HTC, had been steadily updating the OSes on their older phones. Apart from that, once Android became mature as a fully-featured platform, the rapid development that allowed Android to add functionality at a swift pace naturally slowed down. The Android OS was coalescing around three major updates, each named after a sweet: Android 1.5, a.k.a. Cupcake; Android 1.6, or Donut; and Android 2.1, nicknamed Eclair. By the time Android reached 3.0, or Honeycomb—designed from the ground up for tablet computer-like devices with larger screen sizes and improves on features like browsing, multitasking, and widgets—Google began to enforce anti-fragmentation measures to a greater degree.

Google had the foresight of giving itself some controls while crafting the Android developer license so that open-source platform didn't go out of hand. Eventually, Google tightened its control over the source code and began exerting the final say on how its partners could tweak the Android code to create user interface and add services. Moreover, Google gave access of the Honeycomb source code to a few selected partners such as Motorola, Nvidia, and Samsung and delayed its distribution to outside programmers. Now people who questioned Android's open-source credentials, saying that its openness isn't a free ticket, felt vindicated. Among them was Nokia's newly arrived CEO Stephen Elop, who defended his decision to go with Microsoft instead of Google, saying, "The premise of true open software platform may be where Android started, but it's not where Android is going."

Google also intends to make Android more modular, which would enable handsets to update individual components of the Android system even if their carrier had not moved to a newer version of Android. For example, if a new version of Gmail is released, users can decide to upgrade Gmail on their phones. Another approach that Google has taken to try to prevent fragmentation of Android is to adapt a series of standard procedures that must be followed to gain access to the Android Market. These standards define minimal functionality that a handset must have in order for the carrier to ship the Android-base handset in the market.

Android, despite all its brilliance, faces significant challenges. For instance, Google is to confront the same problem in handsets that Microsoft encountered in the PC arena: herding the legions of software developers. The greater diversity of Android platform aside, Google has to carefully balance openness to application developers with a superior user experience and mass-market appeal. Google enforces much less control over content or presentation for apps listed in Android Market. While Google reserves the right to block unauthorized distribution of apps to Android Market store, it doesn't mandate that apps have to look great or follow any strict guidelines. Such lack of restrictions could result in a clear difference between the apps available for Android and those for the iPhone.

Subsequently, it could lead to a wide gulf between the slick, commercial offerings in the App Store and the experimental-like offerings in the Android Market. However, given Android's large installed base, it would be difficult for Google to erect significant restrictions on software. The dilemma for Google is that once its market took off, the latent problems of permissive platform management grew into serious problems. Security, commercial legitimacy, and professional presentation are all factors that Google seems to have thought will solve themselves. Clearly, the Android model is intricate, but by putting OS software into the hands of so many mobile phone makers, Google believes it has created an accelerated form of smartphone evolution in which handsets will diversify and improve at hyperspeed.

Midway in its fight against fragmentation, in August 2011, Google rattled the technology world by making a bid to acquire Motorola Mobility Inc.—the Motorola spinoff that produced smartphones, tablets, and set-top boxes—for US$12.5 billion. The deal put Google in an awkward position and raised all sorts of questions about the Android business model and Google's place in the mobile world. Was Google aiming to unify Android platform by taking a large mobile phone maker under its wings? Would Google eventually position Motorola as a direct competitor to Apple's tightly integrated devices and services, and make Android proprietary at some stage? Google had attempted in the past to enter the hardware business with Nexus One handsets and had only seen a lukewarm response.

Could a web company with software roots prosper in the hardware business closely tied to complex supply chains and rapid product cycles? And, most importantly, would this deal alienate Android partners like Samsung and HTC? For starters, Google vowed to keep Android as an open platform. Motorola Mobility will operate as a separate business within Google and it won't receive any preferential treatment in the Android settings.

As to why Google needed this new challenge, there were two major viewpoints among the industry watchers. According to the first premise, Google wanted to buy the Libertyville, Illinois–based firm for its valuable war chest of nearly 24,000 patents. Google needed to create a legal shield around its Android platform to protect itself and its handset partners in an increasingly litigious technology environment. Earlier in the summer, a consortium including Apple, Microsoft and RIM had outbid Google for nearly 6,000 Nortel patents—many of them relating to wireless technology—which became available after the fallout of Canadian communications equipment maker.

That made it a bold move on Google's part. However, the premise leads to the reckoning that Google will eventually divest Motorola's hardware business while retaining its patents gold. But then there was another group of industry observers which called this move akin to Apple's mid-1990s scheme which allowed other PC makers to create Mac clones while Apple would continue building its own personal computers running the same operating system. Apple's attempt failed and so did the Palm's effort to split itself into separate hardware and software companies during the mid-2000s. So, in the long run, it could be extremely challenging for Google to both license Android and compete with those licensees at the same time. That could also create new opportunities for Microsoft as the sole mobile platform provider not competing against its partners for hardware sales.

The mobile platform wars are likely to continue in 2012 and beyond as people witness an explosion of mobile software for a wide range of smartphones. This is probably the most dynamic time in the smartphone's short but exciting life span. While this chapter mostly dealt with two success stories in the mobile phone software realm, the next chapter delves into why Nokia, which correctly predicted the evolution of the market toward software and Internet services-driven product differentiation as early as 1998, ended up as an also-ran. With younger and energetic competitors in the market, Symbian seemed like an aging actress that should have already stepped out of the spotlight. Next up, why Microsoft, Apple's old nemesis from the PC wars, which clearly saw wireless orchestrating a galaxy of digital devices powered by a variety of Windows software, was being pushed to the sidelines.

SOFTWARE DIARIES

"More and more, we are becoming a software company."
—Jorma Ollila, Nokia's chairman and CEO in a
2001 interview

In February 2011, Nokia's new CEO Stephen Elop sent his staff an apocalyptical memo in which he likened the company to an oil worker standing on a burning oil rig who might jump into icy water to escape the flames. A few days later, when Nokia announced a smartphone alliance with Microsoft, the Finnish firm just looked like that: a desperate man jumping into icy water. A leading mobile phone player in terms of market share had embraced a mobile OS with 3 percent of the smartphone market. Nokia had effectively killed the Symbian—which, despite all its woes, still boasted the largest base of smartphone users—by turning it into a franchise platform and adapting Windows Phone 7, which was barely out of the gate. What agonized many in the industry was also the fact that at the outset it looked more like a business decision than a strategic one. Both Google and Microsoft made offers to Nokia, and in the end, Microsoft outbid Google in terms of engineering assistance, revenue sharing on mobile ads, search, and mapping services.

Adapting Android could have turned Nokia into another me-too handset house. Moreover, Nokia's biggest remaining software product—Navteq, a mobile mapping service for which Nokia had paid US$8.1 billion—would have been brushed under the rug by Google, who had its own popular mapping service. Microsoft would integrate Navteq into Windows Phone 7 software while its Bing search engine would power services in all of Nokia's devices. Nokia's content and applications store, Ovi, would integrate with Microsoft's Marketplace to provide more options to mobile users. It was a bold move for Microsoft in the sense that Nokia's economy of scale in handset business could solve Windows Phone 7's small market-share problem in one swoop if things went according to plan. But for Nokia, it felt like giving up, not like fighting back. It could even prove hugely damaging to Nokia's internal psyche to give up on its own software and turn to one of its competitor's.

The market sentiment had been overly pessimistic and Google's vice president of engineering Vic Gundotra channeled such perception, saying, "Two turkeys do not make an eagle." The transition would take two years, and during this time, Nokia expected to sell 150 million Symbian devices. But who is going to buy a smartphone with an obsolete operating system that, according to the company head, is not competitive enough. The partnership would take time to implement and deliver phones and that could prove Nokia's Achilles' heel. Moreover, though tie-up could initially boost Microsoft's market share with millions of Nokia handsets being ported to Windows Phone 7, it could ultimately also alienate Microsoft from its existing partners. Microsoft's special relationship with Nokia could antagonize its other hardware partners like HTC and Samsung and push them further into the Google camp. These companies also won't be happy seeing the navigation and mapping technology of their chief rival, Nokia's Navteq, in their mobile devices.

Windows Phone 7 was no one's priority until the world's largest handset maker committed to the platform. That made the logic of killing off Symbian to adapt a competing product with a lower market share highly defective. The critics argued that combining two companies' operations with flawed strategies wouldn't necessarily translate into shoring up each other's failings. Moreover, Windows Phone 7 had strict hardware requirements and allowed for little in the way of hardware design innovation, a core area of Nokia's strength. Furthermore, Nokia had made a huge commitment to its developer community and had sunk millions into it. Its Ovi services center wasn't a smash hit like Apple's App Store, but it was making a steady progress. And finally, though Nokia took a beating in the U.S. smartphone market, the American market wasn't everything. Nokia had huge brand awareness around the rest of the planet.

However, two critical aspects of this deal show merits in the longer run. First, although Nokia will pay Microsoft an undisclosed royalty for licensing Windows Phone 7 software, the strategic partnership would save Nokia a substantial reduction in terms of developing and maintaining its own software ecosystem. Second, Nokia could tap Microsoft's software developer network, and together, they could build a viable ecosystem to create a three-horse race that many in the wireless industry might welcome as an alternative to Android–iOS duopoly. The work that Microsoft and Nokia have cut out for them is going to be long and tenuous, and all the while, the smartphone market is moving forward at a relentless speed. The following sections chronicle the mobile software journey of these two companies who were once fierce competitors but subsequently fell into each other's arms.

NOKIA'S SMARTPHONE TROUBLE

Kai Nyman, a former architect for Internet services at Nokia, had known about the Internet-ready, touchscreen prototype way back in early 2000s and had pitched this to the management, which was reluctant to proceed because of concerns over the performance of the operating system. In 2004, the prototype was demonstrated to business customers at Nokia's headquarters in Espoo, Finland, as an example of what was in the company's pipeline. The management didn't pursue development, worrying that the product could be a costly flop. "It was very early days, and no one really knew anything about the touchscreen's potential," recalled Ari Hakkarainen, a manager responsible for marketing in the development team for the Nokia Series 60. "And it was an expensive device to produce, so there was more risk involved for Nokia. So management did the usual. They killed it." Although Nokia introduced the industry's first touchscreen devices in 2003—the 6108 and 3108 phones, which worked with a stylus—it did not perfect the technology to fingertip precision before Apple did.

Hakkarainen also recalled that in 2004, his team developed the early design for a Nokia online applications store. They made a demo and tried to convince middle and upper management, but there was no way. The company rejected an early design for a Nokia online applications store—an innovation that Apple introduced four years later and other handset makers quickly followed the suit. Likewise, Juhani Risku, who worked on user interface (UI) designs for Symbian from 2001 to 2009, remembered hundreds of proposals that his team offered to improve Symbian but could not get even one through. "It was management by committee," Risku said, comparing the company's design approval processes to a "Soviet-style" bureaucracy. Ideas fell victim to infighting among managers with competing agendas, and were rejected as too costly, risky, or insignificant for a market leader that had become synonymous with mobile phones.

Risku later left Nokia to design environmentally sound buildings. His account of Nokia's highly staid corporate culture is a testament to the dilemma shared by Nyman and Hakkarainen for getting Symbian back on track. "There were plenty of years to make Symbian better," said Nyman. "We could have rewritten the whole code several times over. We had the resources and the people. But we didn't do it." Even Apple's co-founder Steve Wozniak didn't hesitate to take a stab at Nokia while commenting about the rivalry between Android and iPhone. He called Nokia a throwback from a previous generation. Wozniak told Dutch newspaper *De Telegraaf* that "Nokia came late to the touchscreen game and now has 'an image problem' that it needs to rectify with a fresh new brand and younger customer base."

So what went wrong for a company that had long excelled at making beautiful mobile phones? First and foremost, Nokia was a hardware-focused organization that never mastered its software strategy. Nokia's software roadmap was in a continuous state of flux if not downright directionless. About a decade after its first software venture took shape in the guise of Symbian, most of Nokia's allies and competitors had joined the Android bandwagon, getting a head-start both in terms of cost and of time-to-market, and Nokia looked like Alice in Wonderland. Nokia's smartphone ambitions were stuck in a space that was marked with endless gamesmanship for winning the hearts and minds for its Symbian prodigy. Critics blamed Nokia's Symbian OS for the company's failure to pull ahead in smartphones, saying it is so clunky that developers have not been willing to write applications for it.

The anecdotal evidence presented early on affirms Symbian's cumbersome legacy software and a hardware-centric model. When Android and iPhone were fighting over the gadget brains in late 2000s, even pared-down models needed eight times as much memory and processing power in Symbian handsets. Apple and Android were relentlessly eroding Nokia's smartphone market share by bringing cool new stuff to mobile phone users. For software guys looking to develop apps for iPhone or Android, all they needed was download the software developers kit (SDK), register, pop open Xcode (for iPhone) or Eclipse (for Android), and they could get started.

But if they wanted to create an app for Symbian, there were just too many choices: they could develop in native Symbian, in some Java variants, in .Net integrated or in Qt. Nokia had so many versions of the Symbian and development environments that apps created for one set of devices often didn't work on another. And that led to fragmentation similar to a degree to what we have witnessed in case of Android in the previous chapter. But it became far more challenging to develop an app for Symbian compared to Android, let alone the iPhone. Take the example of an app that turns consumer photos into postcards. Developing such an app for Symbian could be a long road because the developer will have to build a plug-in that connects the camera to the gallery of photos; the app could take about four to five weeks of work.

On the other hand, in Android environment, it could take five minutes. It's a feature built into the operating system, and all a developer has to do is just turn it on. Technology waits for no one. The market was moving quickly and Nokia urgently needed to deliver an exciting and genuinely differentiated device to regain its reputation as an innovative technology leader and to retain its leadership position in the market. So far, Nokia's excellent distribution and sales strategies had ensured market share on the mobile OS domain, but the company understood too well that status quo is not a given in these quicksand times. Nokia's quest to feature MeeGo in its upcoming premium phones was a clear sign of growing strains and marked a gradual shift away from Symbian.

Symbian continued to decline until Stephen Elop terminated the Symbian program by hooking up with Microsoft. It was a clear admission that Nokia's mobile platform strategy had faltered. Why Nokia fought a losing battle? To find all these answers, it'd be worthwhile to have a closer look at the Symbian story. The following sections will document Symbian's slow death in the wake of more modern user interfaces offered by iPhone and Android devices.

SYMBIAN HISTORY

Symbian's history is a long one. The software platform, riding on top of the digital wireless wave, was set to forge ahead in its quest to build the de facto operating system for the up-and-coming smartphones housing voice, data, and video services.

As Andrew Orlowski narrates in his article titled "Symbian, the Secret History," published in *The Register*, the first major turning point in the Symbian story came with decision to offer four UIs that could cover the whole array of mobile devices. In retrospect, that only confused the nascent smartphone market. Meanwhile, Nokia, then the smartest kid on the block, came to this realization that if it could own the UI, it could own the user experience and, ultimately, the developers. And that's where the value was going to migrate, reckoned Nokia managers. Nokia, executing flawlessly in those days, subsequently led the Symbian partners into this decision that the venture had to be limited to only developing code for a mobile phone operating system.

That would mean abandoning a project to develop software that provides the underlying features that are essential to a smartphone operating system, such as support for graphics, security, and Internet access. Through that application software or user interface, licensees could also change the phones' on-screen menus and other features like graphics. An interesting twist in the Symbian model emerged when Nokia brought that portion of the project, named "Pearl," in-house and then replaced it with its own software—later dubbed as the Series 60—to control such applications as picture messaging, web browsing, and e-mail.

In the article "Facing Big Threat From Microsoft, Nokia Places a Bet," published on May 23, 2002, the *Wall Street Journal* chronicled how Nokia's initial plan was to keep the Series 60 software to itself. The strategy could preserve its profit margins, but at a potentially high cost. It could make it harder for Nokia to stay ahead of Microsoft in the race to establish a software standard. Apple Computer had taken a similar approach with its PC software by keeping it proprietary and wound up as a niche supplier once Microsoft's Windows took off. So Nokia chose the U.S. computer show Comdex in November 2001 to announce that the Series

60 would be available to all comers. The top mobile phone maker was at pains to persuade the rest of the industry that Nokia wasn't secretly plotting to become the next Microsoft. Senior company executives pledged that Nokia would license the software at very low rates and make the source code—which shows exactly how Series 60 was written—available to licensees.

Nokia built the Series 60 framework on top of Symbian operating system. The company added an abstraction layer to the Series 60 by using web runtime tools and Qt, so that the Series 60 software, employed by application developers, could also run on other mobile operating systems. Qt, pronounced "cute," is an open-source application platform that can create software that works on multiple platforms. "We need to agree on a common architecture for the middleware on top of the operating system to make next-generation services and applications happen," Jorma Ollila told the Comdex crowd. Nokia, for its part, was looking to cement its Series 60 software platform on top of the Symbian operating system as a de facto standard.

In the coming months, the company licensed the platform to Siemens, Matsushita, Samsung, and Sendo. When Nokia added itself to the list, the companies that signed to the Series 60 platform collectively controlled more than 60 percent of the handheld market. Nokia made great play on the fact that the Series 60 was significantly more OEM-friendly, and thus it allowed handset makers to customize their offerings. In the meantime, Symbian OS, the base system that helped Nokia succeed with its Series 60 platform, was ready for the next big shake-up.

In 2004, Nokia took control of Symbian through the purchase of Psion's 31.1 percent stake valued at an estimated US$252 million. At this stage, Ericsson CEO Carl-Henric Svanberg urged minority shareholders in Symbian Ltd to use their proportional preemption rights to stop Nokia from gaining a majority stake in the mobile phone operating system supplier. But Nokia went ahead anyway. The future of the Symbian group had been further clouded after the largest shareholder in Psion announced its opposition to the sale of its Symbian stake to Nokia. Earlier, in late 2003, Nokia had bought out Motorola's 19 percent stake, which clearly suggested that it was preparing to take a controlling stake. The quest for the control of Symbian was seen as a move by Nokia to strengthen its position ahead of a looming battle with Microsoft in the use of operating systems for next-generation smartphones.

It was a time of dramatic realignment, given that the telecom bubble had burst, and limitless money that the Symbian alliance promised had dried up. Nokia's Nordic cousin, Ericsson, was experiencing a dramatic fall and had spun out its mobile phone business into a partnership with Sony in 2001. Motorola had also fallen into a funk. Then there were WAP and 3G fiascos lying at the wireless industry doorstep. It was precisely this set of circumstances that helped Nokia—still riding high on a strong volume of handset shipments—to become so good at getting what it wanted from the Symbian alliance.

The standardization by consensus, the hallmark of the old Symbian, was also seen as giving proprietary systems such as Windows Mobile and Palm an advantage in time-to-market and nimbleness. However, the move also had an opposite effect as many licensees of Symbian began to perceive the operating system as Nokia's proprietary technology rather than an open standard. An analyst called the move "the ultimate manifestation of the boy scout effect." So far Symbian had been a tightly controlled ecosystem where fragmentation had not been allowed to happen. But this control point was now in danger of being breached.

By 2006, 100 million Symbian phones were in the market, a figure that more than doubled two year later as the smartphone market contin- ued to grow. Until this time, Symbian remained an undisputed market leader in the smartphone space, largely owing to the strength of Nokia. But Symbian saw the mobile game changing as the iPhone started to rewrite the rules of the wireless game. Charging a royalty for use of its software and requiring its partners to sign up to a license agreement eventually became a barrier, which for a partner was a financial burden as much as an overhead. Google was clearly aiming at that stumbling block through its free operating system software, Android. So the next crucial stage in the evolution of Symbian came in 2008 when Nokia purchased all Symbian assets and started to guide the software down to the path to open source.

In summer 2008, Symbian co-founder Nokia announced that it is buying the 52 percent of the software maker that it didn't already own and would release its mobile operating system under an open-source license. To support the new open-source project, Nokia established the Symbian Foundation, a collection of hardware and software companies that pledged to donate code and resources to Symbian's development. Phone makers Motorola and Sony Ericsson got on board, contributing software from their UIQ project, a touchscreen interface for Symbian. The Japanese carrier NTT DoCoMo promised support while contributing its MOAP (S) interface. Other supporters included AT&T, Samsung, STMicroelectronics, and Texas Instruments.

SOFTWARE CONUNDRUM

Nokia's acquisition of Symbian was hailed by some industry corners as one of several game-changing moves by Nokia. They advocated that by turning Symbian into a non-profit entity—Symbian Foundation—Nokia had ushered into its future the collective power of open-source community. By making Symbian an open-source operating system, Nokia engineered a brilliant preemptive strike against Android, or so they said. A closer look at the making of Symbian Foundation reveals a different story. The Finnish giant, by pursuing a two-pronged strategy, was seen by some industry sections as trying to have it both ways. It was championing an open philosophy, while fostering software unique to Nokia.

The very problem for Symbian's future was that its success was intricately tied to the success of Nokia. If Symbian had historically been bound to the fate of Nokia, part of the unspoken rationale behind the creation of the Symbian Foundation was to go beyond Nokia. Even if, in the short term, this meant getting closer to its former Finnish partner as Nokia bought up all Symbian shares and then bequeathed its code to the Foundation. But time was the essence here as pressure from Android was relentless.

When Nokia announced its decision to acquire the rest of Symbian and offer its Series 60 and Symbian OS for free to the open-source community, the mode of Nokia's transition initially created some confusion. The Symbian Foundation, a not-for-profit organization set up to manage all the assets related to Symbian, would take a phased-in approach. First, it would make available the source code of Series 60 and Symbian OS "for free" to all members of the Foundation in the first half of 2009. During that time, Foundation members—who were asked to pay US$1,500 membership fee to join—could use the Foundation's intellectual property assets, but would not be allowed to redistribute them. By the first half of 2010, the Symbian Foundation was scheduled to be fully ready to go open source. It never came to that.

The truth of the matter was that Symbian needed to untangle copyrights and intellectual property rights to a massive collection of third-party commercial software—originally gathered and implemented by Symbian when it was for-profit commercial entity. Uniting Symbian OS and the Series 60 platform, Motorola's UIQ and DoCoMo's MOAP (S) was likely to take some time—two years to completion was the target mentioned by the Foundation. Symbian also had to do a lot of scrubbing of its existing codes before the Symbian Foundation could completely hand off its software to open-source developers. Integrating Symbian's various software stacks into one unified platform and managing the millions of lines of codes were overwhelming tasks. So the transition could be costly, and Nokia would be vulnerable in that transition which could take around half a decade.

The old Symbian had helped Nokia modestly succeed with its Series 60 platform in the pre-iPhone days. Over the years, various incarnations of the Symbian OS had powered smartphones with desktop-like features such as preemptive multitasking, memory protection, and Unicode support. The biggest leverage that Symbian OS and Series 60 had was the large installed base. But now that Symbian partnership was dissolved, that Symbian was run by a Foundation, and that the system was fully open-source, it was evident that Symbian must do more to succeed than just going open-source. According to software experts, four major areas existed in which the Symbian Foundation needed to fix things: user interface, app-development environment, developer support and the leadership vacuum for the platform. In a nutshell, Nokia had been fighting mostly on a device-by-device basis when the battle had really shifted to the ecosystem level.

Unlike Android and iOS, which could start from a blank sheet of paper, Symbian had to carry the past that created a bit of slowness in the UI experience. Symbian proponents were confident that it wouldn't take much to fix the UI because the UI was just the presentation layer of the operating system. Next, Symbian accumulated a variety of app platforms that had become prevalent over the years. Symbian technologists hoped that Qt could help change that, making it easy to create apps that will work across all past and future Nokia devices. The Qt platform and library, which have been extremely successful in the desktop world, allowing developers to create cool apps, could help new Symbian overhaul its app strategy by creating hardware and software coherence.

That much about UI and app development environment! The developer community at large still saw it as imperative for Nokia to lead development of Symbian if the OS was to succeed, much like what Google has done with Android. There were people in the industry who believed that the Symbian open-source experiment had failed and Nokia needed to scrap the Symbian Foundation and bring the OS software home. "It would eliminate time-wasting Foundation activities like release councils, architecture councils, user interface councils," said Nick Jones, Gartner's London-based analyst on his blog. "What Symbian needs is agility and vision, not committees, and if Symbian is fixable it will be fixed a lot faster under a single leader. Great user interfaces aren't developed by committees."

Jones asserted that if Nokia took Symbian back in-house, assumed control and leadership, and used Qt to create a good UI, Symbian would be a very different entity from what it was in 2010. In some ways, Nokia's decision to acquire Symbian and turn it into a non-profit entity to leverage the collective power of the open-source community in future software development looked more like an Android me-too. Merely opening one's operating system was in itself no guarantee of success on the mobile market. After a decade of trial and error, time wasn't on Symbian's side, so Nokia needed to move fast. Ultimately, Nokia was to face a difficult decision about whether or not to abandon Symbian entirely and rebuild its software strategy all over again.

There had been speculations for months about whether Nokia will adapt another platform before the world's largest handset maker embraced Microsoft's Windows Phone 7. Symbian loyalists said it wouldn't come to any of that: "The strategy is clear and makes sense." But Nokia's history with Symbian had been checkered. Now the blend of Nokia and the Symbian Foundation made the Finish wireless house look like it was between a rock and a hard place. The Symbian Foundation had been in a state of flux since its inception with the ongoing streamlining of the workforce. It was inevitable for Symbian Foundation to go beyond Nokia, but at the same time, Symbian looked bound up with the fate of Nokia. The Series 60 platform was still significant as of 2010, but that leverage could diminish in the coming years if Nokia failed to put the house in order.

The Symbian saga offers some valuable lessons. First, one can't be all things to all people all the time. Second, if merely acquisitions and strategic partnerships mattered, then the alliance of Ericsson and Microsoft to create a compelling wireless e-mail service would have rendered RIM irrelevant. Third, Nokia ended up spreading resources thin over too many different projects. Now how would Nokia come out of this software conundrum? The common perception among mobile industry watchers was that the world's largest cell phone maker would eventually dedicate Symbian to low- and mid-tier feature phones while steering its premium phones toward MeeGo, the OS platform it was co-developing with Intel. That made sense as it could provide Nokia with much wanted room to correct its missteps and reinvigorate its software roadmap.

But mobile software technology probably grew faster than Nokia's ability to handle it. The epic tale of the world's first smartphone platform came to a sad end in February 2011 when Nokia's newish CEO concluded that it would take too long to modernize Symbian, and pulled the trigger on the ten-year-old software, making it a franchise platform. Stephen Elop called Symbian non-competitive and a difficult environment for developers before making Windows Phone 7 the primary OS for Nokia smartphones and integrating its software services into Microsoft's. By ceding control of the software experience to Microsoft, Nokia could stop trying to be a software company and focus on what it's always been good at: hardware. It's ironic that like Nokia, Microsoft had also struggled in the smartphone arena. And just like Nokia, Microsoft had been a victim of its own success. Both companies stayed with their playbooks for too long and didn't change with the times.

MICROSOFT: THE WINNER'S CURSE

If Nokia was playing the role of monolithic incumbent, the IBM of the mobile, it was countered by timeless Microsoft. The colossus of Redmond's formidable strength was its dominance of desktops and it was aiming for a day when Windows run on mobile phones as widely as on the PC. "The mobile world is moving from voice to data, and data is what we understand best," claimed Ben Waldman, Microsoft vice president for mobile devices, while speaking to *Bloomberg Businessweek* in March 2002. But Microsoft's bid to replicate the PC success faced daunting odds. The biggest obstacle to Microsoft's wider aspirations was the mistrust among potential customers, mobile phone makers, who had watched the software giant gobble up most of the profits in PCs. An even larger stumbling block was Microsoft's own oversized ambitions. All this led to persistent fears that Microsoft would try to dominate the software portion of the wireless communications industry.

The PC software company argued that it was better positioned than its rivals to offer an open platform on which third parties can create viable new wireless applications. The top Microsoft executives were confident that familiarity with Windows would help its mobile software become a standard in the new high-performance wireless gizmos. It's important for developers to have access to data they are already familiar within the PC environment, they argued. But could Microsoft extend its dominance from the desktop to the handset? Firstly, all the smart spin couldn't make up for the fact that Microsoft was generally perceived the Johnny-come-lately of wireless. Furthermore, having pocket Word and Excel programs was still not a sure bet; a wealth of functions made it too expensive for mass-market users such as students.

Mobile phone makers just didn't want Microsoft to dominate this space, so they were doing everything they could to prevent Bill Gates from turning the wireless world into a part of his empire. At first, as the leading handset manufacturers saw it, Symbian would ensure that they stayed in control of their destination. And the license fee that would go to Symbian, rather than Microsoft, was a nice way of keeping all in the family. Symbian's licensing model was publicly disclosed and uniform to all licensees: a 10 percent royalty per unit. At one stage in early 2000s, it seemed as if the cellular industry had written off Microsoft's Windows CE operating system, calling it too specific for portable computing devices.

The cellular industry was visibly uncomfortable with Microsoft's move into its territory. None of the wireless heavyweights had fully embraced Microsoft's mobile operating system. The wireless industry had witnessed how companies facing Microsoft met strange fates. They saw that when Microsoft wanted to get into a market, it first crept slowly and then stomped. It had done that to Lotus and to WordPerfect; then, it had overcome Netscape. In this very backdrop, the cell phone industry was proving a far more difficult nut to crack. As a result, handset makers like Nokia were confident that the Redmond, Washington–based software company would remain a niche player in the wireless space. So after having failed to sign up large handset makers, Microsoft instead decided to get around them by going directly to their customers: wireless operators. The software giant began wooing wireless carriers that sold vast majority of mobile phones to consumers.

For operators such as Vodafone the appeal of this approach was that they could customize the phone and brand it with their own logo, thus differentiating themselves from rival operators. For Microsoft, the appeal was that it could get phones to the marketplace without the support of large cell phone makers. That way, Microsoft set in motion a potentially nightmare phone-clone scenario for handset makers. A growing number of handset manufacturers in mainland China and Taiwan could produce no-name generic phones in mass quantities and were willing to live with thinner margins, just like PC makers. Disrupting the supply chain had provided Microsoft a much-needed entry into the mobile phone realm.

An outcome of the recipe Microsoft was cooking hit the streets in late 2002 when a little-known Taiwan firm launched one of the first high-profile hybrid PDA phones running Microsoft's PocketPC software. The SPV phone was a joint venture between Microsoft, High Tech Computer

(HTC) and Orange, a European mobile operator. Taiwan's HTC, which had produced eye-catching iPaq for Compaq, designed and built the hardware; Microsoft provided the software; and Orange agreed to buy the phones. In the long run, the software giant had hoped that the mobile phone industry will develop in the same way as the computer industry did: away from vertically integrated model, in which software is supplied by Microsoft and hardware becomes a commodity produced by firms like HTC.

But the economics were still skewed in favor of large handset makers that produced far bigger volumes. A wireless operator placing order with an original design manufacturer (ODM) like HTC, in contrast, would order a few hundred thousand at best. With smaller economy of scale, this made the handsets more expensive. So the omens from Microsoft's joint venture with HTC were not so good. Sales of the device, known as the XDA in Britain, the MDA in Germany, and the T-Mobile Pocket PC Phone Edition in the United States, had been slow. The phone also showed some serious pains with irritating, and sometimes unfathomable, bugs requiring frequent warm and cold reboots. Although the lackluster response was generally seen as akin to a lack of enthusiasm for PDA-like devices, it highlighted another problem with the ODM approach: the lack of a strong brand. Mobile phones were now fashion items and branding mattered to their users.

By the end of the decade, Microsoft's Windows Mobile OS software was seen as bowing out of the contest. The company didn't fully update the Windows Mobile in a three-year period, from 2007 to 2009, and that's long enough given how fast this market was evolving. Its hardware partners were abandoning it one by one, and Windows Mobile had lost nearly a third of its smartphone market share it had accomplished by 2008. Microsoft's mobile OS history was rooted in PDAs, which were marketed toward enterprise audiences. But the smartphone shifted into the mainstream as a consumer device, and yet Windows Mobile was still largely focused on enterprise features.

Further into this turmoil, in 2010, Microsoft's early attempt to be hip and cool went bust when it threw a lot of resources into its Danger Inc. acquisition and introduced two phones under a new brand. The devices, called Kin One and Kin Two, were built with social networking services such as Facebook and Twitter at their core. The phones had a browser and could access social networking sites through widgets. And they would make it easy to share photos, videos and access social networking feeds, as promised Microsoft's ads. Manufactured by Sharp for Microsoft, and available exclusively on Verizon Wireless, the phones were targeted at teens and social networking addicts. But Microsoft crippled the overall functionality of the device by not allowing apps or games on the phone. Microsoft scraped the device merely two months after its launch. They would be integrated into the company's upcoming mobile operating system: Windows Phone 7.

"We were ahead of this game and now we find ourselves [number] 5 in the market," Microsoft CEO Steve Ballmer said at the 2010 All Things Digital Conference. "We missed a whole cycle." Over time, it became evident to Microsoft's software leadership that the company has been trying to put too much functionality in front of the user at one time. And that resulted in an experience that was a little cluttered and overwhelming for a lot of people. Clearly, despite having invested seriously in software for handsets, Microsoft had yet to come up with a mobile platform that captivated the imagination of the market. So after Windows Mobile OS tanked in popularity, Microsoft set itself for a complete do-over on mobile. In December 2008 came "The Reset," as Microsoft staff members avidly call it—the company halted work on its Windows Mobile OS and started over.

MICROSOFT'S SMARTPHONE CAMEO

The company spent the next six weeks hatching a plan for a Windows phone do-over, and it set a deadline of one year to build and ship a brand new operating system. The outcome was Windows Phone 7, an operating system with a brand-new user interface that looked nothing like its predecessor. Microsoft knew it all too well that its future in the mobile space depends on this reboot, so it torn up its playbook and built a new mobile phone platform from scratch, completely eclipsing the old Windows Mobile offering. It was a "make or break" for Microsoft, as Windows Phone 7 manifested an attempt at righting its previous fumbles in the mobile space.

Windows Phone 7 was targeted at uniting consumers' personal and business needs into a single device to enable them to navigate seamlessly between work and play. Joe Belfiore, who had earlier spearheaded Microsoft's Zune product launch and was the man in charge of its new mobile effort, summed up this target market as the "life maximizer." Microsoft, Apple's old nemesis from the PC wars, was now planning a major consumer push for its phones, which in the past had been aimed mostly at corporate users.

Windows Phone 7 introduced two new concepts—"the hubs" and "live tiles." Similar to the Zune HD's interface, the Windows Phone 7 main screen was organized into six "hubs" containing different sets of features: People, Pictures, Games, Music plus Video, Marketplace, and Office. For music and video, Microsoft integrated its Zune player software into Windows Phone 7, which enabled mobile phone users to sync and play content downloaded through the Zune Marketplace store. Windows Phone 7 devices would also include a built-in FM radio and Zune Social to share their music recommendations with other users.

The "live tiles" replaced the cluttered front screen of the device with large, customizable icons, automatically updated with important information like reminders, appointments, weather, and anything else the user wished to see at a glance. Microsoft mandated an agreement with manufacturers and mobile operators that phones will ship with half of the front screen reserved for carriers' and manufacturers' customized apps, while Microsoft gets the rest of the screen for its default apps like e-mail, calendar, address book, and so on. Any of these apps could be removed from the front screen if customers didn't enjoy them.

As for usability, Microsoft's phones would support about the same touch gestures seen on the iPhone. But despite the similarities to the iPhone, the general hub-based user interface was a major difference from any smartphones on the market. Windows Phone 7 was a solid operating system that was easy to use and had some nice features like Media Center, software that enabled users to stream TV to their PCs and to record shows. Media Center had been around since 2002, but was greatly expanded for Windows Phone 7. Developers could inject their apps into Microsoft's standard hubs, and they had the option to create their own hubs. Clearly, Microsoft deserved credit for doing things that went against the prevailing smartphone norm.

Microsoft had witnessed the crucial role that Apple's marketing and public relations spin machine played in the iPhone's ascent and how Google was attributed for its aggressive marketing of the Android brand. It is critically important to generate excitement around a platform, and both the iPhone and Android were able to create a lot of buzz. Apple had though targeted its marketing firestorm mostly at iPod users, retail consumers, and its own Mac OS X developer base, while Google's buzz had largely been confined to punditry and open-source advocates. Microsoft, on the other hand, had become a punching bag for many tech pundits. The software giant often attracted jibes on its mobile strategy despite its heft and undeniable importance to the tech industry. So now that Microsoft had come up with a complete overhaul of the mobile platform, it spent an estimated US$400 million at the launch of its new operating system, getting the message out that it had learned from its mistakes.

Windows Phone 7 was way ahead of earlier iterations and provided a vastly improved user experience that pleasantly surprised many people when they used it. The interface design was elegant and intuitive. In sharp contrast to the confusing Kin phone, the new operating system was easy to navigate. Moreover, the integration of Microsoft service assets—such as Xbox Live, Bing, Zune, and Office—greatly strengthened the overall value proposition. Although Microsoft would continue its strategy of licensing its OS to manufacturers, this time it would set the rules: all phones running Windows Phone 7 must meet hardware requirements, and every device must pass a series of lab tests conducted by robots designed by Microsoft engineers.

Although Microsoft was late to the game, Windows Phone 7 was aimed at doing right what Google had been doing wrong. The crucial part of Microsoft's strategy was the quality control it imposed onto its hardware partners. Rather than coding an operating system and allowing manufacturers to do whatever they want with it—like Google had been initially doing with Android—Microsoft required its hardware partners to meet rigid criteria in order to run Windows Phone 7. In its prior operating systems, even Microsoft allowed mobile carriers and manufacturers to determine the features they wanted on a phone; they'd issue a list of specific instructions to OS makers like Microsoft. Consequently, mobile phones turned out to be overloaded with features and were unfriendly to users.

The effort to control quality and consistency was seen as crucial in Microsoft's quest to regain some ground in the phone battle. Microsoft had now mandated handset manufacturers to use the same display size and number of camera megapixels to give users a consistent experience across all Windows Phone devices. Microsoft had to move fast to stay in the smartphone game, because the longer it took to get Windows Phone 7 out, the more difficult it would have been for it to regain the ground. And now that Microsoft had made a splash in its bid to differentiate itself from the rest of the market, the challenge was to execute, to continue to tell the rest of the story, and to begin to expand the market share. The company, three years behind the competition, needed as much developer support as possible and as soon as possible. Withholding some of the tools necessary to build a viable app ecosystem at launch, whatever the reason, wouldn't encourage developers to jump over to Windows Phone 7.

One thing that Windows Phone 7 had going for it was that no one was expecting much. Microsoft's last mobile operating system was clunky, antiquated, and subsequently retired. The Kin One and Two, those smartphone-like pair of devices that Microsoft launched for youngsters earlier in 2010 had already become relics. Now, no one doubted that Windows Phone 7 was good, but was it good enough? Could Microsoft create a platform that stands out enough to compete with Android and Apple? Would it be different enough that consumers would come in droves? Would it be pervasive enough that developers would want to invest their limited resources in making apps for it? By spring 2011, iOS had almost 350,000 apps available to consumers through the App Store while Windows Phone 7 had about 8,000 apps.

Although Windows Phone 7 was late and badly trailing iOS and Android, reviews of software have generally been good, and as to the development of critical mass, the volume of handsets that Nokia delivered could certainly be a springboard for Windows Phone 7. The deal gave Microsoft access to one of the world's largest handset distribution networks. Microsoft had shipped more than 2 million licenses during the first quarter of the launch of Windows Phone 7 while Nokia had shipped 28.3 million handsets during the same period. If Nokia had chosen Android, it'd have probably been the end of road for Microsoft's mobile ambitions. Now, at least on paper, Windows Phone 7 had some momentum. Still, the alliance was more of a gamble, a last-ditch effort of sorts for both Microsoft and Nokia to gain a lasting foothold in the booming smartphone market.

Microsoft was desperately trying to make a comeback and its proponents expected Microsoft to redeem itself with Windows Phone 7, as it did with Windows 7 after negative reception of Windows Vista. Microsoft has had a history of eventually dominating the market in which it has made a big push—until it set its eyes on the up-and-coming handheld and smartphone realm. Microsoft's hope with Windows Phone 7 aside, it has a staggering task ahead of it to catch up to and surpass Android and iPhone. The crux of the story boils down to one critical aspect: as open-source codes emerge with graceful interfaces created and managed by the likes of Google, it will become harder to justify paying royalties to Microsoft. But then Microsoft's multilayer tie-up with Nokia brought an abrupt about-face in software giant's business model.

According to media reports, billions of dollars will funnel from Microsoft to Nokia most likely in the form discounted software licenses, marketing cooperation and other non-cash benefits. Sun Microsystems never really found a way to make money on Java amid fears that charging for technology would hinder its adaption. That leaves Microsoft as the one scrambling for a piece of action in the mobile arena. So the question is if the future of computing lies in mobile, will Microsoft survive post-PC era? Despite Microsoft's shrinkage in the mobile OS market, industry watchers say it's not over for Microsoft. The software giant has a lock on integration with Windows PCs and servers. Microsoft's conviction in its Windows brand is very strong, and the firm is constantly trying to do more to leverage what it believes is the good will value of the Windows brand.

It's possible for Google and Apple to both succeed in mobile phones, but it's harder to imagine Google and Microsoft both thriving. Google's business model, which calls for getting manufacturers to use its operating system, is far closer to Microsoft's than to Apple's strategy of making money on the sale of proprietary hardware. But Google offers Android to manufacturers at a price that Microsoft can't beat: free. Microsoft is pitching it to the same companies that have embraced Android, claiming that Windows Phone 7 is more polished and will give phone makers a better tool for competing against Apple. Meanwhile, it is hoping that fragmentation will eventually become the undoing of Android.

THE FIGHT FOR SMARTPHONE BRAINS

Software usually makes up about 20 percent of the cost of a smartphone. So, chapters 6 and 7 presented the bigger picture of the mobile OS landscape. While the previous chapter avidly looked at the iPhone versus Android juggernaut, this chapter was dedicated to a review of early OS pioneers in smartphone space who subsequently lost their way. The fight of mobile OS dominance is akin to the fight over next-generation gadget brains. Apple and Google captivated mobile phone users with sleek touchscreen software and app-phone bragging rights and, in just three short years, captured a large chunk of the market. But while the iPhones and Android handsets commanded much of the mind share, Nokia still had the actual market share. Nokia's global reach provided Windows Phone 7 with a strong ally on the hardware side, and that could ultimately create a three-horse race. So one should ignore them at his or her risk.

Then there is RIM, which often is lumped with Microsoft and Nokia as a wireless pioneer that squandered early advantages and is sinking toward irrelevance. There was a time when RIM commanded 42 percent of the smartphone market share. RIM provided a proprietary multitasking operating system for the BlackBerry, which made heavy use of the many specialized input devices available on the phones, particularly the scroll wheel, trackball, and trackpad. RIM continued to have a lock on the business phone market with its "Crackberry" environment built around wireless e-mail and later web access. RIM's newer operating system, called BlackBerry 6, released in April 2010, boasted a redesigned browser and touchscreen technology that allowed users to zoom in and out of images using two fingers, like on the iPhone.

The company rolled out the BlackBerry Torch in August 2010 based on this new OS amid complaints from both users and developers that BlackBerry's operating system was slow and clunky and had a dearth of fun apps. But then RIM eschewed the recently revamped BlackBerry 6 OS in favor of a completely new platform built by QNX Software Systems for its upcoming professional tablet: PlayBook. RIM bought QNX, a maker of embedded operating systems used in everything from consumer electronics to cars to nuclear reactors, in what industry watchers said was a bid to replace its legacy software criticized as being slow and buggy.

The move initially implied that RIM would have to juggle two distinct operating systems as the company executives didn't say what its plans would be for the BlackBerry operating system. But people familiar with RIM's strategy reckoned that the company will eventually switch its smartphones to the QNX system as well, although that transition is likely to be gradual.

Coming back to Symbian, which had been a victim of first birth and of long-standing perceptions. Symbian's dominance faded as mobile software platform market became more crowded. When Nokia started calling up developers on the platform of Symbian Foundation, there weren't many listeners. Despite the fact that Nokia was still the top smartphone maker in terms of market share, developers were flocking to competing platforms—Android and iPhone OS—as they saw a negative momentum with Nokia. These software types have a very keen insight of how fast or slow a platform is gaining traction. The way Android had knocked down royalty-based models of Microsoft and Palm and had forced the 800-pound gorilla Symbian to toe its line hadn't gone unnoticed. Apple was another matter, however.

One view is that all the operating systems other than Android will largely be flat or decline over the coming years. But the other camp says that no one OS satisfies everyone and that wireless carriers want to see an alternative ecosystem in the market. The evolution curve in the early history of the smartphone's development showed two fundamental patterns. First, software has been evolving slower than hardware. Second, once a technology platform accomplishes the economy of scale, the business dynamics completely change. That tipping point in the smartphone's software evolution cycle was reached when both Android and iPhone demonstrated the phones could be an incredible mobile platform. Now huge battle of mind share in cell phone software was playing out among Apple, Google, and Microsoft.

It's probably too early to tell which mobile platform will win out. Nokia, having the world's largest software firm at its back, has the advantage of a large installed base. Android could benefit from the pure innovation, which is galvanized when developers take a "sky's the limit" approach to building a new operating system. Apple provided a complete, turnkey approach to software sales via its iTunes App Store, which appeals to consumers as well as developers. Reminiscent to its PC history, Apple is playing things close to the vest while Google is out to expand. But in this ongoing corporate battle for market share, the comparison is tricky because Apple sells phones for luscious margins while Google gives away Android to handset makers for free. Apple develops software for devices it sells while Google develops software and lets other companies use the software for the devices they sell.

The floodgates of the twenty-first-century gadget bonanza are just opening. There is broad consensus on the famous industry adage that all mobile phones will be smartphones by end of second decade of twenty-first century. The argument goes like "The winner of the smartphone wars will capture the ordinary cell phones market of 1.3 billion handsets annually. This is going to be the biggest technology prize of consumer electronics, ever." Nokia's tie-up with Microsoft was a stark reminder of the fact that the battle was moving from mobile devices to mobile ecosystems and ecosystems thrive when they reach scale. The ecosystems include not only the hardware and software of the mobile device, but developers, applications, m-commerce, advertising, search, location-based services, social networking, and many other things.

The huge and increasing costs of developing mobile operating systems could result in consolidation for the smartphone OS market. This, along with the arduous work involved in creating an ecosystem around the OS, could potentially force handset vendors to consolidate their efforts around one or two third-party licensable operating systems. Smartphone platforms with strong consumer focus and deep R&D budgets, notably iPhone, Android, and Windows Phone 7 could ultimately win out. HP's failure to gain traction in the smartphone and tablet domains despite its acclaimed webOS platform is a stark reminder that developers just can't get useful business tools to work on them all. That's because not all devices are created equal. Besides different operating systems, some have different-sized screens, and others have different keyboard layouts. Some come with a standard amount of memory.

At this point, it is important to note that operating system is just a start and is only a part of the overall smartphone equation. The applications riding on top of OS software are also a vital part of the value proposition. The mobile handsets got through to the next stage of smartphone evolution only after users began adding applications from third-party developers to take advantage of the seamless Internet access. But before delving into the apps phenomenon, first I look at the hardware beneath these mobile operating systems. The following chapters take a peep on how the smartphone hardware evolved over the years and how the related functional building blocks like imaging and video are taking shape.

THE ANATOMY OF THE SMARTPHONE

"All cell phones will be smartphones eventually."
—Jeff Hawkins, inventor of Palm Pilot handheld
computer and Treo smartphone

On April 3, 1973, Martin Cooper, general manager of Motorola's Communications Systems Division, walked down Sixth Avenue in New York City using the world's first handled mobile phone to call Joel S. Engel, his adversary and head of research at AT&T Bell Labs, where the project had originated fifteen years earlier. A casual call on a portable device weighing about two and a half pounds touched off a decade-long race between Ma Bell and Motorola to bring the first cellular phone to the market. "As I walked down the street while talking on the phone, sophisticated New Yorkers gaped at the sight of someone actually moving around while making a phone call," Cooper recalled on the thirtieth anniversary of the mobile phone. "Remember that in 1973, there weren't cordless telephones, let alone cellular phones. I made numerous calls, including one where I crossed the street while talking to a New York radio reporter—probably one of the more dangerous things I have ever done in my life."

Cooper used a prototype phone connected to AT&T's wired phone system via a base station on the rooftop of a nearby building. About two decades later, when IBM showcased the first smartphone, Simon, at the 1992 Comdex show in Las Vegas, mobile phone industry was in a whirl land of transition from analog to digital handsets. Still, IBM's rare entry in the mobile arena brought the first glimmers of computing DNA in a handset. Simon was released to the public in 1993 and sold by BellSouth; besides being a mobile phone, it contained a calendar, address book, clock, calculator, notepad, and games, and could send and receive fax and e-mail. The feature set it offered was incredibly advanced at the time. It had no physical buttons to dial; instead, consumers used a touchscreen to select phone numbers with a finger or create facsimiles and memos with an optional stylus. Text was entered with a unique, on-screen predictive keyboard.

Simon suffered from production issues and struggled to handle all of its tasks, including calling, faxing, and paging. But Simon's entrance into the mobile phone marketplace didn't go unnoticed, as evidenced by this *Byte* magazine review in December 1994 issue: "Whether or not Simon is your idea of the ultimate (for now) personal communicator depends on how appealing you find the combination of voice calls and e-mail—and maybe on how little you need a laptop. Clearly, Simon won't replace portable PCs, but it's equally clear that it represents a milestone in the evolution of the PDA." In retrospect, this landmark launch by the computing industry maestro marked the origins of the meeting point of the PDA and the cell phone that would eventually shape into making of the smartphone many years later. This novel idea of the early 1990s concealed the blueprint of a brand-new industry whose time had yet to come.

It was probably Simon's powerful influence that both computing and telecom industries remained hung for more than a decade on the notion of a single portable device that would enable users to make phone calls, plan their schedule, check e-mails and surf the Internet. Here, the telecom guys wanted a cellular phone that could double as a PDA; people on the PC side preferred a handheld computer that could also make a phone call. But almost since the inception of the PDA, predictions that the PDA and the cell phone would eventually converge into a single, easy to carry device had been made. With the passage of time, the call for convergence only grew louder and more persistent.

The other early products also fell short in one or more significant ways, and the consumer often bypassed them in favor of separate cell phones and PDAs. The Nokia Communicator line was the first of Nokia's smartphones starting with the Nokia 9000 released in 1996. This distinctive palmtop computer-style smartphone was the result of a collaborative effort of a modestly successful and expensive PDA by HP combined with Nokia's bestselling phone as the early prototype model had these two devices fixed via a hinge. The Nokia Communicator models were also remarkable being the most expensive phone sold by a major brand; sometimes 40 percent more pricey than the next most expensive smartphone by any major manufacturer.

Interestingly, although the Nokia 9290 was arguably the first true smartphone with an open operating system, Nokia continued to refer it and the following models as Communicator. Only Ericsson referred to its product as "smartphone" at that time. In 1997, Ericsson released the concept phone, GS88, the first device labeled as smartphone. Later in 2000, the Ericsson R380 was another phone sold as a smartphone, which had the usual PDA functions. Its large touchscreen was combined with an innovative flip so it could also be used as a normal phone. It was the first commercially available phone with Symbian OS, though it couldn't run native third-party applications.

In the coming years, products emerged that offered a form factor and range of services that found willing buyers. Such successes emboldened the champions of convergence even more. In the late 1990s, Handspring delivered the first widely popular smartphone devices in the U.S. market by marrying its Palm OS–based Visor PDA with a piggybacked GSM phone module, the VisorPhone. Then in 2002, Handspring started marketing its new integrated PDA-phone as the Treo, which featured a tiny, but full keyboard; a decent-sized, bright color screen; and options for connecting to the Internet. The Treo phone offered e-mail, calendar, and contact organizer features and allowed third-party applications that could be downloaded or synched with a computer. Treo was a study in simplicity and ease of use, and it won over many customers due to an ideal design. That same year, RIM released the BlackBerry 5810 handset which was the first phone optimized for wireless e-mail use.

The mobile phone, which started its journey as a brick-like analog device back in 1973 and later turned into a sleek digital gizmo through digital metamorphosis, had come a long way for the smartphone prime time. But this promising new segment was still in a flux and thus remained largely undefined. The idea of the smartphone was borne out of the wireless industry's desire to move beyond voice—into the data realm— during the days of the Internet euphoria. Over time, the cost of processing power and memory would continue to decline, so phones were certain to get smarter. But what was it that the smartphone actually stood for? For some, a smartphone ran a complete operating system, and for others, it was defined by a few advanced features, such as e-mail and Internet. They said smartphones are all about features. So should we call it a smartphone or an app phone or an Internet phone?

WHAT'S IN THE NAME?

While most of us have some idea what smartphone is, the term is mostly ambiguous. The notion of smartphone emerged in 1998 when a group of wireless manufacturers led by cell phone kingpin Nokia bought stakes in British handheld computer maker Psion to adapt its operating system for data-centric phones. The definition of *smartphone* has been evolving since then. In retrospect, the wireless industry co-opted to phrase the term "smartphone" and later went on trying to figure out what it really means. The term *smartphone*, while the device was still in an embryonic stage, emerged as more of a public relations moniker with no engineering definition. Initially, one way to define a smartphone was a device that offered Internet access, had a relatively large color display and provided, among other things, video and still imaging.

In the early going, both smartphones and feature phones were perceived as handheld computers integrated within a mobile phone, but while most feature phones were able to run applications based on platforms such as Java ME and BREW, a smartphone allowed the user to install and run more-advanced applications based on a specific platform. Smartphones ran complete operating system software, providing a platform for application developers, so a more refined perspective led to the notion of smartphones having a full-fledged OS as compared to RTOS software driving the feature phones. And over the years, a broader consensus emerged for a consistent definition of the smartphone as mobile devices running on an operating system for which third-party applications could be written. Applications designed for smartphones could access core functionality within the operating system, and tended to be more powerful and efficient than third-party software on feature phones.

In the hindsight, however, smartphones evolved from feature phones. The lines between feature phones—often perceived as multimedia handsets—and smartphones had been continuously blurring. The definition of "smartphone" was blurring because even the low-end feature phones began offering software-centric PIM functions such as calendar and address book. The wireless industry aggressively integrated Internet and multimedia features into handsets, and consequently, the pressure on feature phones to evolve into smartphones came faster than expected. On the heels of this confusing picture, the meaning of "smart" broadly stressed that users could now make customized and lean-forward usage of Internet and multimedia functions, while in the past, they were limited to lean-back usage of content unilaterally provided by service providers.

Then there was the PDA dimension. Handspring's Treo was the mobile phone integrated into the PDA. Within a few years, the PDA had all but died out as the smartphone emerged to replace it. Now the same thing was happening, but in reverse: smartphones were getting more functions added to them, and the phone part was fading away as just one aspect of the overall communication and application device. These devices could be called smartphones if one liked, but increasingly, the phone part of the device is just becoming another feature, another widget on the home page. Even smartphone marketing campaigns are dropping the emphasis on the "phone" part in favor of all the other things the device could do. The phone dialer is reduced to a little icon on one corner of the phone's display, which is otherwise filled with apps like Facebook, Twitter, and e-mail, among others.

Even the term *mobile phone* is quickly becoming a misnomer. Amid the rise of so-called smartphones that could perform everything from browsing the web to downloading and storing pictures, there is growing concern that what today we refer to as a mobile phone isn't quite the right description for these new do-all gadgets. Moreover, it's no longer just about cellular technology; handset capability is increasingly defined by the benefits brought forward by the integration of multiple wireless network technologies like Wi-Fi, global positioning system (GPS), and Bluetooth. These new, multipurpose handsets now switched among cellular networks, Wi-Fi, and WiMax, a larger-range broadband network. Cell phones were designed mainly to accommodate voice calls; now technology gurus see the name as antiquated.

What users had was no longer a mobile phone but a landmark, genre-defining product that was a fully capable, no-compromise computer in their palms. A testament of this creative ambiguity was the fact that Nokia execs would say we, as a society, have entered the age of multimedia computers while Samsung called this new category of devices the mobile information terminal. All the while, the "smart" fever generated from mobile phone realm swiftly spread to multiple facets of electronics, smart TV being one prime example. The "smart" buzzword has even been used outside the electronics industry in next-generation phrases such as "smart tariff system" and "smart management."

HIGH-OCTANE EVOLUTION

By the late 2000s, industry watchers were witnessing the fastest evolution of a mass technology to date in what was also unmistakably the emergence of the next genre of computing: the multipurpose, endlessly modifiable "app phone." It was, however, evident by the late 1990s that future cell phones would be used for more than just voice communications. Product designers envisioned at that time that these phones would one day hold keys to users' bank accounts and could even remotely lock and unlock the main door of their homes. It wouldn't really be a phone anymore. It would be whatever the user needed it to be: a phone, a camera, a portable PC, and a remote control.

And in those early days, two views predominated when it came to the future roadmap of smartphones. The first one said why someone would need more than a simple phone for making voice calls: voice, not data, they argued, was the killer application that created a mass market for wireless handheld devices. In consumer surveys, a predominant number of cell phone users favored this viewpoint, saying, "a talk-only phone is all they want because it's simple and convenient." But wasn't it the chief executive of IBM who said in the 1940s that the world would ever need two mainframes. And Microsoft co-founder Bill Gates said back in 1981 that no one would ever need more than 640 Kbyte of memory for a PC.

The history of wireless communications offers more specific usage patterns. Only doctors used to have pagers in the early days of this electronic gadget's availability. Then came a time when a majority of children in the U.S. classrooms started carrying pagers, whose beeper would decide much of their course of action in the after-school hours. And remember the early days of cellular when handsets were big and clunky, and only business people were supposed to use them. So the other group in the debate about the smartphone's future advocated that data- and image-enabled feature phones were inevitable. Otherwise, they reasoned, cell phones are at a risk of becoming low-margin consumer products just like landline telephones.

A new generation had grown up with the mobile lifestyle that would make communicative PDA-like cell phone an even more important proposition than they were in early 2000s. There was clear evidence to suggest that people liked devices such as PDAs, MP3 players, and digital cameras; these gadgets had been discretely successful. What would happen in the natural course of evolution was that each of these devices would find a home in cell phone. A cocoon of features started building around conventional mobile phones, mixing a little bit of computing functions and a little bit of multimedia functions to ignite consumer curiosity. A lot also depended on how well wireless operators did in developing new services and promoting them to consumers at reasonable prices. And once the usage models took hold, the dynamics of the smartphone premise gradually began to clear.

Before going any further in the smartphone's coming-of-age story, a small detour into the genesis of the smartphone from the telecom side could help better understand the technology pedigree. Telecommunications began with landline phones; then people started having cordless phones in their homes; and subsequently they had cellular phones. After the inception of cellular handsets, telecommunication industry's dream weavers wanted to devise a killer machine that was the end-all and be-all of the gadget nirvana, but they mostly ignored the technology's natural evolution curve. In retrospect, basic cell phones and PDAs only embodied an evolutionary stage in the making of smart gadgetry. Smartphones would only evolve over time. For instance, the mobile phone's first killer application, text messaging, grabbed consumer attention half a decade after its inception.

The convergence of voice and data on the smartphone platform was an evolutionary process and kept changing shapes. The origins of smartphone could partly be traced in the early days of GSM project when the first-ever digital handset was conceived. That was when the journey toward making of smart gadgetry began with the fusion of a number of auxiliary technologies into digital cellular phones. One such technology was smart cards. In 1985, eleven years after the advent of smart-card technology at its research arm CNET, France Telecom ordered 7 million smart cards for its payphones to tackle coin theft. That provided the basic impetus for smart-card critical mass in Europe; GSM was going to be the next frontier.

The GSM collaborative circles initially saw smart cards as a basic security tool for preventing fraudulent calls. Based on smart-card technology, subscriber identity module (SIM) was conceived to function as an identification, accounting, and security device in the mobile environment. The embedded device stored a mathematical algorithm that encrypted voice and data transmission, making call eavesdropping nearly impossible. Moreover, a subscriber could access mobile services using a SIM card in areas of coverage not offered by his wireless operator. Over the next few years, however, handset manufacturers realized they could explore SIM cards to bring more adaptable and powerful phones. So in the next development phase of GSM, SIM cards helped cellular handsets offer a range of predefined services. The journey continued over the years in different forms and manifestations.

Then, color displays played a key role in facilitating wireless industry's drive in migration from voice to data- and multimedia-intensive services. Years of squinting into tiny monochrome screens had given consumers hunger for big, bright color screens, and consequently, display technology took a quantum leap in early 2000s. Fast-forward to early 2010s, AMOLED technology is bent on creating super screens that promise brighter, less reflective, and more power-efficient displays. The color display episode also illustrated the shift taking place in the next genre of computing. Microprocessor had been the center of the PC universe and consumers reserved the largest chunk of their PC budgets to buy the fastest chips. But in the next-generation computing device—the smartphone—the processor almost became a commodity component, while items like display and touchscreen were the centerpiece. Of all the parts that made up the iPhone, the touchscreen was now the most expensive—around 20 percent of the phone's manufacturing cost. The phone's processor, meanwhile, cost between mere 5 to 8 percent.

CONVERGENCE AT LAST

Clearly, there was increasing consumer demand for a compelling user experience and for access to services and applications for rich mobile Internet experience: high-definition content for video, imaging, audio, gaming, etc. So the mobile devices were constantly getting smarter with easier browsing, easier content and application discovery and acquisition, and more user-friendly interfaces. Now phones had more senses than just "ears" and a "mouth." They had eyes in the form of cameras, as well as GPS to know where they were, and Bluetooth and other connectivity features to connect to all sorts of peripherals and enhanced multimedia capabilities. The market for just about anything that had to do with multimedia or communication was moving to cell phones. That would drive more smartphones to the market.

The prudish success of the communicator formats like Palm's Treo handset during mid-2000s suggested that the market was tiring for handset devices that do not offer integrated wireless communication facilities. And there was little doubt that the technology needed for powerful convergence devices was now within reach. The challenge for product designers was to find combinations that added utility, not just technological dazzle, and to think carefully about the awkward and potentially hazardous effects of these combinations. For instance, when it came to the form factor, if a device didn't fit into a shirt pocket, it would probably going to be tagged as oversized. Earlier, in the landmark transition from analog to digital, clamshell and flip-screen designs emerged as a way of cheating on size limits and cramming more features into a small form factor. But in case of smartphones, the Treo phone was wider than a typical cell phone and, thus, was uncomfortable to hold for some people.

The overarching problem for many would-be smartphones was that they were either too big to fit easily in a pocket or they were too small to do the real work. Even the earlier versions of the iPhone were bulky enough for the blue-jeans crowd. Smartphones were all about features, but despite bulking up on features, smartphones had to take cost, size, and weight into consideration. The new mobile handsets needed to be designed to serve genuine user needs and to serve them well. In retrospect, this crossroads for smartphones essentially embodied issues centric on user interface, which is the way a human interacts with a computer. That the company that solved this conundrum was also the one that had personalized the computer age more than two decades ago wasn't a mere coincidence. The Apple value system and design philosophy that led to Macintosh computers placed a great emphasis on the needs of individuals rather than technology—what Apple designers once called the user experience.

Apple first borrowed the form factor from the PDA, one of the early forms of portable computing, and then implemented touchscreen technology that used almost 100 percent screen area with a bright, crisp look. A virtual keypad would further reinforce this new form factor. Apple also hit the nail on head by designing apps for the smaller phone screens instead of recycling user interfaces otherwise meant to be displayed on larger-screen devices. The irony was that more than a year before the iPhone's launch, Sony Ericsson's 990i smartphone had offered a perfectly functioning touchscreen stylus, Internet access, e-mail, video cam, and removable storage. In fact, touchscreens have been available since the days of cathode-ray tubes (CRTs), but the technology earned consumer acceptance only when Apple adapted it, first in the iPhone and then in the iPad, to solve the tiny button problem. Multitouch controls didn't work very well in the past, because they had to operate on a standard operating system. Apple created a very touch-friendly user interface in its iOS software. So it wasn't just the mobile Internet experience that Apple had reinvented; the company has a history of turning yesterday's geeky dreams into mainstream must-haves.

The evolution of smartphones has been non-stop since their inception in the late 1990s. The demand for advanced mobile devices boasting powerful processors, abundant memory, larger screens, and open operating systems outpaced the rest of the mobile phone market. But evolution was a creative process, and as the iPhone effectively demonstrated, what mostly distinguished smartphones from ordinary phones was their superior user interface. That was imperative because smartphones were to have a large, high-resolution display; a robust HTML browser; rich messaging and media capabilities; a multitasking and multi-threaded OS; and the ability to do over-the-air software updates. The re-engineering of the mobile phone had started, and it was now on its way to become the most common and powerful consumer electronics information device ever made.

THE MATRIX OF NEW PRODUCTS

What is convergence? It's the coming together of two entities to make a third. For instance, microphones and speakerphones merged to form headset. More than a century ago, the cart and the engine came together to give way to automotive revolution. That redefined society by enabling people to go for shopping and vacations, created new infrastructure like roads and highways, and helped establish auxiliary industries such as gas stations. The collapse of speed and travel time brought us new social norms and societal changes as new usage models took hold to create new opportunities both in social and business segments.

Technology gurus had long been calling for convergence of voice, data, and video to make way for another electronics boom. Subsequently, telecom and computing worlds joined hands for a number of high-profile undertakings only to give them up a bit later. Back in the 1980s, we heard dream weavers saying how Integrated Services Digital Network (ISDN) will change the telecom landscape with converged voice and data services. A decade later, the marketers did an adaptation of the convergence theme to create another pipe dream: Asynchronous Transfer Mode (ATM). At about the same time, cable mogul John Malone put forward his vision of 500-channel multimedia universe on cable networks through intelligent set-top boxes handling multiple streams of data.

By the late 1990s, while industry witnessed an unprecedented boom in computing and telecom markets, the mantra of convergence gradually became synonymous with a technological field of dreams. After so many failed ventures, the notion of convergence of the mobile phone and the PDA seemed more like wishful thinking, a misnomer of sorts! Indeed, many had come to assume it would never happen. But market saw convergence as a failed bid mainly because we saw it in the public relations' smoke and mirrors. The fact of the matter was that convergence was already happening; it's just that we didn't know it. Quietly, without many people taking notice of it, real convergence began happening at some modest levels. Text messaging on cellular handsets was practically the first successful example of convergence between voice and data. Then camera phones came out of nowhere, defying all technological norms, and became the epitome of converged devices.

It wasn't that one day industry would wake up with the convergence dream having come true. The convergence between computing and communications was an evolutionary process, and it would keep changing shapes. Computing and communications are still different industries, but there are growing overlaps in many areas. That has led to the creation of common building blocks that the two industries are increasingly looking to share. The convergence started at the device level and the network followed. And once the convergence started to happen on the network services front, multimedia-rich content was the next one to join the party.

In the smartphone arena, the convergence episode began with this relentless pressure to put cell phone and PDA functions onto a single device. The most crucial driver was the premise that it's more convenient to carry a single device. What we saw with the coming together of the phone, the PDA, and the digital camera was micro-convergence, which eventually gave way to macro-convergence at a new level. As the notion of the smartphone moved deeper into the convergence fray, the new opportunities that it could bring were unimaginable. Who would have thought of adding an MP3 player into a cell phone? This made cell phones literally mobile entertainment systems. In fact, music became the first mobile content type in 1998, and by 2010, about 38 percent of all consumer music spending was mobile when ringtones are included. Video games followed, and by 2010, a quarter of video gaming service revenues were generated on mobile platforms.

The iPhone, which boasted the most downloaded apps in games, was now considered to be the first globally successful gaming phone platform to rival the Nintendo and Playstation Portable. But the convergence endeavor in this specific market began many years ago with the first gaming-oriented smartphone: Nokia's N-Gage. The device was Nokia's attempt at gaining market share from handheld gaming players like Game Boy Advance. In the early 2000s, gamers were increasingly carrying around both a cell phone and a Game Boy, the most popular handheld game system. Nokia spotted an opportunity in combining these two devices into a handier unit.

Apart from games, Nokia packed multiplayer over Bluetooth or the Internet, MP3 and Real audio/video playback, and PDA-like features into the system. N-Gage created a plenty of buzz when launched in 2003.

This combination of cell phone and gaming device was supposed to lure gamers away from their portable devices. But despite the large amount of attention gamers gave to N-Gage before it launched, it earned scorn for its odd, curved design, and the fact that users had to hold the phone on its side to place a call. The speaker was in the side edge of the phone, resulting in many mocking it for side-talking.

The original N-Gage was heavily criticized for its clumsy design: to insert a game, users had to remove the phone's plastic cover and access the battery compartment. Later versions fixed many of the problems with the original device, but the damage had been done. Gamers also blamed the lackluster performance on the poor selection of games compared to those available on Nintendo's Game Boy Advance while still costing more than twice as much. Nokia discontinued the phone line and related gaming services in 2010. The success of the iPhone in an area where Nokia's pioneering effort failed yet affirms that convergence is an evolutionary process and that carefully crafted ecosystem plays a crucial role in such innovative attempts.

What's so fascinating about convergence is that it allows multiple facets of technology to be integrated onto highly accommodative platforms like the smartphone. People want to add web browsing, video streaming, music player, GPS, and much more. Some phones would focus on photography and picture messaging; others on playing music and games; yet others on corporate e-mail access. The digital convergence enabled fantastic new services and abilities we never could have imagined. Cell phones were now moving from the realm of "accessory" to "lifestyle tool," thus positioning themselves as a fulcrum point in people's lives. The convergence of smart gadgetry and lifestyle, and the increased focus that people put on social connectivity, meant that phones would become gateways to their digital life.

The preceding examples establish that convergence is not a mere buzzword for fancy gadgets and that there is a big picture to it all. The smartphone engine is driving the mobile industry, which according to some industry observers, would amount to US$5 trillion. Here, the actual mobile device alone is worth around one-fifth of that; there is far more for outsiders to contribute to this enterprise than for the legacy mobile giants. To converge, instead of being additive, would eventually stimulate new applications and thus would fuel new opportunities and ignite pent-up demand for innovative new products. Convergence is in fact the matrix of new products. Cell phones, which now looked like tiny PCs, were evolving into a mobile terminal that would provide access to personal and corporate data from any location. The smartphone had become a symbol of convergence; it had taken convergence to a new level.

INTEGRATION PARAGON

In the early stages, amid all uncertainty, market watchers called the cell phone merged with a built-in camera and an MP3 player a Frankenstein's monster: technically brilliant but disastrous in practice. However, a mere existence of these devices was promising at that time because it demonstrated an early level of standardization. Although not exactly the harbinger of a golden age of consumer electronics, these devices attained two important merits. First, they drew developers, making it easier to find bits and pieces needed for a converged device. Second, they created a sense of confidence in the interoperability and longevity of these products. Subsequently, the genuine integration became achievable. And it became apparent that there wouldn't be a single device catering to all applications. The so-called single-device paradox had led to the development of so many combo devices, but none of them actually did well.

The true notion of the smartphone manifested integration, not the single-device paradox. In their quest to build the ideal mobile communicator, wireless handset designers initially ignored the fact that designing and building any mobile product is an exercise in trade-offs and compromises. They wanted a screen that was as large as possible, but bigger screens meant less portability. They went for an input metaphor that worked while being mobile, but keyboards were too small and keypads too limited, and character and voice recognition were too imprecise. Power and batteries were also a huge challenge. One place where smartphones won't make a major leap was with battery life. In short, smartphone required technological and design development on several fronts simultaneously.

Majority of people wanted a device that served their distinct needs and looked good, and that didn't cost them an arm and a leg. Did that make the smartphone a mere fantasy? Not if the issues of power consumption, connectivity, and user-friendliness were addressed using technological advances and research into consumer needs. Power was the nemesis of mobile devices and a key bottleneck to realizing convergence in the handset. After all, what is the point of being able to surf the web, respond to e-mails, and take pictures on a cell phone if the battery is dead in a few short hours. At this very juncture, the question the pundits asked was whether the concept will die faster than interactive TV, become bigger than the desktop computer, or land somewhere in between.

So what really turned the dream of convergence into a reality for smartphones? The answer may partly lie in the availability of key building blocks in the form of components like chips that became ready to deliver convergence in a way that was technologically and economically viable only in mid-2000s. Remember that Steve Jobs gave the go-ahead for the iPhone project only when Apple engineers assured him that ARM-powered chips could handle the convergence of voice, data, music, and video. The innovations in system-on-chip (SoC), system-in-package (SiP), memory stack, and dense PCB domains allowed far more circuitry to be packed onto a single device platform. Apart from tighter integration at the silicon level, smaller footprints could also be achieved by stacking miniature devices on multiple dies. That would lead to a new architectural framework with integration at its best, providing more horsepower to process data and hence bring higher value to mobility.

The SoC methodology, which meshed multiple functions on a single chip, had done some wonderful things for wireless and portable electronics devices during the early 2000s. Next, for the pressure to reduce component count amid increasing functionality and smaller form factors, SiP was an important new venue available to handset designers, especially for the radio-based portions of 3G phones. Together, these semiconductor technologies marked new milestones for convergence endeavors the industry had been longing for in many years. The inexorable march of Moore's Law commanded that these astonishing designs squeeze so much into a portable handset that it could even turn a smartphone into a superphone. According to this empirical observation, made by Intel co-founder Gordon Moore in 1965, the number of transistors on a chip roughly doubles every two years.

On the hardware front, communications silicon bellwethers like Texas Instruments (TI) were able to combine their advanced process technologies with system know-how to deliver powerful phone-PDA combinations without inflating size and power consumption. Earlier in the 1990s, by devoting vast engineering resources to Nokia and Ericsson for development of platforms based on its chipsets, TI had just about sewn up the mobile handset silicon market. Now it was pushing one-chip handset mantra to accomplish new integration milestones. The single-chip offerings for the smartphone would include all the basic ingredients of a cell phone: radio frequency (RF) circuitry, power management, analog and digital baseband, and application processor as well as embedded memory and software stacks.

The semiconductor industry aimed to squeeze processor and memory devices along with much of radio functionality onto a single piece of silicon. Consequently, a core mobile phone chip promised an improved battery life and almost five times the computing power of the existing handsets. In the basic composition of the smartphone, the flash memory stored information such as telephone numbers and text messages; the baseband chip handled communication and signal processing part; and the application processor managed intelligent call handling and PIM functions such as diary, calendar, and reminders. Combining these functions removed the need for buffering between chips, leading to reduced power consumption and longer battery life. One chip instead of three also meant a smaller footprint and hence more room inside the handset for other components. That would lower the cost of cell phones and free up space for critical new technologies like GPS and Wi-Fi.

This was a class act, one that remained largely hidden from the media limelight. While trade press has been running a non-stop commentary of software flash points like Symbian and Android, it largely ignored the technological feat that came from semiconductor industry's labor in obscurity. Although conventional definition of "smartphone" is closely tied to OS-based devices, when it comes to elevating user experience, there is more to the heart and soul of smartphones than just full-fledged operating systems. The battery manufacturers, for instance, are the unsung heroes of the mobile revolution. Over the years, they greatly extended talk time and standby time while shrinking the physical dimensions of the battery, making possible handsets small enough to be tucked into a pocket.

During this transitional period in personal communications technology, in the mid-2000s, it all looked like the computer industry in the years before the inception of IBM PC: a slew of operating systems, processors, and platforms and a lot of creative confusion about what will be the hot applications. The early personal computers were hardly consumer products, yet they evolved into something with far broader appeal. Nearly two decades after the inception of the PC, industry was contemplating if color displays, graphical user interface and downloadable games, or a Windows-like operating system would enable the transformation of PCs in a small box? Many experts in the early days of making of the smart-phone also wondered if consumer products of the future could be pack-aged into a computing bundle. The promised land of the smartphone also sparked the long-awaited battle between the PC makers and the phone makers.

In the end, the PDA versus phone debate only proved to be brouhaha as both devices were in fact passing through a fast-track evolution. Phones, however, had started from a higher plateau and were more likely to maintain their lead in the mass market. By 2002, mainstream cell phone models were starting to include calculator, voice recorder, and primitive scheduler functions traditionally associated with PDAs. The next big step came with the addition of digital camera: comput-ing met telecom met consumer electronics. The fusion of cameras onto mobile phones was initially seen as a triumph of consumer electronics. But if viewed from an integration standpoint, with a camera on-board, mobile phones were far more "smarter" in a variety of ways.

9 SMARTPHONE TURNS DISRUPTIVE

"Single-lens reflex cameras would soon be obsolete due to camera phones. There will be no need to carry around those heavy lenses."
—Nokia's marketing vice president Anssi Vanjoki at a gathering in Helsinki in April 2010

The notion of the smartphone remained a PDA fixture for the early part of the gadget evolution. However, that common perception about smartphones as some kind of cross between handset and PDA was being constantly challenged by new developments. Then came a breakthrough beyond any industry pundit's imagination. To put fun into phones, the wireless industry began stitching digital cameras and MP3 players onto handsets. With a camera phone, the user simply snapped a photo, selected a name from the address list, and hit the Send button. Camera phones not only shook the world of digital photography; they also changed the rules of the game in the cellular world. As many as 40,000 units of Sony Ericsson SO5051 camera phone were sold on the first day of its launch in Japan in spring 2003.

Although advanced handsets were generally perceived as pocket com-puters, with the advent of camera phone, they began to relate to the consumer electronics industry more than the world of computing. By putting new technologies like digital photography and multimedia messaging into consumers' hands in an easy-to-use form, the new handsets succeeded where handheld computers had failed in general. The irony was that features like color screens, built-in cameras, and music players subsequently encouraged consumers to start using advanced data services. During the years following the arrival of camera-enabled phones, we saw the lines between the so-called feature phones and smartphones blurring in the same way during the crossover between the mobile phone and the PDA. In fact, the cellular world's switch to digital photography became a stepping stone in making multimedia a key ingredient of the smartphone recipe.

The camera phone roots could be traced to the first wireless picture phone prototype—known as Intellect—developed by Daniel A. Henderson in 1993. The pioneering system and device featured still image and nonlive video clip transfers. This wireless neo-videophone was designed to receive pictures and video data—sent from an originator—via a message center used for transmission and display to a wireless device such as a cellular phone. Intellect was essentially a cellular handset with a large black-and-white display that could show still images and video clips downloaded remotely from a computer via a wireless transmitter. The data transfer protocols pioneered in the Intellect design would eventually be deployed in camera phones launched in the early 2000s.

The complete integration of cellular phone, digital camera, and wireless transmission infrastructure would take a few more years to complete. Many cameras and camcorders incorporated communications technologies in the years following the release of Intellect, but none of them actually focused on the integration with the wireless Internet, which was in fact the key to providing instant media sharing with anyone anywhere. Among those early experiments included a version of the device that was known to the world as Apple's Newton. Then there were companies like Kodak and Olympus who showcased several digital cameras with cellular phone transmission capability. Shosaku Kawashima also led a team of developers at Canon to demonstrate a full-fledged combo of digital camera and cellular phone in May 1997.

The camera phone was invented in 1997 by Philippe Kahn, CEO of the cell phone software company Starfish. It all started from his desire to upload photos of his newborn to the web and share them with his friends instantaneously. He thought about how clumsy it was to have to take a digital photo, download it to his laptop, post it to a website, and then e-mail his friends to tell them where to look. He just wanted to snap a picture, hit a button, and have it automatically loaded to the web. When his wife was having labor at the hospital, he fiddled with his laptop, cell phone, and digital camera and wrote a code to make his idea work. Subsequently, on June 11, 1997, Kahn was able to share the first pictures from the maternity ward—where his daughter Sophie was born—with family members, friends, and associates around the world. A sharing infrastructure and an integrated cell phone-camera combo boded the birth of instant visual communications.

Siemens had laid claim to the first full-color-screen handset with its S10 model just a couple years before the release of Sharp Corp's first commercial camera phone complete with infrastructure: the J-SH04. Launched in Japan in 2000, Sharp's J-SH04 handset had an integrated charge-couple device (CCD) sensor and color LCD screen with the Sha-Mail infrastructure developed in collaboration with Kahn's LightSurf venture and marketed by mobile operator J-Phone. *Sha-Mail* meant "picture mail" in Japanese. While Sharp's camera phone coupled to J-Phone's Sha-Mail infrastructure initially focused on instant sharing of pictures, a team at Kyocera led by Kazumi Saburi was spearheading a somewhat similar project. Cell phones both from Kyocera and Sharp had integrated cameras, but the Kyocera system was designed as a peer-to-peer videophone.

The camera feature proved popular right from the start; J-Phone in Japan had more than half of its subscribers using camera phones within two years. The world soon followed the explosive popularity of picture messaging service. In the United States, they said who would want a camera in their cell phones? When news of such combination devices trickled over from Japan, the idea seemed silly and excessive to some people. But in 2002, when Sprint released a Sanyo-made handset as the first American cell phone with a built-in camera—SCP 5300—the public went crazy for it. Sprint, who had also launched the PictureMail infrastructure developed and managed by LightSurf, sold 1 million Sanyo-manufactured camera phones within two years.

CAMERA CALLING

Camera phones, initially seen as a Japanese fad, became a bona fide cultural phenomenon, allowing the average person to quickly and personally share both mundane and earthshaking events with the rest of the world. By 2010, camera phones had become so common that people stopped calling them camera phones anymore. Camera phones were utilized for citizen journalism, political protests, business applications, and even for invasions of privacy and voyeurism. In South Korea and Japan, camera phones must make an audible noise when a photo is taken to prevent up-skirt shots. In 2007, New York City mayor Michael Bloomberg introduced a plan to encourage citizens to capture crimes in progress on their camera phones.

Camera phones could share pictures almost instantly and automatically via a sharing infrastructure integrated with the carrier network, thus negating the need for connecting cables or removable media to transfer pictures. The sharing infrastructure was critical and explains the early successes of J-Phone and DoCoMo in Japan as well as of Sprint and other carriers in the United States. Images were usually saved in the JPEG file format, and the wireless infrastructure managed the sharing.

The amount of megapixels a cell phone camera could capture determined the quality of the image or in other words, resolution. The higher the megapixels were the better was the image quality.

The early camera phones—successfully marketed by J-Phone in Japan—used CCD sensors. But these cell phone cameras were limited in terms of higher price tag and the amount of megapixels they could capture. However, a new technology helped them increase their megapixel capabilities and allowed these phones the ability to zoom in on objects with greater ease-of-use and quality. The advent of complementary metal-oxide semiconductor (CMOS) image sensor became an enabling technology for camera phone mass production as more than 90 percent of camera phones sold in later years used CMOS hardware. CMOS image sensor, or camera-on-a-chip, developed by Eric Fossum and his team in the early 1990s, proved a cornerstone in the realization of the modern camera phone.

The camera phone, like many complex systems, was the result of converging and enabling technologies. The ascent of CMOS is a case in point. Unlike the digital cameras of the 1990s that mostly carried CCD image sensor, a consumer-viable camera in a mobile phone would demand far smaller power and a much higher level of camera electronics integration to permit the miniaturization. Naturally, camera phones preferred CMOS image sensors due to lower power consumption and lower cost as compared to CCD-type cameras, which continued to be used in some high-end models. The lower power consumption made possible by CMOS image sensor also prevented the camera from quickly depleting the phone's battery.

After all, still images were not the only thing that could be shared using a cell phone camera. Some of these camera phones began to be equipped with some kind of video-capture capability as well. Cell phone cameras could record small amounts of video on certain, more advanced video-enabled models. Cell phone video could then be sent in the form of a text message to be shared with others. Moreover, video clips, like pictures, could be stored on the phone. Cell phone video and images could be quite large and took up much of space in the handset memory. So for people who wished to save their cell phone images and videos, the most flexible option was a memory card. Other enhancements followed, and these enhancements were nothing but amazing.

The phone had to be smart enough to blend the sensor's magic with connectivity, and once that was done successfully, assimilation of sensors into smartphones would bring unimaginable possibilities for the wireless future. One of them was the birth of new support industries. For instance, apps became a turning point for camera-enabled smartphones. Users could take pictures and turn them into postcards by utilizing apps to manipulate images, something well beyond the landscape and panoramic settings of point-and-shoot cameras. Smartphone apps allowed users to apply filters—like black-and-white and vintage—to images just with a touch of a finger. For some users different photo apps meant having different cameras onto the phone. And for some users, it was now camera first, and phone, second.

Who would have thought that digital cameras would one day merge into cell phones? No one foresaw that the most dominant and prevalent digital imaging platform would no longer be a still camera but a mobile phone. Suddenly, handset designers came to the realization that nothing was more important than imaging in the next-generation phones. After all, it gave people the capability of having a means of communication and a point-and-shoot camera at the same time. So cell phone evolved rapidly during the 2000s and what used to be a simple gadget—a phone with the sole purpose of making and receiving voice calls—became a full-fledged mobile entertainment system.

The advent of the camera phone forced wireless industry to move beyond the simplistic view of a phone–PDA combo and to revisit the notion of smartphone. The convergence enabled by a little camera with a big sensor was at the heart of new-age wireless, not the so-called phone–PDA combo, so the wireless market planners started seeing the whole smartphone affair with a renewed promise of a converged handset. At this very junction, when camera phones were redefining society in many ways, the wireless industry clearly found itself at an inflection point. In the mobile phone realm, imaging was the third killer application, after voice and SMS; it also marked the mobile phone's fifth disruption.

THE MOBILE PHONE'S FOUR DISRUPTIONS

There was a time when PDAs were nice, even necessary, companions for many business travelers. The advent of the PDA was originally a valiant attempt at lifestyle organization. These portable computers were made to sync to PCs so that information, appointments, and memos would hopefully not be missed. However, the PDA fell short. As this amazing story of technology evolution goes, both cell phone and PDA makers courted multimedia application vendors in their quest to seek valuable niche markets. In the end, though PDAs had the big-screen prerogative for multimedia features, camera phones ended up as more popular. This partly happened because software vendors wanted to be on the phone side due to its mass-market appeal. A lesson that they had learned from the PC industry was that software sells only when people buy a lot of devices.

Moreover, if the market became saturated with the creation of two product categories instead of one, there wouldn't be much room for many profitable hardware players. With the rise in popularity of Internet-connected smartphones, which could perform synchronization with desktop tools such as Microsoft Outlook, PDAs suddenly looked clunky and unnecessary. The argument was that since the cell phone is smaller, lighter, and easier to carry, and since the smartphone is gaining more and more handheld computer functions, there is no longer much need for PDAs. The growth in the smartphone market and power it acquired just in a few years time inevitably meant trouble for the PDA market.

People were now willing to buy a slightly more expensive cell phone that provided PDA functionality and walk away from the dedicated PDA. The writing was on the wall for an industry that pioneered the concept of carrying a digital office in a pocket. Not long after the release of its BlackBerry PDA, RIM gave it a slam in 2002 by integrating wireless e-mail and PDA features into a mobile phone. The PDA-upstart Handspring was working on similar tracks, and consequently, with Treo, another commercially successful smartphone was born. Handspring delivered the first widely popular smartphone device in the U.S. market by marrying its Palm OS-based PDA with a GSM mobile unit in 2003. Finally, when Sony pulled the trigger on PDA business in summer 2004, it was heralded as the beginning of the end for the PDA.

The second disruption story is about pagers, which had been the primary means of getting instant information to people away from their desks during much of the last quarter of twentieth century. Before mobile phones became a must-have item, teenagers, doctors, and executives all had pagers. Companies offering one-way paging services began rapidly gaining value in the late 1970s. During the next decade, paging turned into a mainstream business segment as operators continued to add subscribers while expanding their networks. But by the time the paging industry started weighing its move toward interactive arena, a new generation of cellular phones was about to make its mark in the commercial domain.

Paging companies like RIM—pioneers in wireless messaging—epitomized their fight back with the realization of e-mail pagers. While witnessing an unprecedented growth of cellular phone market, paging industry stalwarts held on to the notion that mobile phone was an altogether different device used for a different purpose. Furthermore, they argued, paging was one tenth of the cost of cellular and used the frequency spectrum very efficiently. But a testament of paging industry's denial was the fact that, back in the 1980s, many U.S. operators in paging industry in fact considered cellular a logical next step in the radio business. That's why most of the applicants for cellular service license then were small mom-and-pop paging companies aiming to serve this contiguous new market.

Over the years, mobile phones got smarter and cheaper, and they began to offer exchange of small text messages as an alternative to paging. Paging subscriber base was a natural target for SMS, which was more than adequate for passing short alphanumeric messages over cellular handsets. Another element that sidelined the pagers was that digital cellular phones were becoming lighter and cheaper every day. Paging industry experienced a slow death at hands of these sleek gadgets. The destruction of paging industry provided an opportunity for the nascent cellular business, though, in the hindsight, when GSM handset designers incorporated SMS as an add-on feature, hardly anybody heard a death knell for pagers.

Another example of disruption at the hands of the cell phone progression is the demise of payphones. The fate of pagers and payphones was somewhat intertwined because a telephone booth was usually the first thing people looked for after receiving a paging message. Now payphones, which related to pagers more than anything, were fast becoming the dinosaur of the telecommunications world. There was an immense drop in payphone usage caused by the surge of mobile phones. Telephone companies around the world started to hold back their plans of expanding payphone networks by the late 1990s. The rest, as is said, is history.

The fourth disruption story relates to personal navigation device, or PND, which once a star of the consumer electronics space, was now being hit by competition from navigation-enabled smartphones and in-dash navigation devices. After hitting a peak in 2008, global PND shipments began to decline in 2009 and continued their downward spiral afterward. The business model of GPS linchpin Garmin Ltd, and that of its main competitor TomTom International BV, came under pressure in 2009 when Google and Nokia started offering free turn-by-turn navigation on mobile phones. While Garmin decided to fight fire with fire and launched its own navigation-focused smartphone, TomTom responded by starting to shift its business mix toward higher-margin, value-added services, making PND a smaller portion of the total revenue.

Garmin announced the original nuvifone with a huge amount of hype behind it, but while it was delayed for around a year, the iPhone came out and stole the thunder from every other smartphone on the market. The incredibly shrinking Garmin then teamed up with Asus to create the Garmin–Asus partnership, but the phones that shipped under this label saw only a lukewarm reception. The problem was that Garmin just didn't have any credibility in the smartphone realm, and it was competing with flagship companies that did. Moreover, GPS units were becoming so cheap that few people saw the benefit of buying a GPS-smartphone hybrid. The unkindest cut of this all was that even Garmin seemed to understand that it was not in the best position.

SENSOR: THE INVISIBLE COMPUTER

Now the point-and-shoot camera—a part of households since the beginning of the twentieth century when George Eastman introduced the Kodak Brownie—was in danger of extinction. Like other single-use devices—the answering machine, the desktop calculator, the Rolodex—it was being marginalized by multipurpose smartphone and its camera, which took better snapshots with each new model. But more critical was the fact that it was much easier to share those pictures with smartphones through messaging, social networks, or e-mail. With a point-and-shoot camera, on the contrary, people had to plug it into a computer and upload the photos; it was just a few more steps they had to take.

When a camera phone was in a user's pocket or bag, every moment could be a potential photo opportunity. Moreover, for casual picture-takers, the convenience of the smartphone meant that they had one less thing to carry. The ubiquity of a 5- or 10-megapixel camera phone in one's pocket was clearly hard to overcome. By 2010, the camera on the phone was used by 65 percent of all mobile phone users. In the hindsight, the incorporation of megapixel cameras in phones was just the beginning; it set the stage for what would transpire on the ever-converging smartphone platform. The fifth disruption was eventually triggered by the ubiquitous image sensor on top of cell phone when it was hooked with always-on mobile Internet and apps marvel to create a perfect storm.

It's ironic that the camera industry correctly forecasted a dramatic growth in consumer adaption of digital cameras during 2000s. The growth was even better than they expected, only it shifted from stand-alone digital cameras to camera phones and that the sales of stand-alone digital cameras mostly stalled and stagnated. By 2003, more camera phones had been sold worldwide than stand-alone digital cameras, and at the end of 2008, camera phone sales had reached 1.9 billion worldwide. Nokia became the world's most-sold digital camera brand in 2004. The Finnish wireless titan sold more camera phones than film-based simple cameras sold by Kodak, which made Nokia the biggest camera manufacturer in the world.

Back in early 2000s, mobile phone industry was desperately in need of good news amid WAP and 3G debacles, so camera-equipped mobile handsets breathed new life into the wounded wireless segment. But the early camera phones were quite puny, and most serious camera freaks considered them a joke. Those neophyte units had poor resolution, featured cameras of poor-quality optics, and carried limited storage. They were hindered by the need for better image sensors, more on-board memory for picture storage, longer battery life, and network improvements to make picture sharing easier and interoperability a reality. No wonder the camera industry initially thought mobile phone–based camera phones would never catch on and would never rival their industry. In retrospect, however, the battle was short-lived and was over in merely three years.

Initially, cell phone cameras were indeed far behind the compact digital camera in terms of quality or features like image stabilization and larger lenses and sensors. But they continued making improvements required in areas like image sensor and lens design. More powerful processors and better sensors helped improve image quality to levels many consumers found acceptable. Since Sharp and J-Phone jointly launched the world's first camera phone in 2000, the device evolved with Moore's Law and continued getting better both in terms of megapixels and of operational capabilities. By 2005, handset makers were offering megapixel camera phones, and by 2007, phones with 5 megapixel cameras had appeared on the market. In 2010, mobile phones had surpassed the 12 megapixel threshold for their built-in cameras.

Meanwhile, the realization of 3G networks provided faster pipes for the transport of multimedia content generated by these phones. In 2004, camera phones alone outsold all types of stand-alone cameras—digital and film-based—and by 2010, camera phones were outselling stand-alone cameras by ten to one. Makers like Canon, Nikon, Minolta, and Konica had never accomplished even 100 million active users for their camera brands. Moreover, all stand-alone digital cameras and film-based cameras ever made during more than 150 years of the camera's existence had not reached a billion shipments. Ironically, it was Nokia now preoccupied with Carl Zeiss optics, Xenon zooms, and raising bar of megapixels; the handset maker had even started offering camera phone tripod mounts. Nokia, who launched its first camera phone back in 2002, had become the world's biggest camera brand within a few years.

The first decade of twenty-first century embodied the golden age of cameras, yet some of the big names in the camera industry failed to survive this enormous opportunity. While specialists like Konica and Minolta quit the camera business altogether, Canon and Nikon quickly shifted their focus from consumer snapshot cameras to premium professional and semi-pro camera systems. The camera phone didn't kill the stand-alone camera; it just took 90 percent of the market from it. The big phone makers like Ericsson, Motorola, and Nokia were not in the business of creating professional cameras. But when the mass market took a radical shift, major camera industry players had to abandon the camera-related businesses and shift to something else like professional imaging, scientific instrumentation, or photocopiers.

It was evident by now that image sensors, not PDA-like devices, embodied the true promise of proactive computing. The ubiquitous camera phones and their relentless pursuit for more megapixels sparked growth in areas like gaming and m-commerce and, thus, boosted wireless industry initiatives conceived on the smartphone platform. But probably the most remarkable accomplishment of camera phones was to lay the foundation for content-rich services. Mobile users could take image or video footage at anytime, anywhere and thus could create their own content. Camera phones became just another convenient way to start a conversation online when mobile users hooked onto Internet services like Facebook and Twitter, where they shared photos and thus documented moments throughout the day. Combine the camera phone's reach with the richness of the mobile Internet and you have a world at hand in which billions of people would be able to consume and even create their own content over these mobile gizmos and thus take the information revolution to a new level.

VIDEO: THE NEXT FRONTIER

First, it was ringtones, then came along camera phones, and now it was video in a variety of manifestations: clips, video streaming, and mobile TV broadcasts. One application that seemed to be crying for video was the camera phone itself. So after the camera phone take-off, it was apparent that video is the next frontier for the smartphone and that the next logical step would be turning camera phones into camcorder phones. Imaging quality, both for still images and for video, was constantly improving amid a relentless technological drive. By 2010, the smartphone had transformed into a multimedia powerhouse, equipped with a back-facing camera that snapped photos and a front-facing camera that recorded videos and performed video chats. These mobile phones boasted excellent cameras and video recorders/players that ultimately aimed to go high-definition.

Still, while many mobile handsets integrated camcorders, they fell short when it came to required resolution and picture quality for video. The trouble with legacy camera phone products was that they provided mobile video telephony by incorporating a small digital camera on the top of the handset, which sent TV-style color image to other handset's screen. So far, video had been jerky because it didn't contain as many frames as a television picture. Setting up the service was difficult and costly, and the video quality was choppy at best. The challenge was to make video chat as easy-to-use for consumers as making a phone call or sending a text message. If video communications required anything more complicated than pressing a single button, most users would likely ignore it.

During 2000s, several Japanese companies tried to revive mobile video-phones with little success. The mobility required handsets to be small and light. So even with a color display, tiny screens were still tiny screens. Moreover, the wireless industry at large had been slow to develop technology to support multimedia services simply because it was a question of creating a value chain through radically different content production, which required different knowledge skills from handset manufacturers and wireless operators. It seemed much like the early days of multimedia PC when some users were doing it just because they could and not because it was really useful for a customer. There was a lot of experimentation going on.

Videophone and related products have a long history that goes back to AT&T's demonstration of an experimental picture phone service between Disneyland and New York's World Fair in 1964. "A logical extension of today's telephone service" boasted a Ma Bell advertisement. AT&T and technology journalists alike predicted that videophones would become standard in homes and office within a decade or so. But in the coming years, video communications failed to reach critical mass amid many setbacks and false promises. Video communication products were expensive and hard to operate, and they kept changing forms and shapes after being provided energy dozes every five to ten years. Half-hearted attempts like desktop videoconferencing brought some value to consumers, but the real promise of video communication came within the reach only after the glory days of the smartphone arrived.

Video calling had been a sci-fi dream in the United Sates for decades, but a reality for years in other parts of the world, such as Japan, where mobile phones have traditionally been more advanced than in the United States. Japan had earlier defied the norm when it came to realization of camera phones, and undeterred again, the wireless industry here continued to test video prospects. The Japanese mobile phone companies were hoping to usher in a new class of services such as interactive video games and quick music downloads. They reckoned that streaming video could become the first popular application for features such as sports highlights and electronic postcards. Despite the technological limitations—smaller screens, jumpy connections, expensive devices, and interminable downloads—plenty of companies in Japan believed that the demand for delivering photos or streaming videos to handsets was there.

As mobile bandwidth expanded and smartphones with increasingly sophisticated cameras hit the market, new services offering live video streaming from mobile devices mushroomed around the world. Competition among these services is fierce, with a hodgepodge of diverse business models reflecting alternative visions of the future of mobile streaming. Few services are making money yet, but they are all racing to upgrade and add new features in an attempt to anticipate and to monetize trends in mobile streaming adaption. The mobile streaming subindustry is still in embryonic stages, and it is not yet clear how or even if the public and enterprises will use mobile streaming on a large scale. For the business sector, getting information where and when it is needed is what could actually drive the mobile adaption of video, not fancier and costlier applications like streaming media.

For instance, video-on-the-move offers an interesting new element to the users and brings new possibilities for vertical applications such as video surveillance. But such multimedia services also depend on user demand and how the future technology shapes up. Probably, the most crucial factor in the success of multimedia content services is to understand consumer behavior. That was evident from the fact that one of the first significant breakthroughs in mobile videophone came from NTT DoCoMo who had mastered this art during the triumph of its i-mode service.

DICK TRACY COMES TO LIFE

For decades, futurists had predicted that videophones would become commonplace, but for decades, it didn't happen. However, experts said that video would eventually become a much more useful communication tool and that there was a real opportunity for video chat—particularly for mobile video communications on cell phones—to take off. Perhaps a more accurate prognosticator was Chester Gould, the author of the *Dick Tracy* comic strip. Decades ago, he painted a picture of instantaneous two-way video communications via a smart watch. Heaps of abandoned products have taught the industry that the wristwatch is too small to bring about effective and rich interaction, though it could eventually make some significant headway into the futuristic wearable computing domain. Cellular phone, however, is a somewhat similar personal accessory that may indeed be the nexus needed for the convergence of wireless communications, the Internet, entertainment, and more.

Then there was this "Apple effect." The ability to understand the visual world in the mobile context in fact began to develop in the post-iPhone arena. If the original iPhone wasn't a videophone, it was evident that there will almost certainly be one eventually. Video calling was the key feature on the iPhone 4, which more than doubled the number of pixels per inch from the last iPhone 3GS version, making its sharpness resemble that of a printed book. The iPhone 4 was probably the first phone to make good video call as well as record video in high-definition. Apple sold 1.7 million iPhone 4 handsets in the first three days of its availability.

Apple gave video chat its biggest boost with the launch of the iPhone 4. The phone included a front-facing camera and software called FaceTime for users to make video calls initially over Wi-Fi networks; the support for 3G video calls would come a year later. Apple presented the FaceTime video-calling feature on the iPhone 4 as a novelty despite the fact that video calling on mobile phones had been around for years. That's because Apple's bet wasn't on traditional video calling, to which many users around the world were now used to, but on a carefully crafted, closed ecosystem. Apple must have studied the history of video calling and cleverly picked up its ideas.

Apple did several things right with FaceTime, the first and most important being that it was free and that users were not limited by a wireless operator. Second, its initial launch was on Wi-Fi, which ensured a certain quality of service and good experience. Third, FaceTime—the engine behind the iPhone 4 video calling software—was an open standard, so apps on other platforms could integrate this video service. Being an open specification, FaceTime was built on standard technologies such as H.264 video compression.

Ultimately, the success of Apple's FaceTime relies on whether software and phone makers will embrace the new open standard—and of course—on how many iPhone 4 devices and its successors the company sells since compatibility is initially limited to the latest Apple iPhone, iPad, and iPod Touch. Another feature that predated iPhone 4 on a select few recent handsets was video recording, but with the iPhone 4, Apple was able to bring this capability to mainstream, complete with video editing function. Capturing and processing video at a full 30 frames-per-second needed the right processor and the right software—both of which became readily available for the smartphones only by late 2000s.

The new feature iMovie let users edit video and add music and photos using a storyboard. A user could take movie clips captured by the iPhone's video camera, mix them with the 5 Mpixel images stored on the device, and integrate music to create a compelling iMovie on-the-fly. Such innovations are turning mobile phone into an ideal instant self-expression device, one which would let people instantly share their lives with friends and family. That may also be a harbinger of smartphones and tablets becoming serious content-creation tools, making them more PC-like than anyone would have imagined in the past.

Computer-like smartphones, faster networks, and consumer readiness meant that the technology was ready for take-off. For starters, front-facing camera provided a perfect venue for video chatting and it could encourage people to use the video calling a lot. However, Apple wasn't the only tech company renewing the conversation over mobile video-chatting platform. The perfect storm was here, and it didn't go unnoticed by Skype Technology SA, the biggest provider of Internet-based international calls, who was raising the bar on mobile video calling by introducing a version of its online video-chat service for cell phones. The programs like iChat and Skype had existed for the PC, but this technology on mobile platform could give users a powerful new way to communicate with people all the time.

Until now, thanks to the iPhone 4, most mobile video calls had happened via Wi-Fi connections at homes and coffee shops. Most wireless carriers didn't offer video calling over 3G networks for risk of being overwhelmed; an average video call consumed five to ten times more bandwidth than a voice call. We're talking about a service that would consume more data than anything else on the network. But that could change as companies like Skype release apps that run on their Internet-friendly networks. In early 2011, Skype opened a collaborative front with Apple as it announced the availability of its video calling service for Apple's wireless devices in a bid to broaden the appeal of mobile video communications.

Microsoft, desperately trying to turn around its sagging fortunes in the mobile market, probably took cues from Skype's Apple liaisons and bought Skype for a whopping US$8.5 billion, just three months after its landmark deal with Nokia. Evidently, among other things aimed for Skype acquisition, Microsoft wanted to tightly integrate the Skype offering into its Windows Phone software, as well as desktop and notebook platforms, and thus present Skype as a viable equivalent to Apple's FaceTime video chat software. Microsoft, however, needed to tread carefully for two reasons. First, Microsoft could foray into mobile phone operators' territory and end up antagonizing them. Second, Skype was based on peer-to-peer communication technology that eats up bandwidth quickly and was considered relatively old. According to media reports, Google courted Skype before Microsoft, but backed off precisely on these grounds.

Would high-speed data networks kick off video calls? Sprint certainly thought so. In 2010, the Android-powered HTC Evo 4G ran on the faster but limited Sprint 4G network and required the Qik video-streaming app for users at both ends of the video call. The Evo had a big 4.3-inch display and a front-facing 1.3 megapixel camera for quality video calls. So, as exemplified by Sprint, wireless carriers will need the extra cash to expand the capacity of their networks, which are already becoming increasingly stretched by the streaming of movies, YouTube clips, and other types of video content. Meanwhile, YouTube, the world's most popular video streaming site owned by Google, had started collaborating with mobile phone manufacturers and operators to help optimize flow of data across overwhelmed cellular networks and thus reduce the impact of video content on these networks. The sixth disruption on the smartphone platform looked just around the corner.

10 MORE TALES OF DISRUPTION

"No tree grows to the sky."

—Wall Street adage

Just when the media were buzzing about Apple's forthcoming iPad launch in 2009, a team of developers at Microsoft was working on a tablet device with multitouch, stylus-friendly screen, and icon-rich user interface under the project name "Courier." According to a story published on Gizmodo by a blogger named The Paperboy, when Robbie Bach, in-charge of Microsoft's entertainment and devices division, presented the idea to CEO Steve Ballmer, his response was less than enthusiastic. The problem was that the Courier tablet would have an operating system of its own, just like the iPhone and the iPad were built on a new operating system, iOS. The Courier craving ended up being folded into the next version of Windows operating system, Windows 8, due in 2012. Back left Microsoft after that.

Steve Jobs was once asked, early on during his first tenure with Apple, what would happen if the company failed to make its latest products successful. He replied, "I guess we'll be just another billion-dollar computer company." Also, Steve Jobs had predicted in his early days of the rivalry with the Redmond, Washington–based company that Microsoft might become the IBM of 1990s. That more or less seemed to have happened a decade later than Jobs's prediction. Interestingly, there was a time during the roaring 1990s when Microsoft was on top of the world and its self-assured top managers would ask themselves, "Are we becoming IBM?" in an apparent gesture of keeping themselves on toes in a relentlessly competitive market. But then they would answer it themselves, saying, "No, we can't be like IBM."

Ironically, a decade later, it wasn't just IBM with whom trade media were drawing parallels. Now there were people who saw Microsoft as mirroring General Motors's position in the auto business. Like Microsoft, GM was an icon in its industry, held a quasi-monopoly, produced eye-popping profits, and was often distracted by antitrust lawsuits. But after living in this kind of environment for over a couple of decades, GM eventually lost its competitiveness. Microsoft was the iconic technology company of our age built to compete and win in the global economy. Now the press was having a field day in narrating Microsoft's journey from being the cash-flush envy of the software world to a case study of mismanagement. The company that had once bulldozed IBM and seemed to hold the future of America on its keyboard was now struggling in the mobile arena where it failed to build a clear strategy.

The metabolism of digital technology became just too fast, and transformative innovations weren't taking nearly as long to play out. Consequently, behemoths like Microsoft became classic examples of how behaving as though they have plenty of time to jump on a new opportunity can backfire. Many industry observers saw Microsoft's problems rooting in its legacy and lack of innovation. Microsoft clearly suffered from the incumbent's curse during a technological transition in which computing moved away from the desktop and onto small mobile devices, and thus away from Microsoft's strengths. In all fairness to Microsoft, it constantly developed the new software products while raising the slogans of transforming its consuming desktop occupation to the new computing paradigm.

But it failed in developing the new business models, leaving the PC kingpin with approximately nothing in the mobile domain despite its cash, power, research, and market might. The software giant was now often described as slow and out-of-touch as the company couldn't shoot straight on anything other than a PC or a laptop. Despite massive investment in Windows CE platform for more than a decade, Microsoft failed in having attracted the same kind of attention that Apple's iPhone got in its first year and a half on the market.

Part of the problem was Microsoft's reactionary instincts in the mobile computing realm. Microsoft introduced Windows CE, the predecessor of Windows Mobile OS, back in 1996 with software tools familiar to Windows developers. It invested a lot of effort in making Windows CE a viable mobile, handheld computer platform in reaction to Apple's 1994 launch of Newton MessagePad, then as a PDA platform in response to the Palm Pilot in 1998, then as a smartphone platform as an answer to the Handspring Treo in 2002, and then as a portable media player Zune in reaction to the iPod in 2004. Finally, Microsoft set itself to build Windows Phone 7 touchscreen handheld devices in reaction to the iPhone and the iPod Touch.

Then there was Microsoft's obsession with being a software conglomerate. Case in point: Microsoft spent fifteen years hemorrhaging money in its Internet division and still had nothing to show for it but annual multibillion-dollar losses and a distant third- or fourth-place market share. It's worth imagining what would have happened if Microsoft had never become distracted by Google's search business or Yahoo!'s Internet service provider (ISP) and ad businesses and, instead, focused all that money and effort on developing the new markets around smartphones. The company had yet to successfully transition into popular consumer domains like tablets and mobile phones, and many insiders saw Microsoft dying a death by a thousand cuts.

MICROSOFT'S MIDLIFE CRISIS

Many software experts acknowledge the fact that Microsoft's Windows Mobile operating system was a good one and that Microsoft had it long before anyone else did. The problem was that the built-in applications were uninspiring, and that set a very low bar for developers who were coming to the platform. Another quandary was segmentation in the hardware ecosystem. Windows Mobile shipped with several different manufacturers' hardware, including HTC, LG, and Samsung. From a developer's perspective, that requires coding an app for several phones with different UI styles, buttons, and screen sizes. On the other hand, the iPhone operates on a closed system, which can only run on Apple hardware, meaning that third-party developers can produce apps and games that work exclusively with the iPhone. Therefore, despite Apple's questionable and controversial approval policy for iPhone apps, developers can code one app that works with 40 million iPhone and iPod Touch devices, which is less time-consuming than developing several versions of one app for a variety of Windows Mobile smartphones.

Then there was this issue regarding Microsoft seeking to exert even more control over its own third-party mobile software store. The company asked its developers to pay fees for every app submitted to the Windows Mobile Marketplace, charged new fees each time an app was rejected, and angered developers by denying apps repeatedly. All this resulted in stagnation of the Windows Mobile Marketplace, an absolutely wrong thing to do especially at a time when sales of Windows Mobile phones were beginning to plummet. By the time the store became viable despite Microsoft's crippling rules and micro-management, there weren't any customers left. A major knock against Windows Mobile wasn't the OS itself, but rather the weakness of the bundled apps included with it.

From 2000 to 2006, Microsoft did a reasonably successful job by toppling Palm as king of the PDA market and later penetrating into the smartphone software line. In retrospect, however, Windows Mobile had been anything but a smash hit and was gradually fading out of relevance in the smartphone domain. The experiences taught Microsoft how critical it was to develop and support an ecosystem of software developers to build apps. So in a strange about-face, the company vowed to pay developers to build applications for its vaunted Windows Phone 7 operating system and thus help narrow the lead of rival products from iOS and Android. The company promised financial incentives ranging from free tools and test handsets to funds for software developments and marketing. In some cases, Microsoft would provide revenue guarantees and would make up the difference if apps don't sell as good as expected.

The OEM-related business issues also cost the Redmond, Washington–based software house dearly. Microsoft took a sever hit when it decided not to make Windows Phone 7 operating system compatible with Windows Mobile. In that transition, Microsoft lost a key ally, the second-largest handset maker supporting the Microsoft platform: Motorola. Sony Ericsson is also said to have made a similar cut to Android after initially considering Windows Phone 7. In another setback, Microsoft lost Hewlett-Packard, the ever loyal OEM partner, as it bought Palm and its webOS operating system. Microsoft's Phone 7 transition from Windows Mobile, without any path for upgrading, also left many app developers out in the cold.

There are no simple answers to why a leviathan like Microsoft struggled while it was having so many plates spinning. This section has chronicled some of the major lapses that industry watchers point to for Microsoft's mobile business being in such disarray. Microsoft—like its former smartphone nemesis Nokia who eventually became its best friend through Windows Phone 7 tie-up—found itself in a constant catch-up mode. Microsoft's failure also presents a classical case of companies with deep pockets that diversify into new turfs amid fears of maturing of their existing, lucrative markets.

The dawn of the iPhone, the ascent of Android, and the failure of Vista were the latest wake-up calls that Microsoft got to heed that its old ways weren't working in this new age. It's about time that Microsoft stops looking at every new opportunity through Windows-colored glasses. The arrival of Windows Phone 7 and Nokia adapting it as its primary smartphone platform are encouraging developments, but the company would still have to grapple with the odd conundrum of turning its desktop success into an asset rather than an albatross. For a technology company, a late product can be as deadly as a bad one, and that might be the situation confronting Windows Phone 7. Nevertheless, Microsoft leadership continues to hope that the company will pull off the same come-from-behind wins that it did in the early days of the PC industry.

NOKIA'S PREDICAMENT

The cell phone was where the consumer action was, but the ability to merely make a phone call was no longer a factor in selling these devices. People now wanted a single phone for their work and personal lives, and the new, do-all gadgets did that bidding. Smartphones did everything from browsing the web to downloading and storing music and pictures. The mobile phone market changed tremendously during the first decade of twenty-first century. It was no longer a mobile phone; it was now more about services and operating systems. That's where troubles started for Nokia, who had a history of innovation in the hardware space. A decade after Nokia had surpassed Motorola in 1999 to become the world's top handset maker, the Finnish company was troubled, confused, and conflicted.

Apple's launch of the iPhone was probably most instrumental episode in changing the mobile phone industry with thousands of applications. And three years after Apple introduced the iPhone, changing the mobile phone industry by making usability of the handset the new basis of competition, the world saw the cell phone king Nokia struggling to develop a smartphone with the same mass appeal. Industry watchers feel that Nokia was caught flat-footed by the iPhone's success and blame its weakness in smartphones for shaving about 70 percent of Nokia's market value during this period. What happened? How did Nokia fall from grace? It's another epic tale of disruption that started with the birth of an industry that this cellular innovator sowed with its own hands.

Nokia was the first one to discover that handsets were too personal to trust to engineers alone. A cell phone was something that people kept close to their body, so it must have an emotional appeal. Just as much fashion and marketing went into a successful handset as good technology did. "Mobile phone is in a way an extension of a person's own personality, likening it to that of a fashionable watch," as a Nokia executive once put it. In the transition from analog to digital age, branding was decisive and Nokia got the message better than anyone else did. To Nokia's credit, it anticipated shift toward software and services much earlier than other handset makers. It launched Ovi in 2007, almost a year before Apple opened its highly successful App Store. A few months later, Nokia bought Navteq, a maker of digital maps, to be able to harness the power of location-based services. Shortly thereafter, Nokia launched "Comes with Music" service, an innovative pairing of a handset with a digital music subscription.

Nokia used to be a stodgy Finnish conglomerate, making everything from rubber boots and cables to lavatory papers and television sets. Even in that form, it had come a long way from its start, in 1865, as a lumber mill on the bank of Emakoski River. But few would dispute the claim that Nokia had been Europe's outstanding business story of the last decade of the twentieth century. Nokia was the most successful European company of the 1990s and politicians lined up to praise Nokia as an example of how Europe could prosper in the twenty-first century. Romano Prodi, president of the European Commission, drew attention to the success of Nokia and its Nordic neighbor, Ericsson, in a speech in 2002. "Their achievement in mobile telephones helped to create two vibrant clusters, around Oulu in Finland and Stockholm in Sweden, which have attracted a large number of startups as well as investment from foreign companies."

Nokia correctly predicted the evolution of the market toward software and Internet services-driven product differentiation as early as 1998. However its execution has been disappointing, resulting in substantial high-end market share loss to Apple, and increasingly, to Android. The poor execution reflected in a number of factors, including Symbian's cumbersome legacy software and a hardware-focused organizational model. Nokia was a hardware maker that wanted to be a software and services company, but its top management grossly underestimated the challenges related to moving from a hardware-driven business model to software settings. It was simply taking too long for the software and services strategy to unfold. Like Sony, Nokia couldn't find a way to shift from hardware to software.

At one stage, Nokia's segmentation strategy, which it had pioneered earlier in 1997, was seen as one of the answers to the company woes. In 2002, the Finnish wireless house split its handset division into nine "mini-Nokias," each concentrating on a different market segment while sharing research, development and manufacturing facilities. Olli-Pekka Kallasvuo, after taking over as Nokia chief in 2006, insisted on remaining in all segments from 25-euro basic phones to gem-studded luxury models. As part of his services drive, he bought the U.S. map company Navteq Corp. for US$8.1 billion, and built up the software organization with toolmaker Trolltech ASA and social networking startups. Kallasvuo drove the development of services such as music downloads and GPS navigation to increase the value of Nokia handsets and retain customers. But these efforts did little to stop Apple, RIM, and Android devices from taking market share from Nokia in smartphones.

The cruel truth was that despite its entire residual market share, Nokia looked like a has-been. Under Kallasvuo, Nokia worried about hanging onto market share rather than creating innovative products that excite customers. The company misread the way the mobile phone industry was merging with computing and social networking. Time and again, it vowed to fight back to this eminent smartphone leadership shift. The key issue facing Nokia was how to tie together its hardware and software into a compelling package, as Apple had managed to do. So in September 2010, when embattled Kallasvuo made way for Stephen Elop—head of Microsoft Corp.'s Office software business unit—for the top job at the world's largest mobile phone maker, hardly anybody in the industry was taken by surprise.

What really stunned the industry was what transpired soon afterward. The Microsoft veteran turned to his former employer for a turnaround plan and replaced Symbian with Windows Phone 7 software as Nokia's principal smartphone platform. Elop also made radical changes in Nokia's management leadership and operational structure, and vowed to bring half of the top management from outside Finland. Nokia would also operate around two distinct business units: smartphones and mass-market basic handsets. Nokia's new software liaison was apparently a win for Microsoft, but it set Nokia on a highly risky and uncertain path with no guarantee of success. In fact, there was no silver bullet for either company given strength of iPhone and Android.

PALM READING

The story of the smartphone won't be complete without Palm. The company that jump-started the nascent handheld computer industry with the Palm Pilot had a smooth sailing in the early going, leaving competitors, such as Microsoft, in the dust. Some bullish observers even predicted that Palm would become the Microsoft of the handheld world. Palm had now its eyes set on the business market and was ready for a fight against everybody's nemesis, Microsoft. But as a result of a poorly executed strategy, Microsoft and others got an opportunity to capitalize on Palm's mistakes. Handheld computers using Microsoft's PocketPC operating system, such as Compaq's iPaq, began to win over business buyers with feature-packed devices that boasted more power and better screens.

The single greatest weakness in the Palm design: its display featured 160-by-160 pixels. This produced coarse, grainy screens that did a mediocre job in displaying text and were pretty much hopeless for graphics. A series of management miscalculations, tough competition and weak market conditions had left Palm reeling. The company was bleeding cash at a time when it should have been spending millions at updating its technology to fend off renewed challenge from one-time neophyte Microsoft. Palm's role in the marketplace was fast changing from lead innovator to an also-ran as it struggled in correcting its missteps. The biggest technological surprise in the offing was the emergence of voice as a key technology for handheld devices. Palm's offshoot Handspring vowed to turn its PDA into a mobile phone. But while Handspring's Visor already had a built-in microphone, and all it needed was a phone module, Palm had yet to parry this thrust into the hybrid-device market.

However, advancements for the Palm handhelds picked up a considerably more rapid pace when the company joined hands with Handspring and started to merge PDAs with mobile phones. Still, its phone products didn't prove more attractive than those from RIM, which had dominated the early smartphone market. The Treo unit turned heads as one of the first smartphones, but it was a relatively bulky product in a market trying to get away from chubby devices. The later versions of Treo were slimmer and more elegant; these handsets had larger keyboard and screen, and the form factor looked more user-friendly than were Palm's old calculator-like designs. However, the new Treo didn't solve Palm's troubles. The crux of the matter was that Palm had an aging software product.

So by necessity, Palm began to think about its next act. It started to develop a new operating system for smartphones, knowing that the process would take up to two years. A Silicon Valley veteran and an acolyte of Steve Jobs was brought in to resuscitate Palm software. Jon Rubinstein, a rail-thin figure who commuted by train to Palm's Sunnyvale office from his home in San Francisco, had spent his formative years as a computer engineer at Hewlett-Packard. Later in 1990, Rubinstein joined NeXT, the workstation computer company Steve Jobs founded when he was kicked out of Apple. Then in 1997, when Apple was wallowing in financial losses and the Mac's appeal was waning, Rubinstein moved to Apple just ahead of Jobs's return, to lead the hardware engineering division at the company. Over the next nine years, Rubinstein and his team of engineers breathed new life into Apple by helping develop the iMac and the iPod.

Rubinstein had a big job ahead of him. Palm, once king of the hand-held world with its Palm Pilot, had fallen from grace when PDAs began to morph into the phone realm, and now the company's attempt at a comeback largely hinged on the Palm's new smartphone dubbed as Pre and its new OS platform webOS. In fact, making the Pre along with webOS represented what could prove to be the company's last best shot at survival against the onslaught of RIM and Apple. Palm's finances were increasingly shaky, and its stock price had reached the rock bottom. A start-over of sorts could give the company a fighting chance against its more powerful rivals.

At the 2009 Consumer Electronics Show (CES), when Palm unveiled the Pre and its brand new mobile OS, it initially got rave reviews and quickly became a sensation on the web. This brought Palm back into the game as some industry watchers regarded Palm's widely-anticipated phone as the first potential iPhone killer. "Mobile is in our DNA," said CEO Ed Colligan, who had a long history with Palm and its various antecedents and spinoffs. The company's second wind with the Palm Pre caught an enormous wave of buzz, but that alone failed to actually result in sustained success for its next attempt at the smartphone bonanza. Palm Pre was highly praised for its competitive design and pricing, but it didn't have the carrier relationships to get the sales. When the dust settled, the Pre represented a win for Palm only in the sense that it demonstrated clear potential for the company's new mobile OS: webOS.

If Palm failed to wow users this time around, it was because of the Pre hardware not the mobile operating system. Palm's new software store was exciting and potentially game-changing as the company built it from the ground-up according to its own vision and idea of the smart-phone. The software was written specifically for mobile devices and could run multiple applications simultaneously. "It is going to rede-fine the center of your access point to the Internet," said Colligan. "It is built on industry standard web tools and if you know HTML, CSS, and Javscript you can develop for this platform." The Pre's Achilles' heel was the hardware.

Palm's unit sales were hovering around 5 million per year, nothing to sneeze at in a market with so much potential. Palm even began licens-ing its Palm OS to Sony to create a new line of Clié handheld devices. The company also lined up an impressive roster of developers and at one point claimed a library of 50,000 apps. And yet Palm's software busi-ness never resulted in much actual success for the company or its users, and subsequently became a historical footnote. Palm ended existing as a phone maker in 2010 after having last three years of consistent loss-making quarters. Although it looked as though it had been mercifully sold to HP, analysts hoped that the industry would see the true merits of the work done by Palm engineers now that HP owned the company.

INNOVATOR'S DILEMMA

A close look at Palm's transformational story reveals an endless maneu-vering, which consequently led to a fatigue somewhat reminiscent of Nokia's Symbian dance. Palm Computing was founded in 1992 to pur-sue computers small enough to slip into a pocket. Amid the financial difficulties looming over the project, Palm founders Jeff Hawkins and Donna Dubinsky found a savior in modem maker U.S. Robotics and successfully launched the Palm Pilot handheld in February 1996. Same year, networking pioneer 3Com acquired U.S. Robotics in a bid to combine its LAN cards business with the strong portfolio of U.S. Robotics's WAN products. Hawkins and Dubinsky persuaded CEO Eric Benhamou about the lack of Palm Pilot handheld's synergy in 3Com's grand vision of creating a one-stop networking shop.

Hawkins, along with Dubinsky and Colligan, left 3Com to found Handspring in 1998 after a dispute with Palm's management over the future direction of Palm Pilot. In 1999, 3Com finally gave up to the market pressure and formed Palm Inc. as a separate company. In 2003, Palm Inc. merged with Handspring to form PalmOne. This merger then split into two companies, PalmOne for hardware and PalmSource for software. In 2005, PalmOne acquired full rights to the Palm name by purchasing the shared rights that PalmSource owned, and changed it to Palm Inc. PalmSource was subsequently acquired by Access Systems in 2005. Next year, Access Systems sold the Palm OS source code back to Palm Inc. What a roller-coaster journey it had in a span of just fourteen years.

Palm holds such an iconic position in the history of mobile computing that its passing into oblivion is tragic. There was a time when Palm could do no wrong. In 2000, the Palm OS was powering 85 percent of all PDAs and advanced phones as the company broadly licensed its operating system to power handheld computing devices. It owned the PDA space when Apple was flatfooted with the Newton, and it pioneered the concept Apple has refined of opening up a mobile operating system to third-party software developers as a means of increasing demand for hardware. Palm anticipated the death of the unconnected PDA market it had dominated with the rollout of the Treo, a respectable player in the enterprise phone segment though vastly overshadowed by the BlackBerry and the Windows Mobile handsets.

Palm gave industry the first portable operating system with real multitasking capabilities. Android has pretty much the same utility—multitasking being high on the list—but Palm was there first, more than a decade ago. Palm's original DNA was also crucial in minimizing the number of steps mobile phone users had to take to browse the web. But webOS was struggling to win developers; it was no longer a major differentiator because there were enough good operating systems in the market. Despite novel features and forward thinking design, Palm's webOS was now sixth or seventh priority on developers' lists. In February 2010, Palm's share of the U.S. smartphone market had fallen to 5.4 percent from 9.4 percent a year earlier.

The early popularity of the Palm operating system was based on genuine merits; Palm had clearly demonstrated it could be quickly adapted to bring wireless capabilities. That had opened the eyes of Motorola and Nokia, sparking them to work with Palm and its operating system solution. And Palm, like Symbian, was based on an open standard. But Palm lost its way over the years. In its early going, Palm's main business challenges were logistics and distribution—getting phones built profitably and sold widely enough to make an impact. So while the PDA king deftly trimmed the product portfolio, managed component costs, and streamlined inventories, both of its serious competitors—Microsoft and RIM—focused on corporate users. Palm basically had the consumer market to itself.

Then, Apple's iPhone and Google's Android OS struck a chord with consumers and software developers, selling tens of millions of devices and taking mobile computing to mainstream. Palm wasn't simply used to such kind of scale. But then it was reckoned that the Palm brand and the webOS platform would be a different story under HP's stewardship. Hewlett-Packard had resources and the wrong product while Palm lacked financial backing. HP not only could bring an impressive base of consumers and business users to webOS, but also could focus its internal software developers on providing a powerful base of apps for home and office use. Moreover, HP's hardware engineers could focus on building interfaces to work around a unified operating system. The hope being that integrated model and large unit volume would eventually encourage third-party developers to commit to webOS.

HP's Windows Mobile business was dying a rapid death, and HP needed to revamp its product line in order to stay in the rapidly evolving portable gadgets business. HP didn't do anything significant after the acquisition, which added to a degree of ambivalence about what it wants out of Palm. What was initially a sensational move became duller by the day as pundits argued that this was the hottest time in smartphones business and that there was no time to waste. HP would eventually commit to migrating all its mobile computing products to webOS and to using the Palm brand across as many of them as possible.

If Google's Android was out to disrupt Microsoft's royalty-based software mammoth, HP mostly likely saw Apple as a real competition. Apple had successfully demonstrated cramming hardware goodness with software magic to create beautiful gadgets, and now HP was the only player other than Apple who boasted hardware and software teams working together to deliver a total product and stellar consumer experience. Moreover, just like Apple, HP commanded a tremendous developer reach. Developers go where the units are and HP as the world's largest PC maker boasted the largest installed base of connected users across the world. In the mobile battlefield, HP started far behind the competition. But if it could successfully deploy webOS on 100 million units with a unified user experience across all devices, this would put the company back in the fight.

All that didn't happen. HP tried to buy its way into the post-PC era by acquiring Palm, but couldn't make it work. In summer 2011, HP became a textbook case study of smartphone disruption, announcing that it would explore spinning off the PC business, shut down its webOS–based smartphone and tablet operations, and focus on software and services. HP took a u-turn later and vowed to hold on to PCs.

THE MOTOROLA STORY

The smartphone is a prime example of a disruptive technology that was dismissed by the industry in early 2000s. Take the case of Motorola resurrection and ultimately the metamorphosis of its hardware business into Google's fold, and one sees a testament of how deep the power of the smartphone archetype goes. Motorola became one of the most admired companies in the world when it led the gold rush to the first generation of cellular handsets. Its StarTAC mobile phone launched in 1996 was to wireless what color was to television. The pioneering device would be remembered in the wireless history books for its small size and for being the first flip handset. The StarTAC mobile phone used a lithium-ion battery and had the first vibrating alert feature on a phone.

Motorola was a market leader in old-fashioned analog mobile phones; when the wireless industry shifted to digital during the 1990s, the company tried to hold back the digital tide and got stuck on developing phones for the existing analog cellular market that was rapidly diminishing. The predominant mobile phone maker started to see its market share shrivel as it misread consumer demand, alienated wireless operators with its arrogance, and tripped over its own feet in developing new products. Industry folks started calling Motorola the "Xerox" of mobile communications. However, the company had a history of reinventing itself through engineering strength to bring the right products in the market.

During the early 2000s, when most cell phones were starting to look the same, Motorola started to break the status quo with Razr, a slim, slab-like clamshell phone with a large screen, a stylish and flat keyboard, a built-in camera, and multimedia capabilities. Cell phones had continued to get thinner and more stylish over the years, but the debut of the Razr took design to a new level. Impressive technical features aside, with its super-slim lines and sleek metallic appearance, it looked so cool that it quickly became a must-have accessory. After launch in 2004, the

Razr sold more than 110 million units and earned the accolade of the "iPod of mobile phones."

The Razr was the ultimate feature phone to have with a VGA camera, 2.2-inch, 176-by-220 pixel internal screen, Bluetooth, and 10 MBytes of internal memory. A design marvel, the all-aluminum, clamshell, dual-screen Razr immediately put the cell phone pioneer back on the wireless industry map. Thin, light and unbelievably sleek, Motorola redefined the stale flip phone and paved the way for a slew of so-called fashion phones that tempted users with glitz and gloss—often at the expense of functionality. The Razr phone was saddled with a weak user interface, suffered from battery and speaker issues, and was known to be susceptible to compound fractures.

Until early 2007, Motorola was the world's second largest shipper of mobile handsets after Nokia, but after the Razr leadership fell off, the company's rank tumbled to eighth place by the close of 2009. Motorola was a small fish in the big sea, again. Regaining leadership is generally a difficult thing to do because the demise tends to originate from a combination of problems rather than one or two issues. But Motorola resurfaced with a flash of innovation when it shifted its focus on the smartphone opportunity while it reduced reliance on feature phones. Feature phones had started to commoditize by mid-2000s while smartphones' thrust into the mainstream was just beginning. Smartphones also offered higher average selling price; Apple effectively leveraged it when it introduced the iPhone in 2007.

In August 2008, Motorola brought the former Qualcomm executive Sunjay Jha to head its struggling portable devices unit, which was spun off as Motorola Mobility later in January 2011. Jha made the strategic move toward smartphones while betting on Google's Android operating system, eventually turning Motorola into an Android smartphone house. The Droid phone, which took the brand from *Star Wars* creator Lucasfilm, turned out to be a star in its own right and put Motorola back on course to become a formidable player in the mobile arena. Motorola's Android-powered Droid was now the smartphone increasingly compared to the iPhone.

According to some in wireless industry, the Droid went beyond the iPhone in two key areas: like the Palm Pre, applications could run in the background, and switching among them was a smooth process. Moreover, as a robust GPS device, it accepted and gave turn-by-turn verbal instructions, making it the closest approximation yet to a total GPS solution that obviated the need for a dedicated vehicle device. Not since the launch of the iPhone had a new product had as dramatic an effect on the competitive landscape of the smartphone business as did Droid. It was the first viable competitor to the iPhone, and probably the only handset in the early going that had challenged the iPhone supremacy. Motorola was out of the shadows, and Apple now looked in its crosshairs.

"This is a new Motorola," confident Motorola executives now told the press. Smartphones had vindicated Motorola. The former ally of Symbian and Microsoft was seeking to reinvent itself in the arms of newcomer Android. For some people Motorola's initiative to go with Android was no-brainer, but it was a key test for Jha. The move could either lead it back to profitability or come back to haunt Motorola.

The next flashpoint in Motorola's smart resurgence came with Xoom, which in many ways embodied things to come on tablet computers. Again, riding on Android's Honeycomb tablet-ready OS version, Xoom was initially seen as the first truly comparable competitor to Apple's hit iPad. The 10.1-inch tablet with capacitive touchscreen and improved resolution of 1,280 × 800 pixels could make voice calls and featured front and back cameras, communication notifications, and application status bars, along with easy movement to airplane and Wi-Fi modes. In the end, the tablet didn't make much commercial headway, mostly because of its higher price tag.

The Android connection unmistakably gave Motorola a shot in the arm, but other Android partners like HTC and Samsung were also catching up fast. Betting on Android was apparently the right move on Motorola's part, but that alone wouldn't fit the bill in this relentless market. Motorola's Android liaison eventually led the mobile phone titan into Google's corporate umbrella in summer 2011, when the search giant launched a bid for Motorola's rich radio heritage. Motorola's wireless story was coming to a close.

THE DISRUPTION GOES ON

In chapter 5, I explained how RIM lost its reigns in the emerging smartphone market by overlooking mobile web and let the iPhone make a clean sweep. But that was only the half of the story; the other half of the story that evolved over time related to the so-called unwired enterprise. In the early going, IT managers dismissed the iPhone as a toy, and many even banned it from their companies outright. The BlackBerry, which looked like a glorified pager and had a small keyboard and always-on wireless data connection, allowed users to send and receive e-mail on the move as though they were sitting in front of their desktop PCs.

The BlackBerry was the first of a new breed of devices taking up the space between handheld computers and mobile phones. A BlackBerry device, unlike a laptop, could be used in taxi. Stable and secure e-mail and messaging services made BlackBerry so beloved by corporate IT departments that the number of BlackBerry subscribers worldwide nearly tripled during 2008 to 2010 period. RIM had moved so aggressively to hang on to business customers that it began offering free server software for small- and medium-sized companies worried about costs. But when the BlackBerry elite movement started in the mid-2000s, most companies didn't allow executives to use their BlackBerrys for personal calls and e-mails, and this forced executives to carry two phones. That's when competitors like Apple took aim at this lucrative market.

Companies like American Airlines and Union Bank initially restricted their employees to BlackBerry handsets and wouldn't support the iPhone for security concerns. But mobile phone is a hugely personal object, so the decision to pick a platform wasn't going to be made by the companies for long. On one hand, the workforce was clamoring for the iPhone experience, and on the other, Apple was methodically adding enterprise features like e-mail, spreadsheets, contact management, and security through hardware and software updates. By 2010, RIM's hold on business segment had weakened while the iPhone, which started as a consumer device, had made significant inroads with business customers. Much of the impetus for widening iPhone adaption in business came from employees themselves. People purchased their own iPhones and used them at work to access corporate data or networks.

The BlackBerry is far from doomed. It's going to be fine for the next few years because it has an enormous installed base and migrating from a key platform usually takes a while. Moreover, the BlackBerry handset is still widely popular in developing markets around the world. That gives RIM a good deal of room to get its technology and marketing acts together. Part of RIM's problem is its inability to stay cool; its attempts at generating a modern appeal through the launch of the Storm and the Torch handsets have largely flopped. RIM is trying to address the coolness and software issues by shifting the OS platform to QNX software. In the final analysis, the crossover between enterprise and consumer worlds has become an Achilles' heel for RIM and could well decide its future in the coming years.

The smartphone shake-up didn't stop at the door of RIM or Microsoft or Nokia. Just a week after the change of guards at Nokia, Nam Yong, CEO of Korean giant LG Electronics Co., resigned amid difficulties in reshaping its handset business toward smartphone shift. The high-profile departure was yet another sign of the turmoil that the technological transition was causing in the mobile business as result of the smartphone-model realization. In its quest to win market share, from 2005 to 2008, LG raced up from sixth to third in global handset sales. But it won market share at the cost of poor margins and, consequently, went from generating profits to making losses. LG had a small presence in smartphones and made its first global launch of a smartphone in September 2010.

Earlier, in 2009, just when smartphone was emerging as a game-changer, LG started on the wrong foot by ignoring the rising star, Android, and embracing the setting sun, Windows Mobile. The fact that even Microsoft was about to dump Windows Mobile software for the new Windows Phone 7 project shows how clueless LG was in the new smartphone game. I have documented many of the problems with Windows Mobile software earlier in this chapter. Worse, LG's first smartphone in collaboration with Microsoft would arrive in the market after three years, in 2012. In the meantime, LG had no choice but to depend on its feature phones, which were being increasingly commoditized.

LG's Korean rival Samsung Electronics also initially missed the opportunity but then it worked feverishly on its smartphone strategy and made sure it won't miss the next cycle. Samsung went on to become one of the most important manufacturers in the Android camp. Android had given Samsung's smartphone business a shot in the arm, and now the Korean electronics giant was looking to replicate that success in the tablet market.

The pen computing frenzy didn't go beyond the early 1990s but it provided an initial impetus for the shift from desktop to portable arena.
(Photo courtesy of Microsoft)

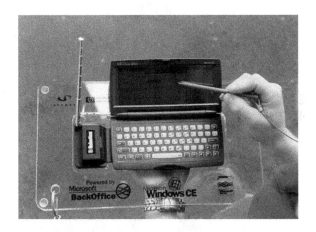

Windows CE, Microsoft's early stab at smartphone software, proved too big for small gizmos. It was a general-purpose OS for a whole range of portable devices, from set-top boxes to handheld computers. (Photo courtesy of Microsoft)

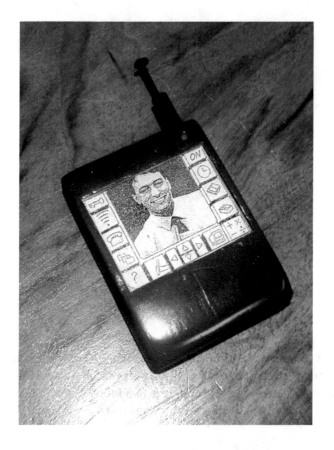

Daniel A. Henderson's pioneering wireless picture phone—Intellect— introduced in 1993 was a precursor to more viable video communication devices like the iPhone 4. (National Museum of American History photo)

The Nokia 9000 Communicator, according to Psion founder David Potter, was a bird that couldn't fly. (Photo courtesy of Nokia)

Sophie is born and so is camera phone. Sophie's dad, Philippe Kahn, integrated camera and photo-sharing software into his mobile phone and sent this picture from maternity ward to family members, friends and associates around the world on June 11, 1997. (Fullpower Technologies Inc. photo)

RIM, founded in 1984, hit the tech business limelight in 2002 when it married its highly focused e-mail device with a mobile phone. The first successful smartphone had come to life with always-on connectivity. (Research In Motion photo)

Danger Inc.'s Hiptop, manufactured by Sharp and marketed by T-Mobile as Sidekick, was the forerunner of Android and among the early mobile phones to offer mobile Internet access. (T-Mobile USA, Inc. photo)

The building that housed ARM in its early years; the British chip designer would eventually rule the mobile world with its processor architecture that powered a predominant majority of gadgets. (Photo courtesy of ARM)

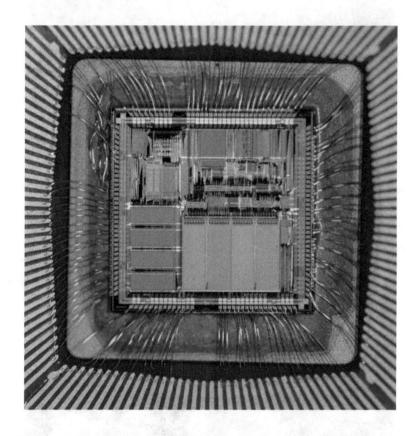

ARM's 1999 projection that 70 percent of its market would come from processors in the mobile phone segment defined its early success. (Photo courtesy of ARM)

Steve Jobs and Stan Sigman at a dinner hosted by the Wireless History
Foundation; Sigman orchestrated AT&T's partnership with Apple for
the launch of the iPhone, which changed the wireless industry forever.
(Coracle Group LLC photo)

SEARCH FOR KILLER APPLICATION

"The cell phone is the single most transformative technology for development."
—Jeffrey Sachs, Columbia University economist and
emerging markets expert

In 2007, soon after Olli-Pekka Kallasvuo took charge of Nokia as the
new CEO, he embarked on an ambitious technology acquisition. Nokia
bought the Chicago-based digital map company Navteq Corp. for hefty
US$8.1 billion to bring navigation out of car and deliver it to pedestrians.
Nokia coined the buzzword "context-aware" Internet while asserting
that it will reshape the Internet. To accomplish that Internet panacea,
the Finnish mobile phone giant was pinning its hopes on operator-
independent, cross-platform phones conceived through development
of new software and services. The company claimed that Map 2.0 would
enable context-aware Internet by combining multimedia features with
the freewheeling Internet and assisted-GPS technology.

Nokia engineers claimed that by adding context—such as time, place,
and people—to the Internet, the mobile web experience would
become something entirely different. Once the context was added to
the network, they contended, the Internet experience would become

more mobile, contextual, and personal than on the desktop. The building blocks necessary to make this happen included GPS, broadband access, a back-end service, and enough processing power and memory residing on mobile phones. Here is one scenario depicting how it would actually work: a user takes pictures with a camera phone, and the GPS coordinates are simultaneously stored in a metadata file; Bluetooth could sniff around and discover who is around him or her. Location, therefore, would no longer be an application; it would become a core fabric of the mobile Internet.

Nokia managers loved to play up fascinating new scenarios at technology press events. Their hyperbole was reminiscent to the early days of mobile commerce talk, which was stimulated by the arrival of WAP-based mobile phones. At that time, during early 2000s, marketing dream weavers conjured up whiz-bang scenarios in which mobile-phone users would resort to all kinds of amazing adventures. One might have heard this: walking down the street, a user approaches a Starbucks coffeehouse and his or her mobile phone starts ringing; on the handset screen pops up a coupon for a $1 latte. Or this: A user strides into a department store and slips into that perfect pair of jeans. A bit pricey! No sweat for his or her mobile phone. The user punches the bar code of the jeans into his or her handset and receives 20 percent discount from an online retailer.

However, these marketing potions carried a fundamental flaw: the shoppers already knew what was happening in every corner of their favorite mall with an almost prescient knowledge. So all these pie-in-the-sky ideas ended up as wishful thinking; broadcasting people every time they pass a McDonald or a Starbucks seemed incredibly unrealistic. What Nokia did in 2007 was recycle this notion by combining two chic technologies of the time—the mobile web and GPS-based location—and started spreading the context-aware Internet gospel. However, after three years and some failed projects, there was little evidence of any tangible payback to Nokia's foray into location-centric premium phones. When the dust settled, Nokia seemed to have fallen to the classical marketing paradox that was all too familiar in the twilight world of the mobile phone and the Internet.

Apple, in a stark contrast to Nokia, carefully rationalized the supporting technology components, worked out a robust product roadmap, and then mobilized its legendary marketing machine. It's official now that the iPhone was initially conceived as a phone that would play music and video from iTunes, but its primary appeal quickly became the millions of head-slapping, useful software applications that ran on it. The iPhone certainly played music, but owners were just as likely to use it to check the weather, book dinner reservations, read a newspaper, get directions, or play a quick game of Taxiball on the subway. The creation of the App Store for the iPhone was a revolution that literally changed the smartphone world overnight and helped the iPhone reach dizzying heights.

The Cupertino, California–based company sent shockwaves across the communication world by readily shifting the consumer expectations for smartphones. The App Store created billions of dollars in new revenue for software makers and catapulted the iPhone into being the largest must-have device on the planet. Suddenly the competition found itself scrambling to make its own version of the App Store, create software tools, and get programmers on-board to begin making applications for their devices. Companies as varied as Nokia, Motorola, and RIM rushed to adapt, also adding touchscreen-based devices to their product lineups.

The combination of smart handheld devices, mobility-enhanced applications, GPS software, and robust wireless Internet experience added up to the promise of mobility. Apple had successfully demonstrated that those companies that effectively execute mobile strategies for commerce, entertainment, and finance will be the winners in the mobile Internet arena. When the movers and shakers behind m-commerce's false start of early 2000s saw the changing tech headwinds, they rejuvenated themselves to bring users banks, shopping malls, enter-tainment centers, and all of their friends to their palm, pocket, or purse. But where was the next killer application? This is one of those perennial questions whose answer often seems obvious in retrospect but is not visible when one is peering into the crystal ball.

The initial hyperbole unmistakably related to m-commerce. In the evo-lution of modern shopping, consumers had very recently progressed from visiting a physical store every time they wanted to buy something to shopping online via a desktop or laptop computer, at least some of the time. The next evolutionary step would presumably have shoppers making more purchases online from mobile devices. Over the years, the smartphone intelligence had been steadily growing in parallel to the device's emerging status of being an important showcase of rich content with handset screens getting bigger and brighter. The sense of shifting sands was further reinforced when 2009 data proclaimed more smartphones than laptop computers being sold.

M-COMMERCE: A BRIEF HISTORY

Less than a decade after the term "wired" surfed into the public conscious-ness, heralded a brave new world of wireless Internet that promised to bring forth unprecedented communications power to businesses and ordinary people alike. At the dawn of the new century, when cell phones were singing the happy tones of the Internet, wireless operators saw writ-ing on the wall and began tailoring a variety of new services according to users' demands. They promised users would now be able check their bank balances, get weather updates and traffic reports, and read news head-lines through text-based data services. Wireless Internet, as it turned out, was just an icing on the cake; it greatly helped stimulate new industries.

Mobile e-commerce, or simply m-commerce, harnessed the ability to make commercial transactions from a mobile phone. Back in 1999, in the heady days of dotcom euphoria, it all looked so easy: to make money from the mobile Internet, simply create a mobile version of what worked so well on the fixed-line Internet. The notion of m-commerce quickly became the face of the mobile Internet in a way that was reminiscent to the rise of e-commerce soon after the wired Internet took shape in the commercial arena in the mid-1990s. Then, people would call e-commerce the second chapter of the Internet. Mobile commerce went one step further by promising the ability to purchase goods anywhere through a wireless, Internet-enabled device.

The vogue at the time was for business-to-consumer e-commerce, and m-commerce drew a lot of strength from this thriving premise because of the ubiquity of mobile phones. Surely, anything that could be sold over the conventional Internet to PC users could be sold over cellular networks to mobile subscribers. And because mobile users have their phone with them at all times, they might be expected to do more shopping than stationary customers. With the rapidly maturing concept of the mobile Internet, the early backers reasoned, m-commerce could serve as a powerful platform to host new services in collaboration with Amazons and Yahoo!s of the Internet world.

But there was an interesting twist in the formation of m-commerce prodigy. Although the Internet euphoria spread across the whole globe, when it came to business realties, it was pretty much an American phenomenon. From the super-ISP America Online to the Internet hardware king Cisco to the UNIX server powerhouse Sun Microsystems, almost every predominant company on the Internet scene was of American origin. Hence, when commercial Internet extended its reach to the big-time e-commerce bonanza, it seemed like just another American show. In the ongoing wireless revolution, however, Europe and, to some extent, Asia called the shots. In countries like Japan, where the personal computer penetration was low due to cultural reasons, mobile phones could well become a potent tool to access the plethora of nifty consumer services.

Inevitably, the companies in these parts of the world thought up of m-commerce as an answer to America's e-commerce hegemony. But although European and Japanese footprints were initially predominant in the m-commerce roadmap through projects like WAP and i-mode, when it came to actually doing the Internet, America seemed to know the way forward. It became evident that the Internet-savvy America had still an important role to play in this nascent marketplace when Amazon. com launched its m-commerce efforts in late 1999. The Seattle upstart had invented one-click ordering that let the buyers store credit card number and address after the purchase. Next up, the online retailer, through its "Amazon Anywhere" initiative, started assembling partnerships with a number of cell phone operators in the United States.

The arrangements typically called for Amazon's website to be given a prominent placement on the screens of mobile phones, critical in those early days because of difficulty users had in navigating the web by punching on a phone keypad. In March 2000, just before the dotcom bubble burst, Amazon chief Jeff Bezos predicted that by 2010, all of his firm's customers would use wireless devices to make purchases. Describing m-commerce as "the most fantastic thing that a time-starved world has ever seen," he predicted that it would change the way people shop, since they would be able to make impulse purchases anywhere, at anytime. Within five to ten years, he claimed, "almost all of e-commerce will be on wireless gadgets." In an industry not yet humbled by the dotcom collapse, Bezos wasn't alone in seeing the wireless web as the next sensation.

Market analysts queued up to make rosy forecasts of m-commerce revenues. With such a bonanza conceivably around the corner, it's no wonder that wireless operators paid so much for the 3G licenses. But when the Nasdaq crashed and the dotcoms started going under, the wireless world came to a rude awakening that making money was hard enough even on the conventional Internet, where technology was rather mature. Once-fashionable dotcoms had failed to bring a workable business model and were now burning in style. The dotcom fiasco had shown to the world the ugly side of the Internet. Now the prospect of buying things on cell phones, with their tiny screens and keypads, suddenly looked far-fetched. The new surveys showed that consumers found the reality of m-commerce hugely disappointing.

In the hindsight, the way America Online, Yahoo!, and Amazon approached the wireless Internet was to take what they had on the wired web and simply put that on the handset. Another problem with the mobile Internet was that there were too many clicks. An early trial found that it took over forty minutes to order a book by cell phone. Likewise, booking a hotel room on a mobile hotel-reservation system required thirty-seven clicks. Then there were those dreaded lessons from the e-commerce chapter. For instance, its ad-based revenue model had proved untenable, leaving many in the wireless industry wondering whether consumers would be the ones to pay for m-services. Banner ads, pop-ups and the like were also deemed failures in the wireless domain. On the heels of this confusing picture, online merchants started placing a low priority on developing and marketing m-commerce services. The user base also remained small because only a few websites were equipped to accept m-commerce transactions.

ENGINE OF M-COMMERCE

So what was the reality of m-commerce: online shopping done on the run? Was it just another reflection of PC-based Internet commerce or the next Internet joke? In a way, m-commerce was just any wireless data activity that made money for a company along the value chain. So when a mobile-phone user made a restaurant reservation, the amount that the user later paid using cash or credit card could be counted as m-commerce revenue. Under this view, m-commerce, in collaboration with the mobile Internet drive, could potentially bring a myriad of new activities for the wireless industry. For a common user, it was all matter of value. Once the wireless industry worked out this piece of puzzle, m-commerce and other wireless data services could come of age much faster than anticipated.

How the promise of m-commerce would descend successfully to the cyberspace was still a question mark in the early 2000s. But as wireless devices got faster, smarter, and cheaper, the hope floated that more effective software platform to power the m-commerce products would subsequently emerge. The broad consensus was that it won't be before 2005 that the world sees the three As of m-commerce—anything, anytime, anywhere. In the meantime, m-commerce architects had to find ways to complement the desktop web but avoid replicating it. There was little point in making an impulse purchase of a book or a CD if it will then have to be delivered by post. Rather than spend ages pecking at a phone keypad, why not wait until one get home and order in comfort from a PC.

The mobile Internet was different from its fixed-line counterpart in three important aspects. First, a mobile phone was a far more personal device than was a PC. It was likely to be used by only one person, who would probably have the phone with him or her for most of his or her waking hours. Second, mobile phone operators could broadly determine what menus and services appeared on their handset screens. The ability to set the default portal was a big advantage for operators, because it allowed them to act as gatekeepers. Last, and most important, people knew that using mobile phones cost money, and there was a mechanism for the network operator to charge them for that use. Sending an e-mail over the Internet from a PC was essentially free; sending a text message from a cell phone cost about 10 cents. So users were more likely to pay a mobile premium to do things while on the move.

Beyond the stark contrasts of wired and wireless Internets, another significant challenge for the wireless industry was to make m-commerce a seamless social experience. The traditional Internet had existed for many, many years and was only used by a handful of scientists because it was too complicated. Only after people got reasonably easy way to browse the web did the Internet explode and became a driving force in the economy. The same could happen to the mobile Internet and eventually to mobile commerce. The logical conclusion: make it easy.

Then there was this risk that wireless operators might be tempted to set up "walled gardens" of services and contents, hence restricting users to a handful of approved services that would enable operators to capture a much larger chunk of the expected bonanza in data revenues. In the 1990s, online services such as America Online, CompuServe, and Prodigy operated on the walled-garden principle; but as soon as one of them offered unfettered Internet access, the others had no choice but to follow the suit. The walled-garden model could turn out to be just as unsustainable on the mobile Internet because it annoyed users.

In fact, in the early stages, mobile carriers generally limited the number of sites offered to customers by signing exclusive agreements with content providers, or selecting a package of sites from a content aggregator—a third-party portal or a WAP provider. But subscribers didn't want restrictions on whom they could do business with or which sites they could visit. This strategy backfired as majority of users found the m-commerce experience unsatisfying. Mobile phone companies subsequently removed these restrictions. M-commerce was initially written off also because the early versions of Internet-enabled phones, the so-called WAP phones, didn't deliver what they promised. When WAP failed to live up to expectations, there was a backlash for m-commerce. In many ways, WAP became the acid test for m-commerce viability.

Mobile Internet access providers, in this case, the wireless operators, generally charged by usage, either for every minute spent online or for every byte downloaded. This meant that they made money on transporting data, so it made sense to offer users the widest choice of content possible to encourage them to run up transport charges. That's how i-mode worked; the vast majority of NTT DoCoMo's data revenues came from transport, not the sale of content. The Japanese mobile phone company generally offered a selection of approved services through its own chosen portal, but also gave subscribers the option of going elsewhere. That's what America Online did with its dial-up Internet service; it offered services such as instant messaging, chat rooms, and e-mail, as well as access to the web.

Wireless industry, predominantly made up of traditional telecom service providers, faced an important crossroads. The telecommunications model was predominantly based on a closed architecture. A telecom operating company typically owned the network; provided the transport; supplied the services, such as caller ID and calls waiting; and billed the subscriber directly. But was this model applicable to the convergence of mobility and the Internet? Probably not! The ability to control the entire value chain didn't exist in many industries. Now the convergence of voice and data was testing this model because original expectations of the mobile Internet were largely based on the experience mirroring the wired Internet.

In the open architecture of the Internet, service providers bought the raw bandwidth from a backbone carrier such as UUNet, while end users paid for their connection media: dial-up telephone line, DSL, or cable. Once connected, they could subscribe to and pay for any service and complete transaction at their discretion. The consumer accepted this model, in which the transmission of data was separate from the provision of service. Overcoming these expectations was proving a challenge for wireless operators. To the question of how wireless operators would overcome this historical odd, the answer was probably in the value chain they needed to refine for a new user base.

Although consultancy firms had predicted massive m-commerce revenue streams, there were big hurdles—standards, security and presentation—before these vast revenues could be tapped. Vendors were making all the right noises, but because of the complexities of initiating the technology, m-commerce marketplace simply didn't seem ready in the early 2000s. The key m-commerce players had more than a few months of experience to draw on it. Then there were all these whacky ideas like web-enabled ice cream delivery. While it sounded fascinating to have a refrigerator send users an e-mail to let them know they are out of milk and then send a message to Webvan to bring more, in reality, this only came out as one of those "oh-golly" scenarios.

The wireline Internet had taken off because it offered users instant gratification to get what they wanted, when they wanted it. Next, mobility started giving users the additional space to get answers when and where they wanted them. But while the Internet strategy had indeed become indispensable for every wireless company, the problem was that the public seemed unenthused. The slew of companies that once emerged hoping to make it easier to translate the riches of the web into the wireless world missed one important step. No one bothered to ask consumers what they wanted. Mobile phone users at large were not sure what the wireless Internet was because the medium itself was still in its embryonic stage.

Back in early 2000s, for the wireless Internet, while much had converged, much also remained to be converged. Many of the mobile consumers' concerns, such as security and privacy, and difficulty with navigation, were reminiscent of worries of the early days of e-commerce, and could eventually be overcome. Still, there were broader problems with using handheld devices for shopping. Compared with PCs, which had large color screens, handheld devices of early 2000s were hopeless for browsing. Scrolling through lists was cumbersome, and features and prices were hard to compare. But the difference between these two worlds also presented a window of opportunity. People carried mobile phones with them everywhere, so it was only a matter of time before all necessary components converged onto the mobile phone, making it even more invaluable.

MOBILE INTERNET 2.0

The mobile Internet's troubled youth was a form of natural progression in the evolution of m-commerce. Mobile-commerce generated a huge buzz when introduced during the late 1990s, but with the exception of NTT DoCoMo's i-mode offering, it had been a major disappointment. Due to network and handset shortcomings and limited content available on WAP and other data services, potential customers were not enthused and failed to sign on in large numbers. Mobile commerce had gone through a lot in its brief existence and the insight this learning curve garnered from failures of wireless operators, content providers, and WAP service companies proved to be immensely valuable. For instance, wireless industry began to ponder on why had m-commerce been a hit in Japan only?

One of the primary reasons of WAP failure was that services were launched before they were ready. Moreover, mobile phone networks were inadequate for handling the e-commerce-like shopping experience that was promised to subscribers. However, the basic problem with mobile data in general, and m-commerce in particular, was that there was no clear business model. M-commerce was a bright idea whose time could come once the wireless industry came over its apathy for embracing data services. That would involve the transformation of the existing telecom practices into entirely new business models and it was an uphill task. But if done well, mobile phones, for instance, could emerge as a popular vehicle of payment, at least for small items. Since wireless operators were used to handling large numbers of small transactions, their billing systems could handle such transactions at around a tenth of the cost of a bank or a credit-card company.

Unlike on the fixed-line Internet, people were prepared to pay for content and services they really wanted. But they preferred to pay lumpy subscription fees rather than a small charge for every morsel of information they accessed. Once past the initial stumbling blocks, mobile commerce, being rife with opportunities, could kick-start in several fast-growing niches: for example, provision of services to stock traders and others who need instant information. Mobile commerce could also transform stores into virtual showrooms. Handhelds with bar code scanners would let users purchase and customize a product and have it waiting for them when they get home. A person looking for a new apartment or a house, while driving in a particular neighborhood, sees a "for rent" or "for sale" sign with a code listed on the signboard. He or she could key in the code into cell phone to receive short video clips of the interior and decide about arranging a visit.

In many ways, the mobile Internet in 2005 was at the same stage of development as its wired counterpart was in 1995 when the web was in its infancy. Then, retailers asked if this new channel was for real. Could it be used as a competitive channel? There were hundreds of startups, and nobody really knew which technologies and business models would win, or what consumers or corporate users wanted. But this cycle of boom and gloom on e-commerce trail also meant that there were plenty of lessons to be learned from the mistakes made on the fixed-line Internet. While which model would prove most successful remained to be seen, there was certainly money sloshing around on the mobile Internet.

By late 2000s, most of the pieces that had to be in place to make m-commerce successful were already in place. Text messaging had provided m-commerce with initial traction and necessary learning curve, but SMS had mostly been successful as an appetizer. That's probably because text messaging was primarily a marketing vehicle; when it came to making purchases, the action was on mobile websites. Only fully transactional m-commerce sites would enable consumers to shop more or less as they would on an e-commerce site, except on a pared-down version. It was also imperative that m-commerce sites provide shoppers access to the same number of products as e-commerce sites.

The search for the "killer application" went on until the iconic iPhone came, bringing a revolution in the way people look and use their mobile phones. The iPhone was the first mass market mobile device that made the Internet fun and easy to use, especially on websites optimized for mobile phones. After the iPhone launch, m-commerce started moving away from SMS-driven systems and into the actual business applications domain. SMS had significant security vulnerabilities and congestion problems, even though it was widely available and accessible.

Retailers had heard the m-commerce mantra before. A lot of retailers that had earlier invested resources and effort in the mobile market were disappointed in results, which generally stemmed from consumers' poor experiences on conventional mobile phones. Also, in the early stages, mobile-optimized sites were quite dumbed-down; they were basic versions of websites that reminded users of clunky web pages of the early 1990s. And the fully capable mobile Internet devices were not there yet. The iPhone changed all that. Once phones and the mobile Internet converged on powerful new platforms, like the iPhone, more consumers started using the mobile web as a viable communication tool. Devices like the iPhone boasted significant computing power and made accessing the Internet from a handset far easier than with mainstream feature phones.

The iPhone set the bar for a true mobile browsing experience in which sites rendered the way they would on a PC. Now consumer expectations went up: they had good experiences with some mobile sites, and now they were expecting standard websites of other companies to offer a good mobile experience as well. Eventually, the post-iPhone era saw a flurry of retailers creating text messaging-based marketing programs, m-commerce sites and downloadable mobile applications to reach out to consumers who wanted to do more with their phones.

There was a difference this time around, and it was called the smartphone. These slick new gadgets with great visual experiences and added functionality like GPS location awareness changed everything. Just as broadband made shopping online far more attractive, so too, did these powerful phones by changing the game of m-commerce. eBay, for example, reported a whopping US$380 million in sales through its iPhone app and m-commerce site for the first nine months of 2009. Now retailers were lining up to learn the rules of the m-game.

Wireless companies initially touted m-commerce as a mobile extension of e-commerce instead of portraying it as a unique, value-added mobile service. So they found users flocking to the wired web and ignoring the mobile alternative. Customizing m-commerce to multiple consumer tastes, as shown by DoCoMo's i-mode service, launched nearly eight years before the arrival of iPhone, offered better chance of success. The i-mode service allowed Japanese mobile users to use their phones for most of their everyday transactions. They could purchase a plane ticket and use the e-statement stored on the phone to check into the flight. They could locate nearby convenience stores or vending machines using mobile GPS and then make the purchase from the device itself.

Like its remarkable mobile Internet execution, Apple seems to have learned all the right lessons from the success of i-mode in the m-commerce realm. Like DoCoMo, Apple carefully cultivated an entire ecosystem made up of thousands of applications, which in turn made it possible for people to go beyond the mobile web to get things done. The company that had put MP3 players into the hands of the masses and had changed the way consumers interacted with music was now courting software developers to build an industry around its iPhone platform. By 2010, the iPhone was at the center of a huge m-commerce maelstrom with thousands of alluring apps. An increasing number of consumers armed with smartphones like iPhone were relying on online product reviews and recommendations while in the aisles of brick-and-mortar retail outlets.

THE FOURTH SALES CHANNEL

In retrospect, the 2001 version of Amazon's m-commerce site was bare bones compared with 2009's Amazon Anywhere, which according to the company reports, has been able to yield significant traffic gains. Over the years, the e-retailer added features and functions based on customer e-mails and comments that came in through the Feedback buttons. By 2010, Amazon was selling over US$1 billion annually via mobile devices. What the world's largest e-retailer had learned through a decade of experience was that how it sold on the wired web couldn't be ported straight to the mobile world. That crucial inflection point hidden in the twilight zone of wired and wireless Internets for many years was finally becoming evident to the industry by 2010.

Now retailers were nervously looking at the m-commerce bandwagon on its way to joining stores, catalogs and websites as a mainstream retailing channel. M-commerce was on way to becoming the fourth sales channel. In the hindsight, wireless Internet and its alter ego m-commerce were not a wishful thinking at all. It's just that wireless industry in general and retailers in specific needed to think through their mobile strategy from the ground up. The m-commerce affair was a lot like e-mail: in the early 1990s, e-mail wasn't hot but later became the killer app of the Internet and caused the demise of fax. Almost everything could be communicated via e-mail. Just like e-mail, the infrastructure for m-commerce needed to be in place, and once it was there, people wouldn't think twice about joining the m-commerce bonanza.

M-commerce is still in its infancy because mobile Internet-enabled phone users are more likely to employ their devices to get weather forecasts, read news, find movie times, and bank online than buy products. The i-mode case study suggests that users may be prepared to pay a small amount to receive news, weather, sports scores, horoscope, and so on. But though subscription revenue associated with these services is tiny, the real money could be in transport. Mobile operators will most likely make money on m-commerce and location-based services, if only through the associated transport revenues. Access to free content, such as train timetables, would also enhance transport revenues.

However, there is a group in the wireless industry who believes that the real value proposition for wireless operators might not be in the ability to own spectrum and provide the transport, but in their ability to differentiate by offering unique services that end users find valuable or entertaining. But there comes the rub: wireless operators are still largely focused on voice, which accounts for the bulk of their revenue. Voice could remain the epicenter of wireless communications for the foreseeable future because everyone needs to speak, but moving forward to the new data paradigm, new forms of communications will evolve.

The innovative new services built around location, m-commerce and personalization could harness a new business foundation for mobile phone companies. People are starting to use their mobile phones to send and receive data in increasingly interactive manners. Although mobile sales could be slow initially, a big bubble could eventually come through the air pipe and deliver a major wake-up call, showing everyone just how important this sales channel is.

Wireless industry had first started pondering m-commerce for transaction-based services, but then messaging came up first with rich content to follow. Next up, there were image, video and media streaming technologies that promised a new world of multimedia possibilities. Sharper color screens on mobile phones have made wireless data services more appealing. The collection of these features and services would get to the core of what m-commerce and wireless Internet had been promising all along. The combination of personalization, location, and willingness to pay made all kinds of business models possible. That could turn m-commerce into a powerful new engine for smartphone take-up while entertainment is likely to become the significant secondary source of smartphone revenues.

People won't buy smartphones so that their kids could play games. Smartphones took off when companies like Apple and RIM found ways to enable people to do useful things on their phones. For that, as the old PC industry mantra goes, a sustainable architecture is critical for creating useful new applications. This chapter chronicled the evolution of e-commerce as a core smartphone application and m-commerce's relationship with its prime vehicle: mobile Internet. Beyond the indispensible wireless Internet, however, there are two other intrinsic elements of the m-commerce equation. The next chapter encompasses both SMS and location-based services as two primary enablers of m-commerce along with the third: mobile Internet.

SMS AND GPS LIAISONS

"I'll txt u in 10mins when I know wh/ restrnt."

In January 2001, when the people of the Philippines threw out their elected president Joseph Estrada on charges of corruption and incompetence, texting, as Filipinos would call it, was credited for this popular uprising. Short message service (SMS) became a key tool to mobilize "People Power" rallies that eventually ousted the former film star from the presidential office. Small text messages were ahead of radio, television, and newspapers, and for the first time in the history of media, common people became a proactive part of the information mainstream by merely using their mobile phones. Breaking news would reach out to handset screens even before the journalists could process it for presenting on their respective media channels.

A year earlier, mobile phone users in Manila had left behind Helsinki in producing the largest number of text messages in a single day. The Helsinki connection was reminder of the fact that it were Finish teens who had popularized this technology during the mid-1990s. When Finland's leading wireless operator Sonera recorded 1 million short messages in a single day in 1997, teenagers were credited for the explosion of this new kind of data service. Every generation needs a technology that it can adapt as its own to communicate. The gadget-crazy teens were the innovators in the text-messaging space. From Finland to the Philippines, they were the most pervasive users of cell phones, and for them, SMS was the best part of the package because it was face saving as well as money saving. The truth is that text messaging became so popular also because it was cheaper to send a message than to actually talk on the mobile phone.

To create a message, a user would punch it by hand into the phone using the keypad as a miniature device. SMS was initially perceived so difficult to use that only young people could overcome the man–machine interface impediment and use the service. It is an industry truism that children most readily learn the necessary new styles and habits. So they instantly acquired the most esoteric typing skills required for short messaging over mobile phones. The fact that the entry barriers to learning the service were so high actually became an advantage for teens as this meant that it would take a while for parents and teachers to figure out the functioning of SMS. In Finland, during the early days of text-messaging craze, teachers had to confiscate mobile phones from high school students to prevent any chances of cheating.

The first SMS text message was sent from a computer to a mobile phone in 1992 in Britain; the message was sent to Vodafone director Richard Jarvis saying "Happy Christmas." The first person-to-person SMS from handset to handset was sent in Finland next year. Eventually, simple person-to-person messaging fueled the volume growth as people told each other how they were feeling and what they were doing. The scene of watching teenagers beaming text messages to one another in a Helsinki park was enough to convince even the most cynical observer that one day wireless data could transform the way people communicate. Before the emergence of SMS, sending data over cellular networks was never an easy proposition.

Although cellular systems supported fax and file transfer features, reliability and cost were always an issue. Things started changing with the advent of GSM, which allowed people to send and receive messages of up to 160 characters on their mobile phones. People became remarkably enthusiastic about using phones to send short messages. The love for text messaging spread over the globe and the SMS coup de grace left many analysts wondering. Given that mobile handsets were hampered by slow transmission rates, small screens, and clumsy keyboards, the popularity of short messages was a blessing, and it boded well with the future of wireless data.

GSM designers, while searching for add-on features to make cellular services more compelling, augmented the short messaging capability for handset users. Early contribution in the development of SMS came from Friedhelm Hillebrand of Deutche Telecom and Bernard Ghillebaert of France Telecom. Back in the early 1990s, on the arrival of GSM handsets, packed with a number of upscale telecom features such as caller ID and call forwarding, both users and network operators hardly paid any attention to the SMS function. The commercial Internet was still in its infancy, and data were nowhere on wireless industry's radar screen. Eventually, short message service caused the demise of paging and laid the groundwork for a new arena in wireless cosmos. It became the stepping stone of a new industry: mobile data.

People on their way to work, people on their way home, and people just out and about banged greetings, jokes, and poems at incredible rates. Text messages combined the pleasure of reading and writing with instantness and a handy little gadget. The technology became a modern art when people started turning text messaging into a creative force. Because the size of a handset screen was limited and an SMS message could hold only 160 characters, mobile phone users came up with rather interesting ways to express their thoughts in a constantly evolving SMS lingo. Textual artists—a new breed of romantics—had started establishing this innovative new lingo. A whole new alphabet emerged because text messages took longer time for data entry and were quite abrupt in composition as people attempted to say as much as possible with a keystroke. Abbreviations such as "C U L8er" sprang up for saving time and for coolness.

By the early 2000s, in commuter trains in Asian cities like Manila, there were as many mobile phone users as there were in any big European city, yet these phones were mostly silent. Instead of speaking into these sleek gadgets, users were thumbing on them to constantly exchange text messages. SMS was the triumph of consumers—a grassroots revolution that the mobile phone industry had practically nothing to do with. The great leap forward of SMS was in a stark contrast to the top-down technology- and industry-led approach adapted in other non-voice services like WAP. Surely, the wireless industry could learn a lot from SMS in its quest to create other non-voice services.

SMS FOR STARTERS

Text messages eventually became part of daily life as chatty teenagers and commuting workers sent and received quick tidbits of gossip over their cell phones. The emergence of text messaging once more highlighted the historical reality manifested by early forms of wireless communications: economics and usability. The wide take-up of short messaging had proven that when a technology is simple and useful, people would use their phones for more than just conversation. SMS was an accidental success that took nearly everyone in the mobile phone industry by surprise. There was hardly any promotion or mention of SMS on mobile phone companies' part until after it started to be a success. Wireless operators were even slow to show innovation when it came to billing for these services.

Unlike next-generation networks—whether 2.5G or 3G—which promised faster Internet access but required billions of dollars to build, SMS-based services could be done on the cheap over existing cellular networks. The messages required data rates as low as 2.4 Kbit/s and were so small that they were effortlessly transported over GSM-based 2G networks. Also, over the years, SMS technology evolved from simple text messaging to more sophisticated data services such as games and ringtone downloads. Nokia introduced smart messaging protocol built on binary SMS rather than the standard text SMS and thus allowed users to download ringtones and logos and to pay for them as part of their phone bill, generally by using a premium-rate number to access the service. The earliest mobile phone successes have been these simple applications that allowed consumers to download ringtones, thereby personalizing their phones.

In 2008, over 4.1 trillion text messages were sent worldwide, generating revenue of US$8.1 billion. The country that ranked number one in SMS adaption—way ahead of most developed nations—was Philippines, where the average Filipino's income was about one-tenth of an American's. Text messaging was at the root of the wireless explosion that this heavily agricultural country had witnessed during the first two years of the twenty-first century. In 1996, Philippines had less than 500,000 mobile phone users. Within five years, however, the count had reached 7 million that also made almost twice as many mobile phones as landline telephones. By far the cheapest form of communication, SMS quickly spread to the all cross-sections of society in countries such as Philippines where wireless operators offered clever pricing strategies to get users hooked.

More than 12 billion messages were recorded during the year 2000 in Europe alone. It's pretty ironic that while the blizzard of SMS was manna from data heaven for Europe, most folks in America initially found the idea of typing a short message on a phone utterly bizarre. People in Europe and much of Asia couldn't live without short message service, whereas in America they couldn't have it. Short messaging service became a staple of Europe and Asia mostly because mobile networks there were predominantly based on the GSM standard. Conversely, it was a non-starter in the United States because the region had a soup of different mobile phone standards, and GSM was not the dominant one. As a result, SMS became a telling illustration of the wireless gulf between America and Europe.

Not that wireless messaging was non-existent across the Atlantic: Americans were still using their pagers to send and receive messages. Next, people in the United States got hooked to Internet-based instant messaging services like Yahoo! Messenger. In many ways, SMS was like instant messaging, except that users didn't necessarily have to be in front of a computer or load a particular messenger service. Moreover, SMS continued to evolve into a more interactive form over the years and thus spawned a new generation of instant messaging services on its own. Although regular text messages were limited to 160 characters, costing an average of 10 cents each to send, text messaging could be used to do more than just send a quasi-telegram to other people.

The United States lagged behind Asia and Europe in adapting the texting habit, but once GSM-based American wireless operators noticed a leveling off in their customers' use of talk minutes, they introduced texting options and spent more on marketing them. In 2003, AT&T signed on to sponsor *American Idol*, a television show that had viewers voted by texting to the designated numbers. *American Idol* put texting on the U.S. map and unleashed a new wave of text interaction with television. One of the crucial factors in the successful union of television and text messaging was the availability of special four-, five-, or six-digit numbers called shortcodes. While reality TV shows like *American Idol* allowed viewers to cast votes through text messages, there was more to TV-texting assortment than mere voting. News and current affair shows encouraged viewers to send in comments; game shows allowed viewers to compete; music shows took requests by text message; and broadcasters operated the on-screen chat rooms.

The introduction of prepaid mobile tariffs in which people could pay for their airtime in advance and thereby control their mobile phone expenditure was also the catalyst that accelerated the renewed take-up of SMS.

When hundreds of millions of text messages were zapped from handset to handset each day during the early 2000s, the service made about 10 percent of operators' overall revenue. Now the advent of premium text was proving very lucrative as mobile operators charged special rates for messages to particular numbers. The TV-related text messaging now accounted for an appreciable share of wireless operators' data revenues. Operators usually took 40 to 50 percent of the revenue from each message, with the rest divided among the broadcaster, the program maker, and the company providing the message-processing system. Text message revenues now became a vital element of the business model for many television shows.

Short message service was one of the simplest and most useful means of communications and the secret of its success lay in the business model. Now that one of the fastest-growing uses of text messaging was interaction with television, the success of TV-related texting became a stark reminder of how easily an elaborate technology can be unexpectedly overtaken by a simpler, lower-tech approach. The ubiquity of SMS on mobile phones was transformational and the testament of the fact that textophiles in the United States were starting to see voice mail as a waste of time. Text messaging also provided impetus to other wireless initiatives like m-commerce and imaging, just like what the Internet had done for a myriad of wired and wireless communication services.

Another ringing endorsement of the SMS power was Twitter, a microblogging digital media channel that let people share 140-character or less text messages over the Internet. Instead of calling each other and distracting people during a specific session or having to text seven different people, individuals could stay in contact over Twitter. In a way, websites like Twitter with postings of no more than 140 characters were creating and reinforcing the habit of communicating in micro-bursts pioneered by SMS. And just like SMS, these sites were pumping out sheer volume. Many Twitter and Facebook devotees even created settings to alert them, via text message, every time a tweet or message was earmarked for them.

M-COM WITHOUT NET

Two hours after earthquake struck Haiti in January 2007, a texting dona-tion campaign, "Text HAITI to 90999," was up and running. After three days, the effort had raised US$5 million for the American Red Cross, and "Text" and "90999" were in the top-ten trending topics on Twitter. Nine months later, more than US$40 million had been donated by people sending as little as US$5 to US$10 from their cell phones. The millions of dollars the Red Cross raised and collected via text messaging dem-onstrated how the public had turned a corner with m-commerce. The Red Cross, a well-respected organization, gave a clean way for people to contribute using a mobile phone. It was a major validation about the possibilities of text as well as mobile transactions. Another lesson was that simpler m-commerce technology could also be effective.

In 2008, US$300,000 in text donations went to just over one hun-dred charities, and within two years, mobile giving in the United States reached US$50 million for as many as five hundred organizations. The mobile-giving industry had the potential to change the face of global philanthropy. The Haiti campaign had proven the possibilities of mobile giving—particularly in response to a high-profile crisis. As easy as it is to mail an envelope or visit a website, the operators behind the mobile giving were now betting that the mobile would be even more effective at capturing the short attention spans of potential givers. Mobile giving does have drawbacks though, including US$10 donation limits and the fact that it could take up to ninety days to deliver funds because phone bills must be paid first. On the contrary, online donors don't have dollar limits, and their donations could be turned around in as little as two days.

Nevertheless, the Haiti earthquake proved a turning point for SMS-centric communication as this incident demonstrated that text messag-ing was a great medium to connect with the people and the role it could play in innovative campaigns. Text messaging, being the most com-monly used data application on mobile phones, had become a major catalyst in the broad realization of wireless data services. Initially, text messaging was overlooked as a key enabler of m-commerce, both as a method to initiate transactions and as a trigger for wireless data use. That was mainly because despite the popularity of text messaging, revenues generated by SMS services remained relatively small. Secondly, m-commerce without the Internet sounded like an absurdity to many retailers. But after the Haiti earthquake in 2007, SMS became a business reality and a giant leap for m-commerce viability.

Text messaging offered something that m-commerce sites, mobile apps, and even e-commerce sites and e-mail couldn't offer: access to consumers without the Internet connection. Text messages travel over the same wireless networks as mobile phone calls, and require no Internet access. So when it comes to getting as close as possible to customers, tiny text messages hit the bull's-eye. Amazon.com accomplished a pioneering milestone in text message retailing when it introduced TextBuyIt service in spring 2008. QVC Inc. followed in the fall that year with QVC Text Ordering feature.

The use of the mobile technology as a payment gateway had started in Helsinki in 1997 when a company owned by Coca-Cola installed two mobile-optimized vending machines. These machines accepted payment via text messages. Same year, Merita Bank of Finland launched the first handset-based banking service using SMS. In 1999, the Philippines launched the first commercial mobile payments systems on the platforms provided by the country's two large mobile operators: Globe and Smart. Over the course of years, the idea spread and m-commerce manifestations ranging from mobile banking to mobile credit cards to mobile payments became widespread in Asia and Africa, and in selected European markets.

Other SMS-centric data services enabled by handsets included mobile music, downloadable logos and pictures, gaming, gambling, and advertising. In 1998, the first sale of digital content as a download to mobile phones was made possible when the first downloadable commercial ringtones were released by the Finnish wireless operator Radiolinja. The first mobile news service, delivered via SMS, was also launched in Finland later in 2000. Mobile news services expanded in the coming years with many organizations providing "on-demand" news services through text messages.

The evolving business model made use of the premium-rate text messages as a means of charging for one-off lumps of contribution or content such as ringtones, logos, or horoscope. Users send a text message to a special number, have the content delivered in the form of a text-message reply, and are charged accordingly. This made SMS a communication channel that consumer companies couldn't ignore. The text-based m-commerce-like services spread rapidly in early 2000s when countries like Austria and Norway launched mobile parking payments. In Vienna, a majority of people paid car parking fees using their mobile phones. Ten minutes before a user's allotted time is about to expire, he or she gets a text alert and the option to extend parking minutes.

SMS was a great communication tool because it commanded an effective integration. For the wireless industry in search for the next big thing, text messaging was probably the closest thing to a killer application. Short message service not only brought a social revolution through a very natural form of interaction, it also offered new ways to use text messaging as a marketing tool. For instance, real-estate agents could send buyers real-time, on-site images from the property, along with text on the brief profile of the place. Then there was text advertizing that allowed companies to do brand building through SMS. All these diverse commercial leanings combined with the proliferation of text messaging could spur growth in m-commerce.

Text messaging had become so pervasive in places like the United States that mobile subscribers sent and received more text messages in a month than they did phone calls: as per data gathered in second quarter of 2008, an average of 357 messages per month compared with 204 phone calls. However, the most compelling use of SMS-based business services was seen in emerging markets like India and Kenya, where a growing number of consumers were using even the most basic cell phones to order food and flowers, do their banking, pay bills, make charitable donations, and buy airline, bus, rail, or movie tickets. M-Pesa service, launched in Kenya in 2007, allowed people to send and receive cash through mobile phones, thus replacing banks in ordinary people's lives. The service became so popular that, by 2011, a quarter of Kenyan GDP passed through it. In India, Hindus even made advance bookings at temples via mobile phone to reserve the offering of prayers during busy holiday periods.

Nokia, which had largely missed the smartphone movement but was still king of the hill in basic phones, saw an opportunity. The European wireless titan was now gearing the text-based applications toward countries like India that have a plethora of lower-cost phones and a preference for exchanging information through text messages instead of the mobile web. Short message service was low-cost, easy-to-use and didn't require a phone with service plan, and thus helped build services with a very low barrier to adaption. According to a *Forbes* report filed by Elizabeth Woyke in November 2009, Nokia introduced a set of mobile programs called "Life Tools" that provided agricultural information and educational material to people in rural areas. In October 2010, the Finnish handset giant introduced two mobile applications that let Nokia phone users create chat groups and buy and sell products using text messages.

Unlike Life Tools, which supplied farmers with prices and other information, these applications targeted urban users and encouraged people to communicate with each other, not just consume content pushed to their phones. Take the example of We Meet service, a social-networking application that could be used by families, groups of friends, or small businesses. Users could create chat groups from their contact lists and communicate by sending text messages. The application threaded the messages in chronological order, making it easy to follow the conversation. The effect was something like an instant messaging conversation, but at the fraction of the cost and on devices with no data plans.

These SMS-based services were designed to be location-aware. But instead of pricey GPS technology, typically found only in high-end phones, it tracked people's location via cellular towers. When people move, the location updates. It wasn't exactly a digital map, but it served the same function. The service allowed the phone's contact list to add notes about clients and accounts, and linked them tightly to the device's calendar. That way, a merchant could make a note about a future payment in the contact list and have it automatically saved to the phone's calendar as a reminder. Another service, called MoMart for "mobile mart," comprised product listings delivered by text message.

Interested buyers would subscribe to the service and specify the goods they wanted; the program would then push matches directly to their phones. Listings could be text-only or could include an image embedded into the message. They could also be targeted to particular areas using cell-tower location technology, enabling buyers and sellers to meet in person. Nokia managers liked to compare this digital marketplace to a Craigslist or eBay for India. These services might not look flashy, but they are still smart. Moreover, these services are customer-driven, not just technology-driven. These simple but innovative services also craftily integrated two of the most promising mobile phone applications: text messaging and location.

GPS: COMING OF AGE

During the summer of 1999, the North Atlantic Treaty Organization (NATO) forces raided Serbia with precision munitions amid growing threat of mass murder and genocide in Kosovo. It was a unique event in the history of warfare when a side won the whole battle in the airfields without losing a single soldier. Pilots, using sophisticated precision software, digitally planned their missions beforehand. What lay at the heart of the NATO bombing of Serbia and the subsequent victory was "digital bombing," a smart satellite doctrine based on the global positioning system (GPS) technology. The GPS technology had also played a crucial role in the rescue of a U.S. Air Force fighter pilot after his plane was shot down over Bosnia. The allied forces had previously used this technology during the first Gulf War. This powerful navigation tool was credited for much of the success the U.S. troops had in operation Desert Storm during the Gulf War.

GPS, once the epitome of military communications, eventually became one of the most exciting commercial prospects in the smartphone realm. Over the years, the promise of m-commerce, originally built around the wireless Internet paradigm, branched into areas far beyond the scope of simple transactions over mobile phones. Location-based services became one of the first major outgrowths of the m-commerce prodigy. Though initially anticipated for mentally handicapped and elderly people, these location services expanded to new niche areas, from tracking valuables to recovery of stolen vehicles. By the year 2000, personal locator services were available to 72 percent of Japan's population. Japan's Pioneer Electronics introduced the first commercial GPS device for car-based navigation in 1990. GPS later became part of operations for airlines, ships, and rescue squad vehicles, while hikers and boat people solely depended on GPS devices to guide themselves home.

There was really nothing new about GPS: the technology was reinvented from the old. After satellite communications were established, scientists and engineers started to look for different ways of utilizing this fascinating space marvel. Radio navigation systems had been developed during the World War II for aircraft operations, which subsequently evolved into Loran satellite system. In 1958, the U.S. Navy began working on Loran satellite to develop a system "Transit" for indicating the position of a receiver on the ground. Two years later, the navy launched Transit-1B system to demonstrate the feasibility of using satellites for navigational aids. A receiver on a ship used the measured shift of satellite's radio signal, along with known characteristics of the satellite orbits, to calculate the ship position.

A practical system was born out of the need of the U.S. troops to pinpoint their locations during the Vietnam War. However, this system had a limited accuracy and was difficult to use due to its bulky terminal size. So, in the mid-1970s, the U.S. Department of Defense began a project to upgrade the navigation devices built around this concept for classified military use. The solution they developed required two dozen satellites, atomic clocks, microwave radio transmitters and some heavy-duty number-crunching hardware. A more portable unit could now pinpoint an object's exact location anywhere on the globe by receiving signals from a network of satellites in an orbit and triangulate them to determine latitude and longitude. The military called it Navstar, after the satellite constellation it used, but the industry and users ignored this nomenclature, and technology became known to the world as GPS.

The operational system contained twenty-one satellites in three orbital planes, with three spare satellites. The GPS collection of twenty-four satellites orbited twelve thousand miles above the Earth. These satellites constantly transmitted their precise time and position in space. With GPS, a receiver on ground or in the air could calculate its position using time signals from the satellites. The calculation itself was based on a kind of triangulation—a math technique used to locate an object based on its distance from three points. So signals from three satellites were necessary, although in practice a fourth satellite was used to improve the accuracy of the other three signals. The result was that a GPS receiver could produce highly accurate coordinates of latitude, longitude, and altitude.

GPS receivers could listen in on the information received from between three to four satellites and determine the precise location of the receiver, as well as how fast and in what direction it was moving. The satellite systems transmitted spread-spectrum signals in two frequency bands denoted as L1 and L2. The signals were then modulated as pseudo-random noise codes, and to acquire these signals, the GPS receiver generated a replica of satellites' pseudo-random noise codes. The navigation message could be demodulated only if the replica matched and synchronized with the pseudo-random noise codes. Otherwise, the GPS signal appeared to the receiver simply as noise.

The U.S. Air Force played a crucial role in nurturing the GPS technology by incorporating features like accurate digital maps and satellite photographs. As a result, the pilots were able to spot the key target areas and hit them effectively. Precision-guided munitions, dubbed "smart bombs," increasingly used GPS to hone in on a fixed target such as a military installation or an airfield. Although primarily developed for military operations, this cutting-edge satellite technology was eventually allowed for civilian applications. The first interagency testing of GPS receivers was conducted in California in 1984. By July 1995, using Navstar constellation, GPS was fully operational across the country. Now auto navigation systems became commercially available in the United States.

The former Soviet Union had developed its own GPS at the height of cold war—Global Navigation Satellite System (GLONASS)—as an answer to American Navstar. Both GPS and GLONASS were to be used by an increasing number of civilians, from aviators and sailors to car drivers. GLONASS signals were also used by a number of Western GPS receivers as a complement or back-up to GPS. In the private sector, former NASA engineer Allen B. Salmasi founded Omninet in 1984 to track down truck fleets using military satellite systems. An Iranian émigré, Salmasi had spent the late 1970s and early 1980s working on satellite communications for NASA's Jet Propulsion Laboratory. After the U.S. government urged the agency to privatize part of the GPS technology, he used US$5 million of his family's wealth to start Omninet, a satellite navigation system Salmasi marketed to companies eager to monitor the location of their trucks. After struggling for four years to expand, he merged his business with another small telecommunications startup to form a new venture that ultimately became Qualcomm.

Garmin was another star of the commercial GPS pioneering era and was founded by Garry Burrell and Min Kao at a dinner in 1989. Both had worked for King Radio: a legendary manufacturer of aviation radios. In fact, Burrell had lured Kao, a native of Taiwan, to King Radio from defense contractor Magnavox, where Kao had been developing military navigation systems using the GPS. Now Kao took Burrell to Taipei to raise money for their new company; they were able to get hold of US$4 million, which also included their personal savings though they didn't have to rely on venture capital. The duo eventually hired a dozen engineers and set up an office in Lenexa, Kansas, naming their new company as ProNav. The first product was GPS 100, a dashboard-mounted GPS receiver aimed at marine market that sold for about US$2,500. In 1991, a competitor named NavPro took the GPS pioneer to the court, and subsequently, its name was changed to Garmin, which was a combination of the two founders' first names.

GPS was inherently designed by the U.S. Army as a military system and was still being used for that purpose. It had never been fully adapted for consumer uses. Although the U.S. government offered the infrastructure to the world, it didn't allow equivalent level of service to business users. To protect military interests, Pentagon impaired signals released by the satellites in order to limit the accuracy of system for non-military use. For national security reasons, the U.S. government had required that the signals be scrambled, making them accurate only to within about a hundred meters. Moreover, the operators provided no guarantee of service or liability cover, and if there was a political crisis, GPS could be switched off without any warning to its users. Another technical limitation of GPS was that transmission was sometimes unreliable.

Still, things still kept moving for the technology as GPS was incrementally advanced beyond the scope of a navigation tool that the U.S. military used. Then Pentagon vowed to stop impairing GPS signals within a year's time; with that requirement no longer in place in May 2000, the devices became accurate within one meter, which in turn would help the technology take off in a variety of areas. With the end of "selective availability," civilian users were able to pinpoint locations with at least ten times more accuracy. The U.S. space agency also announced plans to maintain the satellite network at no cost for commercial users anywhere in the world. The location concept, once out of military playfields, began making new headways in customized location services. Applications were developed to make the effective use of technology in a number of niche areas where new pocket-sized devices promised enormous, untapped commercial potential.

PHONE'S YEARNING FOR LOCATION

Benefon, one of GSM pioneers, shifted its focus to marrying cellular with GPS in response to the European Union's Mobile Rescue Phone (MORE) project during the mid-1990s. The result of this ambitious effort was the launch of the Benefon Esc! phone in late 1999 and Benefon Track in 2000. The Esc! phone was splash-proof and featured a large, grayscale LCD. It allowed users to load maps onto the phone to trace their position and movement, and even to call or send their coordinates via SMS to a list of set numbers by setting an "Emergency Key." Interestingly, it also featured a "Friend Find" service, whereby users with Esc! handsets could track each other's locations directly on their handset display. It was evident by the late 1990s that by harnessing the power of location services in wireless handsets, GPS could radically alter the smartphone makeup.

But GPS still had a long way to go before it could become part of the mobile Internet anatomy in a technically and economically feasible way.

In fact, in the early stages, many in the industry had envisaged GPS to become a general-purpose gadget because it was mostly seen as an expensive novelty. That would change with the gradual integration of GPS in the existing wireless phone infrastructure services. However, for that to actually happen, the industry had to overcome a few major stumbling blocks. For a start, GPS was a line-of-sight satellite technology while cellular was not. Then there were problems regarding indoor reception of GPS signals, which was inherent in satellite communications. A user couldn't rely on the phone's GPS to get around inside buildings.

A server-assisted approach was introduced in 1998 to overcome inherent GPS limitations and facilitate the technology into the commercial wireless arena. The technique, also known as differential GPS, allowed a stationary computer server in a certain coverage area to assist mobile phones to acquire the GPS signals. These stations enhanced GPS services by carrying weak signals from satellite receivers and passing them to a locator device or a mobile phone. Industry players at large supported this approach as a way to bring pinpoint accuracy to GPS devices. Moreover, handset manufacturers like Ericsson and Nokia were initially reluctant to embed GPS circuitry into mobile phones, citing time-to-market issues and the added cost. A more crucial challenge related to lowering power consumption in the GPS circuitry in order to integrate it into mobile devices. The complexity and footprint of GPS chip, as well as the need for a separate antenna, further complicated a successful integration onto the mobile phone platform.

Furthermore, GPS maps for smartphones generally required a fairly high-speed wireless Internet connection, consumed significant processor resources, and were not highly optimized for driving-like situations. The combination of satellite and server-assisted versions of location delivery initially failed to solve the dilemma: mobile phone users generally couldn't get a GPS signal at home, and differential GPS was often not reliable enough to pinpoint people on the road. So the dream of having GPS in the heel still looked distant. It was about the time when the wireless engineers started looking for alternative solutions. Then there came this marvel of system integration that crystallized a new direction for cellular networks' liaison with server-assisted location services. In May 2000, J-Phone, the first wireless operator to release a phone with a built-in camera, launched the world's first location-based mapping service that displayed interactive maps within a web micro-browser. GPS meets location meets mobile Internet!

Just like the ascending of the mobile Internet and m-commerce, the Land of the Rising Sun once more provided the pioneering grounds for the location-finding settings. Japan's second-largest mobile phone operator launched the J-Navi service, letting users in Tokyo enter a phone number, address or landmark, and then search the area within a 500-meter radius. This made it possible to find the subway station nearest to a particular shop, or a particular kind of restaurant within walking distance of an office building. Most important, users of the service could download a full-color map. At the time of its launch, J-Navi was expected to handle around 100,000 hits per day, but on its third day of operation, it already had 1.6 million users. Searching was free, but users paid for the data transport costs, so in practice it cost about 4 cents for a location search.

Location-based services were now the new buzzword. Initially, the most popular services were those that enabled users to find the nearest restaurants, railway stations, car parks and patrol stations. Other uses of the technology included giving directions to pedestrians and car drivers, and providing localized weather forecasts. Thanks to the U.S. government's decision to open up the GPS to the general public, as well as lightning-fast advances in wireless technology, it was starting to become technologically and economically feasible to track people down at all hours. Sophisticated location-tracking technology was rapidly finding its way into mobile phones, cars, trucks, and boats.

The commercial possibilities for the location technology were beginning to cause a lot of excitement. One helpful factor had been an

American legislation that required mobile phone companies to be able to pinpoint the position of any mobile phone from which an emergency call was made, prompting operators to add positioning technology to their networks and handsets. The Federal Communications Commission (FCC) asked wireless service providers to be able to identify the exact location of most callers by the year 2002, so that 911 calls could be instantly tracked. The U.S. regulatory body mandated that all wireless handsets include some type of locating capabilities so emergency calls would automatically include information about where to dispatch rescue services.

Location-based services could be deployed in two ways. The first, with a chip placed in each handset, used the same GPS technology that tracked ships and vehicles for years. The second used intelligence in the wireless network to communicate with cell sites to determine the subscriber's location. Initially, there was a lack of consensus among American wireless operators on which way to proceed, and the FCC provided no definitive direction. Wireless infrastructure makers, such as Ericsson, were increasingly looking to the latter solution based on location-finding radio triangulation systems. The network-based locating methods involved triangulating the radio emission of the mobile phone or made use of radio fingerprinting to identify the most likely position of the radiating source. These local positioning services employed a software process that used the geo-physical location of the mobile unit as part of the algorithm for generating presentation content.

Such location services, however, won't come without the rigors of significant technology innovations. To start with, while such systems could locate a user within a particular cell, they did so without pinpoint accuracy and were capable of delivering accuracy within fifty meters in a cellular network. In urban areas, where cells are quite small, such location services naturally fell short. In the meantime, specialized chipmakers continued to improve the accuracy and availability of the GPS technology. The new circuitry was also able to gradually trim the GPS power for use in cellular phones. By late 2000s, smartphones had GPS systems on board, and with location-aware Internet services, they were helping people to get from point A to point B. The GPS-powered applications now helped people to find family, friends, and colleagues, as well as connect them with services in areas of their vicinity.

By 2010, three different kinds of location-based systems were in operation: dashboard navigation systems, smartphones with GPS features and electronic tags that helped people zip through toll stations. And, reminiscent to the camera phone saga, GPS functionality on smartphones was now eating stand-alone portable navigation devices (PNDs), which consumers didn't carry all the time, but they had their phones with them almost during all waking hours. If they needed directions, they would just crank up location software, type in the address, and they were done. The GPS-enabled handsets weren't as responsive as PNDs because they took a few minutes to fire up, but they got people to where they wanted to go, and it was still very convenient. People wanted to interact in real-time and real-place, and the dream of such fully connected, location-aware devices had finally come to fruition in the form of smartphones.

Location data had been available for quite some time, and they were massively underused. And when location data were used, they were used in a very simple way, and there's a lot more users could do with all the data generated. The up-and-coming smartphone industry could take advantage of location as an information source and as a way of creating a richer experience, which is one way to overcome the limitations of the smaller screen. A new breed of startups has started to frantically work on the technologies that would enable additional services such as location-finding in emergency, location-based "push" advertising and location-based service listings. These upstarts are offering location-based services in the realm of traffic and weather reports, driving directions, travel and entertainment information, and restaurant recommendations. Many of these firms have established partnerships with wireless operators, who control access to the location information; others provide operators support for location-enabling platforms off-the-shelf.

THE RISE OF THE CONTEXT

For years, technology had been able to pick up a person's location through GPS satellites and deliver up-to-the-minute traffic and mapping data. But it was only after the advent of smartphones that electronics was able to put a person's location into a larger context. The so-called context-aware software was now poised to fundamentally change the way people related to and reacted to portable computing devices. Context-aware computing—the technology that harnessed the information to provide useful tools to the smartphone user—offered at-a-glance background information about a business associate or prospective client, or it could aid travelers in finding a place to eat or with locating where they parked their car.

Mobile phone companies could leverage their close relationships with subscribers to develop highly customized services that were intimately associated with the location information. That way they could capture two potential revenue streams. First, they could charge users who request location services. Over time, however, the bigger prize could come by meshing these mobile-based location services into the world of m-commerce. So every time a wireless company steered a moviegoer to a particular theater, it could extract a small bounty. In the hindsight, location-based advertising—which became synonymous with the notorious fable of walking by a Starbucks and immediately being alerted to a discount—was only a piece of the mobile marketing puzzle. Just knowing where people are was still an incredibly powerful phenomenon and by combining it with context-aware technology, the wireless industry could create a wide array of localized and rich content and thus make way for lucrative business opportunities.

Making use of GPS, cell tower triangulation, or other means to utilize the user's presence could potentially lead to innovative services as well as targeted advertising. The mobile web also has the potential to be a lucrative field if it could integrate location. In fact, Google had originally created the mobile operating system Android to capitalize on the lucrative future of the mobile web. All Android phones would know where they were at all times, either by tapping into on-board GPS or by cross-referencing cellular towers using a Google database.

For instance, take Google's initiative of free turn-by-turn navigation for smartphones that brought a lot of hope to the smartphone industry. When used with a smartphone, the software sends coordinates to a server over the phone's wireless Internet connection to grab mapping data. Maps are also stored on the handset's SD memory card in some cases. That way, directions keep coming when a user can't get cellular reception, so long as he or she is still getting a GPS signal. Google's Maps Navigation application was probably one of the best in the lot; it took up almost no space on the phone because everything was in the cloud.

The search giant also has access to a sea of data on people and their web searching habits and thus is uniquely positioned to take advantage of context and monetize the information that it has. All these factors put Google at the forefront of companies poised to benefit from the surge in location-aware services. It's ironic that wireless industry veteran Nokia envisioned a context-aware Internet for smartphones in the mid-2000s, and that mobile newcomer Google actually executed that vision a couple of years later.

By 2010, location-based services were increasingly integrated into the much bigger m-commerce bonanza mostly because they complemented mobile Internet services. Smartphones were now situationally and contextually aware and thus could present information to consumers in the form of innovative new features. And that made them more of assistants or even companions than mere phones. For instance,

Groupon, the largest provider of deal coupons online, vowed to alert mobile users to nearby discounts using location data from the smartphone app Loopt. Likewise, some companies began using location data to build better maps or analyze traffic patterns. But there were challenges, too.

The prospects of location-aware technology terrified privacy advocators, who feared that the sensitive location data bytes could be abused by employers, insurers, creditors, or even stalkers. It also troubled civil libertarians who worried that another critical wall was falling between the individual and the probing eyes of the outside world. In 2011, Apple triggered privacy alarms with the news that the iPhone kept a database of a handset's location. The truth of the matter was that both Apple and Google went further than that; for instance, an Android phone recorded location data every few seconds and transmitted them back to Google several times an hour. Both Apple and Google collected location data from mobile devices to ensure efficient switching between cellular networks and Wi-Fi hotspots. Google also used location data to provide traffic information on Google Maps.

Although the information won't be available to everyone, it could potentially be sold to advertisers and could be made available to law-enforcement agencies, hackers, lawyers in divorce cases and other civil lawsuits, and to nosy employees of companies that build location-tracking systems. In countries with repressive political systems, location-tracking records could also be made available to secret police forces. The real question was whether a system could be developed that would benefit from the technology while protecting consumers from constant surveillance of the Big Brother. Well aware that privacy concerns could be a land mine, many of the companies aiming to capitalize on location services now began developing safeguards to protect their subscribers. The purpose of these services is to be of value to customers, and we are certainly not going to do anything they don't want, they said.

It was apparent by now that the intertwining fortunes of SMS and location services presented the smartphone with the killer applications that it had been longing for since its inception in the late 1990s. Both were unique in their character but still mutually complementary in nature. Similarly, both text messaging and location awareness were leading to several other innovations, and in some cases, they even enjoyed a close affinity. Take the example of customizable GPS-based applications like Geoloqi that let people set reminders, rules and notes based on their geographic location. Once the app is downloaded, a user could set SMS notes that are received only when he entered a specific location.

The truth is that SMS and location services were two crucial milestones that marked round one in the smartphone's quest for recognition. Once smartphones were past that tipping point, the rest of the applications mostly used the device's fundamental building blocks—web browser and camera phone—and encompassed photo albums, restaurant guides, music players, audio streamers, calculators, news and book reader, games and so on. After smartphones reached their full bloom in 2010, all the apps people possibly needed became available to them either free or at very low cost. In terms of type of applications, anything that had to do with entertainment, music, and video initially did quite well. While mobile games were a stand-out in the iPhone world, Google and its Android software won prominence in apps like the one that offered voice-activated mobile search or the GPS-based turn-by-turn driving directions. Location was for real, after all.

13 THE APPS REVOLUTION

"Even when the Internet was first starting out, computer users preferred apps over browser tools. The reason
is speed; an app is much faster."
　　　—Adam E. Ornstein, founder of VideoTagger software
provider Electric Happiness

On smartphones, the phone itself became just another app. One could
also call it an app phone instead of a smartphone because everything was now an app on this new class of sleek gadgets. Smartphone
apps are software programs that allow users to perform specific tasks
like reading an e-book, playing a game, getting sports scores, or searching for a restaurant without requiring a live Internet connection. Trade
media and industry at large endorsed the apps juggernaut with a sense
of astonishment. The British Broadcasting Company (BBC) declared that
apps are to be as big as the Internet. Before apps came along, the most
common activity of mobile phone users was snapping photos; even
as an e-mail device, a smartphone was quite limited. Industry observers had mostly seen smartphones as limited productivity tools until
apps changed that myth by helping people actually become more
productive.

It's ironic that at the time of the iPhone launch, people questioned whether the device qualified to be included in the smartphone club. Their argument was rooted in the fact that the iPhone lacked the ability to officially use third-party applications. But then a process called "jailbreaking" emerged quickly to facilitate unofficial third-party applications on the iPhone platform. Later, in July 2008, prior to the release of 3G version of the iPhone, Apple launched the App Store for both free and fee-based apps delivered directly to handset over cellular or Wi-Fi network without using a PC for downloading. Again, at that time, the irony was that the launch of the iPhone software development kit (SDK) was mostly seen as a move on Apple's part to conform to the demands of third-party software developers.

Apple, which hadn't been in the mobile phone business until 2007, caught the mobile industry off guard with its very hugely successful App Store. At the time of App Store's official release in 2008 as part of the iPhone 3G launch, it only contained a modest five hundred applications, but still managed to have 10 million downloads in its first weekend. The App Store quickly became the social hub of the mobile world, where people chatted and played games with friends, caught movie trailers, shopped for everyday things and even got a little work done. Call it a coming of age. It's worthwhile to note that in the early days of apps popularity the iPhone mostly had juvenile games. A turning point came in 2009 when social networking apps became equally popular on the PC and the iPhone. In fact, the app-equipped smartphones are seen as a cornerstone in the fast-track evolution of social networking.

Apple not only launched the most compelling mobile experience for web access, it also defined the smartphone battle by pioneering the biggest, most active ecosystem of software developers. The App Store made it brainlessly easy to install, upgrade, and purchase new applications. But more importantly, any software developer could build an app and sell it to the worldwide audience of iPhone users at the mild entry cost of owning a Mac and signing up for the US$99 iPhone Developer

Program. App developers would get 70 percent of revenues, Apple 30 percent. To many developers, the opportunity was unparalleled. These software developers, many of them tiny startups, could write a single version of their app and reach roughly 100 million Apple customers.

It was all about easy access to Apple's services. Apple made this stuff look easy, when historically it has been anything but. In fact, Android's precursor, Danger, produced what may be the first functional app store, but it couldn't attract enough third-party attention to become a viable operation. Moreover, long before Apple started selling trendy applications to iPhone users, there were online stores where consumers could download apps for their cell phones. With Microsoft's Windows Mobile, for instance, even though third-party programs had been available from day one, it was not much obvious to people because Microsoft always sold them through third parties. Back in 2008, when Apple bragged about its 1,100 apps, Microsoft actually had 18,000 Windows Mobile apps, but people didn't see them. They were never in people's face as iTunes was.

Apple's celebrated iTunes database, which it had originally developed as an application for the iPod, would eventually become a fundamental part of the future of the iPhone and its highly successful apps. Back in 2003, iTunes had morphed into a music store, and two years later, it started carrying video, too. By 2008, Apple was the world's biggest retailer of music, supplanting retail giant Walmart. Consumers around the world had purchased 8 billion tracks from iTunes, and it looked very hard for any company to truly replicate the iTunes ecosystem with its 200 million customers with credit cards. The App Store opened up new possibilities for direct-to-consumer distribution and created a rabid user base. It reached 1 million unique downloads in less than half the time that iTunes music downloads took to reach that mark. A brand-new software business had taken on a life of its own.

It wasn't until 2007 that m-commerce really began to grow, helped in large part by the advent of the iPhone, which generated enormous excitement for m-commerce despite holding only a small chunk of the global market. Apps moved m-commerce off the web and onto a more secure mobile Internet platform. They cut through the clutter of domain-name servers and uncalibrated information sources, taking users straight to the content they valued. Top consumer app categories included social networking, utility, entertainment, games, and lifestyle applications. Key business applications related to productivity, document management, sales force automation, messaging and communications, and a three-way tie between finance, retail, and healthcare applications.

With apps, Apple had succeeded in recasting the mobile handset while cleverly extending the price-point past the point of purchase. Now a legion of app developers was looking for ways to make the smartphone more usable. App downloads sold like hot cakes to millions of smartphone users who preferred the one-click convenience of secure paid services on iTunes over the web's arcane URLs, where spammers, identity thieves, cons, and malware lurked. By 2010, a decade after the inception of m-commerce, foundation for an infrastructure of advanced services was in place and new opportunities were opening up with the dynamics of the app economy becoming ever clearer. Electronic commerce firms like Amazon and eBay allowed a user to buy something right from the iPhone. Amazon became the Walmart in the user's hand. A consumer could purchase on the Amazon website from an iPhone app as easily as from the Internet.

APPLE'S APPS MOMENT

The runaway success of apps turned the humble smartphone into a key enabler for a brand-new mobile experience. Software, or more specifically, an enabled ecosystem of software developers, who write great applications, provided the foundation in this amazing story of American craftsmanship. By pulling off this remarkable technological feat, Apple left its competition in a quandary, where they were already racing to respond to the earlier iPhone innovations like the mobile Internet and touchscreen. It's ironic that Apple was building an empire on a decade of innovation from the likes of Nokia, Palm, and RIM. From the start, its working model for apps authors bore a close similarity with NTT DoCoMo's first viable mobile Internet service, i-mode, which offered a revolutionary revenue-sharing plan where DoCoMo kept 9 percent of the fee users paid for content and returned 91 percent to the content owner.

But the iconic success of the iPhone and the App Store is broadly credited to the holistic approach from a computing expert. Steve Jobs envisioned an ecosystem based on a broader value network, one where his platforms would coalesce the best energies of hundreds of thousands of developers and customers seeking new experiences. He knew that the real key to turning a big idea into the right idea was to create value through a sustainable ecosystem. To be fair, App Store's success wasn't merely due to the availability of familiar software tools facilitating new applications. If that alone were the key to Apple's success, then a much larger reach and a far more entrenched ecosystem of third-party developers behind Sun's Java ME or Microsoft's Windows Mobile would have resulted in similar success for those platforms.

Apple also did a fabulous job of educating consumers on why they need apps and how these sleek pieces of software can enhance their lives. The company had trained its audience to go right to iTunes, and now that the iPhone had come to life, they go to iTunes and they have this big thing in their face: App Store. Apple's captive iTunes audience and the company's unrivaled marketing machine are widely credited for the company's astonishing take on the apps.

Although Apple executives got a pleasant surprise in the amount and type of interest the store received, considerable thought had backed up every element of Apple's strategy and very little was left open to experimentation. The Cupertino, California–based company carefully staged the iPhone's deployment to only deliver what it could reasonably achieve. Apple shipped the iPhone for a full year before launching its software tools and store, building up an installed user base of about 5 million users during this period. Once it opened the store, it was able to unleash a flood of new software and new buyers, creating a tidal wave of positive coverage during this time of early euphoria. The company then continued to rigidly follow its own strategic plans, largely ignoring pundit demands that it do things differently.

At launch time, Apple didn't announce all of its plans in advance, didn't attempt to instantly achieve feature parity with existing smartphone platforms, and didn't allow third parties to set expectations or minimum standards for its platform's software. Instead, the company frequently surprised users with positive news of new features, deflected comparisons by focusing on the platform's strengths, and carefully guarded how its App Store library developed. And rather than ceding control of the App Store to third-party software makers, Apple maintained a firm grip over its development environment to avoid allowing the store to develop a reputation of being crass or seedy or sloppy or unfinished or a free-fall feature bonanza.

Once mobile apps began driving the broad m-commerce agenda, Apple's i-envy went beyond loyal customer base and made inroads into the business market. In a classic about-face, all major smartphone makers were now running crazy to catch up with Apple's relentless innovation streak. Google, Apple's archrival in the mobile space, immediately responded by launching Android Market a month after the arrival of App Store. Then RIM launched its app store BlackBerry App World in April 2009. Nokia followed with its Ovi Store a month later, and Palm came up with Palm App Catalog in June 2009. Finally, Microsoft launched its Windows Marketplace for Mobile in October 2009. These companies obviously wanted to be on par with the Apple's App Store, and from a technology standpoint, it was within reach. But the App Store's unique business model entrenched into Apple's larger ecosystem remained a key reason in itself why rivals were to have such a hard time in closing Apple's lead.

Take the example of RIM's BlackBerry, which had long been the standard smartphone for businesses, largely because RIM made it easy for corporate technology departments to manage and secure its devices. In 2007, the original iPhone was short on security and management features that are important to corporations. But Apple took a major step in 2008 to make the iPhone more business-friendly by adding the ability to connect to Microsoft's Exchange e-mail program and to remotely erase an iPhone's content in case the handset was lost or stolen. More and more companies also began adapting the iPhone because of the device's ability to access spreadsheets and contact-management software.

On the other hand, the BlackBerry apps were pretty anemic, and given the device's keypad-centric make-up, BlackBerry users didn't clamor for apps anyway. The tight security that made BlackBerry so attractive to corporate users also complicated matters for consumer-apps makers. BlackBerry's small screen and operating system software limited its media playing appeal as well. Interest in apps among BlackBerry users had been perceived so limited that application developers saw it hard to make money from apps. As a result, BlackBerry loyalists mostly missed the apps revolution. And when they saw their friends effortlessly using their iPhones to find a restaurant and tweeting from sports events, many of them eventually decided to join in the fun. In spring 2011, just a month before the launch of its PlayBook tablet computer, RIM allowed Android applications on its upcoming device and thus made up for the space it had lost on the BlackBerry front.

The iPhone was more than a mere appliance for sending e-mails. The device, with its sleek touchscreen and ability to run thousands of Internet-connected apps, games and utilities, could be used for nearly any purpose, business or personal. Now the very essence of the smartphone was that most people wanted to use a single device to handle both their personal and professional lives. That's what Apple was really good at, and that's where RIM's utilitarian BlackBerry was playing catch-up. In 2010, RIM's famously addictive gadget, which popularized e-mail on the go, was still the most popular corporate device with 70 percent of IT departments supporting the gadget. But Apple was steadily gobbling up BlackBerry's multibillion-dollar pie with about 29 percent of businesses supporting the iPhone, up from 17 percent in 2009 and none in 2007 when the iPhone was launched, according to Forrester Research.

Apps also marked the next chapter of Nokia's software miseries. In a perfect irony, more than a year before the launch of Apple's App Store, the Finnish company had introduced Ovi as a bundle of Internet services that let users keep calendars, access files, download music, back up contacts, and other activities. The store featured games, applications, podcasts, and videos for smartphones running Nokia's Symbian operating system. Still, Nokia found itself flat-footed in the new apps economy. When Nokia finally aimed the Ovi Store to pit it squarely against Apple's App Store, Nokia's online store was clearly lagging behind in paid downloads. Nokia's Ovi Store for mobile apps also got off to a rocky start as users faced problems accessing the store and downloading the programs. The Ovi interface was challenging to navigate and its cluttered offering of services like "Comes with Music" was confusing.

Ovi, a critical linchpin in Nokia's quest to wrestle with new-age smartphone rivals like Apple and Google, found itself in a vicious cycle soon after its launch. The apps phenomenon mostly related to the U.S. smartphone market while much of the support for Symbian came from European developers. The developers in the United States rushed to write programs for the iPhone and for Android, but shunned Symbian. With Nokia's low market share in the U.S. smartphone domain, there was no point for a developer to invest time and money to create an app for the Ovi Store. Creating apps for different platforms often requires learning a new programming language. Moreover, to make a return on investment, developers generally focus on the platforms that provide the most access to users and revenue. Apple's iPhone and Google's Android were evidently fitting the bill.

SENSOR PHONE: THE FOURTH IT WAVE

Nokia put out some of the first GPS phones and then came Apple's iPhone, a masterpiece with gesture-based multitouch interface, an elegant web browser and sophisticated music and video playback capability. The unique combination of the mobile Internet connectivity, GPS, and the ubiquity of handheld devices would open the floodgates of new, cost-effective ways to access information. To Apple's credit, it reinvigorated mobile handsets through innovative user interface, touchscreen and powerful add-on apps, and to Google's credit, it provided a similarly powerful solution, which was open source. Otherwise, the first two generations of iPhone competitors were blatant failures because of the poor usability created by their uncoordinated user experience. It's not enough to have a nice touchscreen or an app store. The integration of hardware, user interface and apps functionality has to come together to form a supportive whole. Individual apps could succeed only by nicely fitting into a platform.

Apple realized that the mobile phone was becoming the default gateway for people to experience content: games, movies and the web itself. But existing phones with their small screens, cramped keypads, and legacy operating systems were doing a poor job on this count. The iPhone solved these problems by combining a large, high-resolution glass touchscreen with a button-less user interface. Moreover, Apple introduced the element of emotion into user interface by mimicking the way humans work in the real world—sliding a switch, flicking through pages, allowing objects to be held or manipulated in a tangible manner. By using real-world metaphors, touch interfaces brought the physicality of interaction to the fore and made technology simpler and easier to relate to for end users. Once touchscreen devices did away with constraints archetypal to contemporary gadgets, even kids and older people could comfortably use mobile phones. The iPhone was the first truly functional touchscreen phone.

Apple harnessed another revolution by equipping the iPhone with an accelerometer to switch its display automatically from portrait to landscape orientation. The accelerometer's ability to respond to a user's motions turned previously pedestrian operations into game-like experiences. The truth is that the ever-expanding ecosystem of smartphone apps owes a great deal to micro electro-mechanical system (MEMS) sensors. The addition of sensors to the computing-cell phone combo would enable new platforms and enhance the magic of connectivity. Smartphones got smarter because of all the sensors being added to them; having a mere Internet connection didn't make a phone smart. Apple was able to create novel apps by exploiting the accelerometer for gaming, health monitoring, sports training and countless other uses thought up by legions of third-party software developers. By using sensor fusion, these software developers could take information from all these sensors and create apps that had never been thought of.

Take healthcare as an example, where increasing adaption of electronic health records was making it easier to incorporate digital information from smartphones into doctor's visits. Apps available on the iPhone would let doctors remotely diagnose patients, tapping cell phone sensors as diagnostic devices. With just an hour's notice, a service called 3GDoctor let patients consult with a doctor via a 3G videophone connection. An automated assistant collected the medical history and information on the symptoms, including, for example, an image of a wound or an audio sample of a cough. A doctor remotely evaluated the information, texted the patient to ask questions or request further images or audio, then called the patient to discuss the diagnosis and make recommendations for treatment using the face-to-face capability of videophones that had both front- and rear-facing cameras.

Now many smartphones came equipped with sensors to record movements, sense proximity to other people with phones, and detect light levels. Smartphones, with their always-on Internet access, were on way to becoming the world's premier wireless sensor network. The race to add accelerometers to the early generation of smartphones eventually turned into a pursuit for incorporating MEMS-based gyroscopes and next in line would be the addition of barometric pressure, humidity, and temperature sensors. Smartphone-based wireless sensor network capable of monitoring air quality, for instance, could become a higher-resolution network for toxic gas detection in a packed stadium. Apparently, wireless industry had only scratched the surface of the applications to which sensors could be put to use.

Now, just like the apps envy, smartphone makers were scrambling to match Apple's sensor complement. The iPhone had opened the eyes of handset vendors to how a sensor—like accelerometer—could harness motion. For instance, a motion sensor in the iPhone provided users a simple way to read wider web pages simply by allowing them to turn the handset sideways to rotate the display. Now mobile phone makers didn't have to identify the next killer application, they just needed to add the sensors—the invisible computers—and apps would find the best way to use them. For instance, combining motion sensors with location-based technologies and payment apps could create a new set of opportunities for people to interact with phones. Smartphone was inherently a sensor now.

Thousands of smartphone apps took advantage of user's location data to forecast traffic congestion and rate restaurants. The next generation of apps would take advantage of unique sensors on devices and networks intertwined with 3D cameras, multiple microphones, and increasingly precise GPS to take smartphones to a new level of user experiences.

For instance, accelerometers gave users only a crude measure of motion, but with the addition of gyros and other sensors, they could perform true motion processing. Adding this extra layer of sensing on top of the mobile Internet could eventually facilitate the creation of fourth IT wave. The IT industry has passed through the initial three information technology waves—mainframe and mini computers, personal computers, and portable networked computers—and now it's on the verge of the fourth wave, that of "IT everywhere."

THE APP AS AN AD

If the premise of m-commerce was for real, it won't be long before everything in the business world will be linked to the wireless Internet and will be remotely controlled via mobile phones. This very epitome of m-commerce in turn brought forth an exciting and enticing idea to the mobile frontier: wireless ads. A couple of years after the m-commerce hype took hold, the notion of wireless advertising surfaced as a way to reach people wherever they were and in a way that was both personal and relevant. For instance, people would walk down the street to a chorus of beeps and rings as coupons and ads from nearby shops arrive at their wireless inboxes. The proponents said that such manifestation of wireless technology could revamp century-old advertising business by bringing powerful features like interactivity and pay-for-performance. In many ways, America was initially persuading m-commerce experiment mostly from a wireless advertising perspective.

In the year 2001, we saw a modest degree of trials of interactive branding ads. The supporters of wireless ads had reckoned that personalized push technology, rather than simple web surfing, would keep mobile phone users informed on the road. That premise in fact led analysts to predict in the early 2000s that m-commerce would become a trillion-dollar business in the next three to four years. Industry at large had also accepted the public-relations hyperbole of handset manufacturers and network operators on how a myriad of innovative ad services could be delivered over the mobile devices. It sounded like a great idea for advertisers and wireless operators, but in the end, it turned into a nightmare for consumers, many of whom were already fending off a growing torrent of junk e-mails on their PCs.

At this juncture of the m-commerce evolution, the whole idea of wireless ads looked more like a paradox because shopping, by definition, was about wandering. Mobile phone companies at the forefront of wireless ad campaigns were evidently making some of the most egregious errors. They were not only harming their subscribers but also credibility of the entire wireless data movement by embracing such an inept form of advertising. Fast-forward to late 2000s: Apple had reshaped the mobile Internet world and, within that periphery, m-commerce. Once the iPhone accompanied with the power of apps got into the game, the whole landscape changed. The smartphone emerged as a powerful new advertising platform and now industry watchers began to wonder if the App Store could do for mobile ads what iTunes did for music and video.

"The iPhone was the first mobile device with a good web browser, and more such devices will follow," said Google's CEO Eric Schmidt. "Advertising will then become very personal. In a few years, mobile advertising will generate more revenue than advertising on the normal web." But if mobile devices take over as the computing platforms for consumers, then Google's web advertising channel, and the heart of its revenue, will be gutted. It doesn't take much of a crystal ball to see where Apple was going and it was not a pretty picture for Google. If the enigma surrounding Android and how the new economy bellwether would make money out of its mobile foray could be unraveled, the key to that had to be somewhere at this very junction in the m-commerce trail. A proof of how crucial mobile advertising space was in Google's scheme of things came in November 2009 when the company announced the acquisition of AdMob, the leading mobile advertising company.

Until 2010, the mobile advertising market was a small fraction of the US$25 billion U.S. online advertising market. Apple's cachet with big brands helped legitimize the entire market and spurred marketers to boost their spending in interactive mobile ads category. But even with Apple's successful purpose-built network designed to support the apps that ran on its smart devices, it could be a long way from creating the sort of broad-based advertising platform that defined Google's forte on wired web. So when Google outbid Apple to purchase AdMob for US$750 million, Apple followed by buying one of AdMob's competitors, Quattro Wireless. At this very note, it is important to set the record straight: Google didn't actually buy Android to compete with Apple. However, once the search giant saw what Apple was doing, Google shifted its plans to create iPhone–like devices instead of Windows Mobile–like, plain smartphones.

The smartphone market was a prize in itself; what both Apple and Google were after was the wireless ad market that the leadership in the smartphone business could facilitate. The integration of a mobile ad network into Google's larger infrastructure was a testament to the web firm's stalwart quest for revenue stream beyond its core search business. Apple, on the other hand, was promoting rich media ads, which kept users within an application instead of transporting them somewhere else. These ads let consumers play a mini-game or interact with the ad without having to leave or the close the app they were using. "We have figured out how to do interactive video content without ever taking you out of the apps," Steve Jobs told the audience at the launch of the iAd platform. "We think people are going to be a lot more interested in clicking on these things."

In 2010, Apple stepped up its rivalry with Google by adding its own advertising system to the next version of iOS that powered the iPhone, the iPad, and other mobile gadgets. Apple's new mobile operating system would include advertising capability dubbed as iAd. Apple's iAd mobile advertising platform let advertisers make inventive messages appear inside apps. Apple will sell the ads, with developers who create the apps getting 60 percent of the revenue of any mobile ad, and Apple taking the remainder. Jobs emphasized the new ad platform's immersive, interactive, and emotional aspects, which enabled developers and advertisers to include animations, video, and rich graphics inherently integrated with the iPhone. So, for instance, users could click on an ad to add a new wallpaper image to their iPhones. Ads were built using HTML5—a not-so-indirect jab at Adobe's Flash platform, which was widely used to provide interactivity and animation in web ads but was not supported on the iOS software.

It was now widely acknowledged that Apple's entry into the market through the acquisition of Quattro and subsequently the inception of iAd had put a spotlight on mobile advertising. Application developers, a key part of this equation, also acknowledged that iAd platform for the iPhone and iPad applications had provided them with a new channel to monetize their software products. Add to that 4 billion iPhone app downloads, 3.5 million iPad app downloads, and 64 percent of smartphone mobile browser usage share and the iAd opportunity would sell itself. Moreover, the iAd's richer ad units and higher click-through rates were complemented by the iPhone and iPad ecosystem, in which to buy a certain item took a single click using a credit card linked to an iTunes account.

The cost-per-click rates on smartphones are generally three times higher than what they are for desktop ads. But these mobile ads are geographically targeted to specific set of people in specific demographics, so smartphones could bring higher click rates than advertising to people in general. Tailoring content to specific sets of people rather than the population at large through mobile ads could also prove a powerful phenomenon and over the years could change the media landscape. There is clear evidence that American consumers are gradually changing the way they shopped, thanks in large part to the apps that reshaped the discovery process with more tempting online offers, easier ways to compare prices, and innovative solutions for attracting customers to stores. Moreover, with location-based apps, retailers are always watching users.

Smartphones, the industry's biggest user-growth engine, haven't been well served by advertisers yet. But now that mobile has presented itself as one of the strongest opportunities in business history, monetizing mobile has become a key agenda item. Marketers had been slow to buy mobile ads in the early going, largely because consumers were not visiting mobile websites in large numbers and the process of creating mobile ad campaigns was technically and logistically challenging. Mobile advertising is still unproved on a large scale and the business model is in an early stage of development. But now that platforms like iAd and AdMob have kick-started this nascent market, marketing pros have begun paying attention to mobile advertising. The increasing commoditization in wireless handsets and networks also makes it imperative that ads become more important as a source of revenue.

FROM CHATTING TO SHOPPING

With apps, the smartphone suddenly became the epicenter of the software world. With an app phone in a user's hand, which was like a mini computer, a revolution was knocking at the door and it was like the PC saga all over again. Take the case of gaming, which exploded on the iPhone in 2009, especially with fantastic shooters such as Alive-4ever; with quality game-play and graphics it looked like something seen on an Xbox 360. Then there were apps like UberCab, which allowed consumers to call a taxi. These early apps either served as a utility tool or provided some kind of entertainment. But as it turned out, there were bigger opportunities hidden in apps.

The next year, in 2010, smartphones changed the retail business forever. The days when consumers merely crawled through newspaper ads and trekked out to brick-and-mortar stores were gradually coming to an end. The power of the smartphone and the tablet just gave them more options than ever. For instance, Amazon's Price Check, a free iPhone app, let users check prices of CDs, DVDs, books, and video games on the fly by scanning products, snapping a photo, saying the product name, or typing in the name, brand, or model numbers. The service made it very easy for users to order the item via the app. In the hindsight, it might not be a mere coincidence that in 2010 we also saw conventional stores struggling. During this year, Barnes and Noble put itself up for sale and Blockbuster filed for bankruptcy. Apple, who had built up a portfolio of 200 million credit card accounts through its iTunes store, had caused enormous disruption at the brick-and-mortar chains.

Then, 2011 was predicted as the year of the rise of the mobile transaction. M-commerce advocates said that cash would become a thing of the past and that the future of digital money is on smartphones. A phone is a lot smarter than a card and retailers welcomed the flexibility and rich experience it offered at the point of sale. Square Inc., a Silicon Valley startup launched by Twitter co-founder Jack Dorsey in 2009, helped fuel interest in mobile payments by turning the iPhone into a cash register. A matchbook-sized credit card reader plugged into the iPhone and automated the entire process. Customers swiped their credit card onto the iPhone-attached card reader, signed the iPhone touchscreen with a finger or stylus, and got an e-mail or a text with a receipt. After the payment was processed, money was deposited into merchant's bank account. Square's customers included food trucks, hairdressers, small retailers, restaurants, and taxicabs, and the service lured them due to its simple software and its flat 2.75 percent transaction fee.

The next frontier in turning the smartphone into some kind electronic wallet centers on a technology called near field communications (NFC), which allowed data transmission over very short distances. An NFC chip embedded into a phone would allow a user to make a wireless payment just by waving the handset in front of NFC-enabled readers at subway stations, drug stores, and taxicabs. NFC technology was born in the early 2000s through a partnership between Sony and Philips Electronics, and had its roots in transport and convenience store segments. The information stored on the magnetic strip on a credit or debit card would now reside on the NFC part of the mobile handset, and that would inevitably turn phones into a digital wallet. NFC would serve as glue in a retail payment ecosystem that offered a central repository for bank account information, coupons, loyalty points, and membership cards in just a single tap with a phone.

NFC was also the common thread in the battle for the control of digital wallet among four major stakeholders: smartphone makers, apps developers, credit card companies, and wireless operators. Mobile phone players such as Apple, Google, and RIM wanted to put NFC chips directly into handsets and upgrade the respective phone operating systems accordingly. Once payment credentials are directly stored on handsets, third-party app developers and handset makers could jointly create payment solutions, facilitate transactions and redemption of coupons and discounts, and gain insight into consumer spending behavior.

Wireless carriers wanted the NFC part to be embedded into phone's SIM card, which in turn, would make it easier to switch to any handset that's enabled for NFC transactions. In fact, it was simply a matter of competing visions, and just like phone makers, wireless operators wanted to get hold of the consumer account data and bind users to their own services. In fall 2010, the three large wireless operators in the United States—AT&T, T-Mobile, and Verizon—chalked out their own battle plan through an m-commerce venture dubbed as Isis. They also partnered with a payment processing firm, Discover Financial Services Inc., and a card issuer in Barclaycard to bring relevant expertise into this venture. Eventually, in 2011, Isis was able to lure all major credit card companies—American Express, MasterCard and Visa—into the fold.

Smartphones equipped with powerful new apps are unmistakably rein- vigorating the long-held promise of m-banking services. A growing number of businesses are eyeing mobile payments as an alternative to credit cards. Such a move would presumably reduce costs to merchants, who are typically charged 3 percent of the purchase price when cus- tomers pay with a credit card.

Credit cards emerged in the early 1950s, first as dining cards in Manhattan, and then spread across the United States during the 1960s, introducing the era of electronic money. Credit card companies had benefited as people abandoned cash and paper checks for cards and electronic payments. Visa and MasterCard accounted for US$2.45 trillion of consumer-spending on credit and debit cards in the United States in 2010. Now smartphones could potentially displace some of the 1 billion credit and debit cards in American wallets. Amid shifting tides, Visa and MasterCard began exploring what could be the future of commerce: quick and efficient payment processing through mobile technology.

Seeing the forces of disruption at their doorsteps, credit card compa- nies spent millions of dollars to issue new cards with contactless NFC technology and vowed to speed up the check-out process. Credit card companies had been toying with the NFC concept for quite some time by embedding the technology into credit cards and thus allowing their customers to wave a card in front of a reader instead of using the tradi- tional swipe method. As to the future, senior executives from credit card companies argue that NFC is merely a wireless mechanism to process a transaction, and ultimately it would come down to creating a viable payment infrastructure that takes care of how money moves, ensures that transactions work every time, and preserves legal rights for con- sumers, merchants, and banks.

Industry experts following the mobile payments market hoped that NFC could become an industry standard in 2012. But it could take a while in the buildup of a robust payment network through smartphones.

While NFC was clearly primed to become the underlying technology for mobile payment systems, the battle for the control of the digital wallet wouldn't be over anytime soon. The m-commerce game had just started.

PRIVATE APP STORES

Beyond the web of m-commerce possibilities, IT establishments wanted employees to access apps related to customer care, sales management, field service, logistics and supply chains, human resources, and financials on mobile devices. After business e-mail was up and running on smartphones, as the next logical step, companies now wanted to make corporate information available as apps. The innovation from application developers in the business domain focused on tapping into vast stores of data from corporate databases and making these data available in the form of highly customized enterprise mobile apps. By 2010, some companies were already using apps for everything from scheduling conference rooms to approving purchase orders and from accessing marketing materials to logging onto the internal social network. Oracle offered six different iPhone apps that did everything from making sales forecasts to helping clients better organize data.

Then there were content-centric apps like special readers for *The New York Times*, written to optimize reading experience, and such apps were starting to develop their own business models. Bloomberg and other stock services offered apps that kept users apprised of their portfolio and relevant business news feeds in real-time. Eventually, we could see many such apps from anyone with a web presence who wanted to service users that are willing to rely more and more on their phones to access content and services. Mobile radio is also coming of age with the delivery of rich content to smartphones, which are now emerging as a major audience and revenue driver for audio streaming business. By 2015, mobile radio apps able to stream programs onto smartphones could generate as much ad revenue for traditional radio companies as will streaming on PCs.

The scale and promise of apps looked as vast as the Internet. Case in point: a Dutch company, Waleli, has developed a way for people to answer their front door, even if they are in another country. The doorbell and intercom system would send a message to user's phone when activated. After the user enters the PIN correctly, he could talk to the guest over the intercom, or even unlock the door to let him in. Then there are all sorts of interesting ideas brewing in the healthcare space. An increasing number of medical firms have made telemedicine services available through the iPhone. The advent of the iPhone has also brought a new hope for mobile gaming.

So much is at stack not only for promising new venues such as m-commerce but also for the two fundamental building blocks of the smartphone: the mobile Internet and the operating system platform. As explained in the previous chapters, the future of mobile Internet is now torn between two powerful forces, mobile web browser and apps, and there is this predominant view that apps would ultimately define which OS platforms would dominate in the future smartphones. Apple's App Store and Google's Android Market are widely considered to be the two most viable mobile software platforms. Conversely, there are people who believe that apps are going to die away in favor of simple mobile websites. They argue that apps would wither away because it cost too much to develop programs for multiple systems from Apple, Google, RIM, and so on.

Nevertheless, apps had marked an inflection point for smartphones, and as of 2011, they were only growing in popularity. From digital money to e-readers to medical tests, app phones touched almost every aspect of business and personal lives. The digital stores from Apple and Google are likely to remain primary delivery vehicles for games and rich media apps in the coming years. Although faster processors and cheaper memory chips with ever-expanding capacity have been the key ena-blers in their realization, these apps, especially involving multimedia- and graphics-centric apps like games and media streaming, also require fatter network pipes and ubiquitous connections. The next couple of chapters encompass the networks that not only facilitated, but to some extent also rewrote the smartphone story.

THE NET EFFECT

"Unwiring the consumer is the next logical advance in consumer electronics."
—Intel CEO Craig Barrett during his CES keynote in January 2003

In June 2010, AT&T delivered the new smartphone economy a wake-up call when it decided to scrap unlimited data plans for its customers. When the iPhone launched in 2007, AT&T charged US$20 a month for unlimited data usage. AT&T subscribers would no longer have the option to pay a flat fee for unlimited data, which they had since the birth of iPhone-driven smartphone world. AT&T's abrupt reversal to metered data usage prompted a backlash from bloggers and consumers alike. AT&T, on the other hand, faced a dreadful dilemma. The second-largest U.S. wireless company had been battling network congestion as it tried to manage soaring demand unleashed by the data-hungry devices like the iPhone.

The industry watchers had predicted that bandwidth demands on mobile phone networks would grow exponentially as more people get smartphones and as they adapt smartphone-like devices such as the iPad, which are billed as suited for a host of bandwidth-intensive applications like streaming video. Until now, AT&T had largely been able to manage that growth by making billions of dollars of network improvements across the United States. However, several times during 2009, AT&T had been reportedly hit by network outages in cities like San Francisco and New York, places with high concentrations of people using iPhones and other smartphone devices. The connectivity and speed issues were plaguing AT&T's iPhone user base, and consequently, much more bandwidth was needed.

Another case study comes from Google's unveiling of the Nexus One handset in January 2009; Google attempted to separate the phone from its competitors by calling it a "superphone." But soon after the launch, users began complaining about spotty data network reception and non-existent customer services, which made the user experience disappointingly mediocre. The Nexus One debacle was testament of the fact that handhelds are no smarter than the systems built to operate them. The overwhelmed AT&T network episode further made the case that smartphones could be data hogs and that wireless networks needed to carefully calibrate data usage in order to survive the stress and strain on the overall network performance.

Since the arrival of the iPhone, industry experts have seen incredible strides in mobile hardware and software, and it is getting to the point where they are hesitating to call these microcomputers phones. Their capabilities went far beyond what we would expect from simple voice telephony. As to the smartphone's existence before the iPhone, industry experts pointed to the fact that the network resources to cater for something like a BlackBerry were radically different from and much less than those needed for the iPhone. The BlackBerry mainly catered to e-mail where the traffic was compressed and it was only one session per user. Moreover, BlackBerry devices were very operator-friendly. The iPhone, on the other hand, was a computing platform on which the user could put applications like Google Maps, which could open up as many as three hundred sessions through the Internet.

The media attention in the mobile phone arena had long been focused on the battle between super-smartphones. But there was a parallel campaign underway among wireless carriers to roll out new, faster data networks to link these sophisticated handheld computers to the Internet at greater speeds and to increase the capacity to handle all the data their owners were downloading. The goal was to overcome capacity limitations and make wireless Internet access on the street as fast as the access people got in homes and offices.

But will there be sufficient network bandwidth to support the dramatic increase in mobile data usage? Mobile phone operators were evidently having difficulties in keeping up with data traffic demand amid a mind-boggling trend: data traffic had exceeded voice traffic in the post-iPhone era. At this juncture, it was also predicted that within a few years, the mobile Internet would surpass the cellular phone system as a far better communications network. The futurists said that cellular connectivity in the iPhone-like devices could subsequently serve the same function as the VHS slot found in many DVD players: legacy support.

In fact, even in the early stages, the iPhone had started causing the troubling problems of dropped calls and digital brown-outs. Smartphones drove data traffic over mobile networks to new levels, thereby requiring new technologies to keep up with the frenetic demand. The transition had also caused incredibly fast changes in wireless operators' networks and business models. They found themselves in the data business, but not the way they planned to be in the data business. They had in fact planned for Internet Protocol (IP) networks to deal with data but not on the scale they were experiencing in the post-iPhone settings. Now, apart from the ongoing evolution of the cellular standards to higher data rates and efficiency, there was hope that Wi-Fi would play an important role in helping to offload data traffic from overworked cellular networks.

AT&T had set the bar for smartphones and proved a handset could conceive data services on wireless platforms. By doing so, it created the same kind of game-changing event that the iPhone had accomplished in terms of creating new user demand on the device front. Prior to the emergence of smartphones, wireless operator's loyalty was tied more to the network coverage. In the early days of cellular, there wasn't much difference between most voice-only handsets. There were size and fashion considerations, or more interest in devices that offered a wider range of compatible accessories. But phones were just phones. Then came the smartphone, and with it came more device differentiations and a greater range of handset capabilities.

The iPhone's ascendance to fame had one thing written all over it: all services of the network are represented by the phone. So AT&T's iPhone liaison forced competitors such as Verizon to accept the idea that the handset is an instrument of network service, not a mere accessory for fulfilling of it. Wireless carriers like Verizon were now willing to integrate the handset into their service offerings rather than simply using it as a mechanism for getting the consumer onto the wireless Internet. The smartphone represented a monumental shift for the wireless industry, and with it, the handset had transformed from being a carrier- or vendor-controlled device to a computing platform.

Wireless carriers were used to build their own agenda into cell phone designs, but Apple's union with AT&T had changed that model. The iPhone cracked open the carrier-centric structure of the wireless industry and unlocked a host of benefits for consumers, developers, manufacturers, and for carriers themselves. For wireless operators, as the number of radios grew, networks began to matter a lot more than before. Wireless carriers in the United States, who saw AT&T eating into their customer base, started scrambling to find a competitive device and they appeared willing to give up some authority to get it. The AT&T episode had shown that having good handset hardware and software mattered, and having good networks mattered, too. After the iPhone affair and the intertwined AT&T's data snag, the industry was unanimous in its verdict: it's the network.

AGE OF WIRELESS NETWORKING

Wireless got into the networking game once technologists became confident that data could travel across the airwaves. Still, the beginning of wireless data was accidental as well as circumstantial. Wireless data chronicles reveal that though the concept has long been the staple of science fiction stories, its genesis could actually be traced to the 1970 experiments of Norman Abramson at the University of Hawaii, when he developed a 4.8 Kbit/s packet-radio network. Aloha, meaning hello in Hawaiian, connected an IBM 360 mainframe in the main campus on the island of Oahu to card readers and terminals dispersed among different islands and ships at sea. Eventually, Abramson's work became the foundation for packet radio as well as Ethernet-based LAN systems.

The next act in the history of wireless data came in 1979 when scientists at the Advanced Research Projects Agency (ARPA) started looking for variable transmission venues to confirm the viability of what would later become the Internet. The Internet's mentor body funded the "Packet Radio Network" project in which experiments were carried out by establishing communication links between mobile vans. The outcome was a multiple-hop wireless network with an extremely low channel frequency and bulky terminals. These were the early days for the notion of wireless data, which would become a reality only after about two decades with improvements in bandwidth, processing power, miniaturization, and power consumption.

During this period, corporate America got a starter in mobile data with systems meant for two-way messaging services for fleet delivery. The initial impetus broadly came from transport, security, and field service sectors, which had traditionally been using dispatch systems based on land mobile radio (LMR) technology. Their mobile workers, who wanted corporate access while away from the office, were keen to use wireless data services. In the 1970s, for instance, FedEx built a U.S.-wide wireless data network to keep track of its packages using a proprietary radio technology. During this period, it became clear that packet radio was highly suitable for the applications characterized by relatively short bursts of data, such as messaging, point-of-sale (POS) transactions, database queries and fleet dispatch.

However, despite forecasts suggesting a massive swing toward wireless data, the larger take-off was just not happening. That's because, historically, mobile communications at large had focused on handling simple telephony services. The first-generation cellular systems were designed purely for voice communications, so while mobile phone usage grew at exponential rates, wireless data remained in low profile. Several factors also hampered the use of wireless data applications in the public arena. Low data rates, expensive airtime, and unfriendly services that were difficult to configure and run were the major constraints. Another important factor was the terminal evolution.

But though wireless data market remained in doldrums, there were some prominent efforts to carve out important niches in this area. Advanced Radio Data Information Service (ARDIS) was the first major packet data-messaging system created in 1983 as a private network for fifteen thousand IBM service technicians in the United States. It comprised of radio base stations covering fifteen to twenty miles, network controllers and network control centers, and supported 19.2 Kbit/s in 25 kHz channels. ARDIS eventually became a public mobile data service as a joint venture between IBM and Motorola. IBM later sold its interests to Motorola in 1994 and ARDIS became a subsidiary of Motorola.

Earlier in 1989, Hutchison Telecom of Hong Kong had launched an ARDIS-based mobile data service in a joint venture with Motorola. But after failing to get subscribers, in 1994, Hutchison sold its 70 percent stakes to Motorola, who eventually formed a subsidiary company Motorola Air Communications that launched wireless data service under the Max brand name next year.

Meanwhile, recognizing the need for two-way wireless data communications, engineers at Swedish Telecom—which later became Telia—started work in collaboration with Ericsson. The Swedish telecom gearmaker, like neighboring Nokia, had flirted with the computer business for quite a while during the 1980s. Eventually, Ericsson and Telia conceived wireless data technology as a private mobile alarm system for field personnel. The two companies built a low-speed data network at 1.2 Kbit/s with a text-based voice dispatch overlay. Commercial Mobitex operation started in 1986, and the technology was significantly enhanced four years later for use in the United States, Europe, and Australia.

Mobitex terminals were typically standard notebook PCs equipped with a radio modem, or it was a dedicated mobile terminal. The user interface was packet-based, data-only system incorporating cellular architecture, multichannel frequency reuse and store-and-forward capability at data rates of up to 8 Kbit/s. Mobitex used hierarchical structure and was the first wireless data technology to offer seamless roaming and battery-saving features. It became a de facto standard for packet-switched mobile data networks after RAM Mobile Data, a partnership between RAM Broadcasting and BellSouth, introduced a Mobitex-based service in the United States in 1991.

Ironically, while Mobitex had been exported to the United States, American ARDIS technology was being deployed in Europe. Both Mobitex and ARDIS were public wireless data services based on proprietary technologies. They were built around the principle of specialized mobile radio, and they used portable radio terminals that were expensive and heavy. But their major drawback was low-data throughput that typically ranged from 2.4 Kbit/s to 4.8 Kbit/s. Although later upgraded to 19.2 Kbit/s, data throughput practically remained below 8 Kbit/s due to overhead related to radio channel protocol and error-correction procedures.

CELLULAR TWILIGHT

For a start, though dial-up modems could work on mobile phone lines, major cellular operators in the United States felt the need of a service specifically designed for data applications. Their collaboration with IBM resulted in the Cellular Digital Packet Data (CDPD) system in 1992. A packet data overlay to the existing analog cellular infrastructures such as AMPS, it used idle voice channels to send short data messages by hopping among available frequencies. CDPD used a forward error correction scheme to combat the interference and fading of cellular channels, so the impact of voice traffic on CDPD was supposed to be negligible. The CDPD spec was the first open wireless data standard against proprietary ARDIS and Mobitex systems and thus brought a new data communication model. It also introduced the IP element into mobile communications for the first time.

Although the first CDPD specification was released in July 1993, initial rollouts were not complete until the mid-1990s. American wireless pioneer McCaw Cellular was a leading supporter of the wireless data protocol. Ameritech, AT&T Wireless, and GTE were also among the first to deploy CDPD in the United States. Using CDPD technology in 1994, Carnegie Mellon University set up one of the largest wireless data networks—Wireless Andrew. CDPD was suitable for applications based on small burst of data, such as e-mail, database queries, and credit card authorization. Its support for IP also went to its advantage at the time of the Internet boom.

On the downside, however, CDPD required a modem with a mobile terminal for the user and an upgrade at the radio base station for the operator. Notebook and PDA users could run data applications at 19.2 Kbit/s using a combination of a PC Card and a wireless modem. But the actual throughput was reduced to 10 Kbit/s to 13 Kbit/s due to system overhead. Consequently, most of the mobile phone companies became reluctant to deploy it, and despite some market penetration during the mid-1990s, CDPD ended up as an underused interim solution.

Wireless data was repeatedly predicted a boom market but applications such as ARDIS, CDPD, and Mobitex only served small niches in vertical markets. Many products had come and gone in their quest of exploiting the wireless data promise. There were multiple devices, multiple approaches to build networks, and multiple ways of delivering and aggregating content over wireless data systems. At one stage, it seemed as if there was no genuine need for wireless data except niche applications such as vehicle dispatch. The opponents said that wireless data were a drag and that mobile phone users didn't want much data; they just wanted tailored information in the form of well-thought-out applications.

The idea had long been a disappointing stepchild to the industry with often-stated "year of the wireless data" slogan perpetually coming in the following year. Although wireless data systems like CDPD and Mobitex had fulfilled their role in nurturing the natural technology advancement, the market still needed a breakthrough, a killer application of some sort, to go any further from here. Only mass-market appeal could save the promise of this potentially great idea. Then, the humble text message emerged as a turning point, providing the much-needed breakthrough from a corridor nobody in the wireless industry was originally looking at. Wireless data, which had long been a dream of technologists, got its first practical manifestation from a source no one had envisioned through all these years.

GSM AT CROSSROADS

Wireless data were screaming for applications as they needed some kind of kick-start, which SMS would subsequently provide. On one hand, SMS caused the demise of paging, and on other hand, it was laying the ground for a new era in wireless data applications. With SMS, wireless data landed in the masses for the first time, although there was nothing smart about it. Before SMS, originally built to exchange small text messages across GSM networks, sending data over cellular platforms was never an easy proposition. Although cellular systems supported facsimile and file transfer, reliability was always an issue.

Things started to change with the advent of digital cellular, which significantly reduced the airtime for wireless data traffic, thus helping it to make inroads through viable applications like SMS. With it began wireless industry's quest to redefine itself by marrying data with the voice-centric cellular systems and to make data an integral part of cellular services. It's ironic that at the time GSM was developed—in the late 1980s—many in the mobile communications industry felt that the standard was being over-specified. Critics said that too much capability had been designed into it and that most of it would never be used. By the mid-1990s, just a couple of years after the GSM launch, more than 300 million subscribers had been won, and GSM was not finished yet. The core standard, now well tried and tested, was constantly evolving.

When the work on the standard eventually entered in the third major phase—generally referred to as Phase 2 Plus—the key element of this project catered to data communications that spanned both traditional circuit switching and emerging packet data. The Phase 2 was primarily different from the Phase 1 in having added a number of supplementary services. The functions were organized as a set of independent tasks so that each could be introduced with little or no impact on the others. That approach favored an evolutionary growth path since such independence implied that one network element could be modified without affecting other parts of the system. This proved a decisive turning point in the future growth of the European mobile technology.

The aim of the Phase 2 Plus initiative was to gradually introduce important changes while maintaining upward compatibility with the GSM network. Wireless data—as the GSM designers found out while working on the Phase 2—was the next logical step in the evolution of mobile communications. So the Phase 2 Plus gradually turned into a campaign to preserve GSM competitiveness through introduction of higher-speed data services. The project took a novel approach to incorporate data services into two modes: High-Speed Circuit Switched Data (HSCSD) and General Packet Radio Service (GPRS).

HSCSD worked by aggregating a number of speech channels into a single pipe comprising a bandwidth of 64 Kbit/s or beyond. Aimed for mission-critical applications due to its support for guaranteed data rate, HSCSD was suitable for constant bit-rate applications such as videoconferencing and telemetry. However, the need to reserve bandwidth for a certain time could become a liability for the overall system. So even before the realization, HSCSD seemed marginalized by the other application, GPRS. Given the IP roots of GPRS and the eminent wireless Internet revolution, the doom of HSCSD became obvious even before it made a foray in the commercial realm. The fate of HSCSD also marked the death of conventional circuit switching technology within wireless data domain.

Its counterpart, GPRS, was seen as bringing a new lease of life to the dying dream of mobile Internet while it could also facilitate new services in a range of areas from m-commerce to remote surveillance. As a GSM extension, GPRS used the same channel, same modulation scheme and the same network backbone as HSCSD to offer data-over-cellular services. But it utilized packet transfer and routing mechanism, and optimized airtime by allocating temporary resources for bursty data applications such as Internet access. Inevitably, IP interface and its support for connectionless services attracted special favor in wake of the rising tide of the Internet usage. GPRS promised to handle data rates flexibly, according to network availability, from normal 9.6 Kbit/s to as much as 144 Kbit/s, and allowed simultaneous voice calls while sending and receiving data.

GPRS STEALS THE SHOW

In many ways, GPRS was a recycling of the older American technology, CDPD, in the digital arena—a network overlay requiring additional cards at base stations that would separate data packets from circuit-switched voice. However, CDPD had its roots in the analog AMPS world, whereas

GPRS belonged to GSM digital cellular technology and was thus perceived to offer better solutions for mobile data applications. This broad distinction was seen as the key factor in GPRS network's ability to push the U.S. wireless data protocol out of picture altogether. CDPD, nevertheless, continued to play a modest role in providing niche wireless data services in some of the American markets.

The work on GPRS started in 1994, and air interface protocol was proposed after an effort that spanned a period of two years. Ericsson signed the world's first GPRS contract with T-Mobile of Germany in 1999. The network, composed of a switching node and a router, directed data packets to the Internet or any other packet-switched data network. GPRS was going to support almost all the major data communications protocols, including IP, which would enable mobile subscribers to connect directly to any data source. Also, GPRS was designed to work within the existing GSM infrastructure with additional packet switching nodes, so that the network coverage could be introduced quickly and easily.

Packet data brought improved spectral efficiency because radio resources in it could be shared simultaneously. Users could remain connected to the accessed data network as long as they wished, but they would be charged only for the actual amount of data transmitted. People no longer had to go through the frustrating dial-up process. Once a connection was made, a user could put the phone down or even make a phone call and then come back to it exactly where he or she had left off. Thanks to GPRS, there was an enormous increase in SMS traffic, while e-mail became an exciting prospect in wireless communications. GPRS seemed to be a practical solution to historical problems of wireless data—cost, bit rate, and user-friendliness.

But the bigger milestone was that GPRS spec tactfully built itself on top of a successful technology, GSM, while it also adapted to the rapidly changing market needs. GPRS—now widely quoted as the 2.5G wireless network—represented the first implementation of packet switching within GSM, which was essentially a circuit switching technology. The timing could not have been more perfect for GPRS as it quickly became synonymous with the celebrated premise of wireless Internet. GPRS, with its always-on-connectivity to the mobile Internet, became a key landmark in the path towards accomplishing full broadband mobile Internet connection. The technology that brought much needed packet data capability to mobile communications was now seen as a stepping stone toward the 3G panacea.

Originally, GPRS promised a minimal hardware upgrade to the existing GSM networks and was merely seen as a software upgrade to GSM. But though its GSM roots enormously helped the technology to achieve widescale industry recognition, the soft launch of GPRS subsequently became an uphill task for the European wireless establishment. There was about ten times as much software involved in a GPRS handset as in a GSM phone. So moving to GPRS could be as much as doing GSM all over again. Nobody was surprised when GPRS was off to a slow start. History would repeat itself when GPRS handsets were nowhere to be seen like the early days of GSM. Following the golden rule of mobile telephony, the handsets were late.

And once these GPRS-enabled phones arrived in the market, they were relatively unaffordable because handsets had only recently acquired multimedia features like color displays and the ability to run downloaded software. This time around, however, the wireless industry was more circumspect in preaching the GPRS gospel. Both wireless operators and manufacturers knew that they faced a potential backlash if they over-promoted GPRS. So the logical first steps were taken without a lot of hype. For a start, they were counting on GPRS to turn around the fate of WAP debacle as accessing the web was quicker with GPRS and the always-on connection was a big advantage for e-mail.

GPRS offered higher transmission speed and constant connectivity with minimal use of network resources. And with no dial-up, no drop-off, and speeds of 20 Kbps to 40 Kbps, it provided a testing ground for the wireless data services. The GSM industry had envisioned an Internet revival through GPRS networks because an always-on connection coupled with data speeds two to three times faster than before could turn wireless Internet into a viable experience. It also allowed wireless operators to demonstrate that they could market data services. Both in terms of network dynamics like bandwidth and new, souped-up applications, GPRS seemed a prefect precursor to the coming-of-age 3G panorama. It was official now: GPRS was the gateway to 3G.

MOBILE DATA IN TRANSITION

Back in 1999, surfing the Internet on a mobile phone emerged as a very cool idea. But a couple of years later, waiting several minutes to peer at a few lines of text on a small monochrome screen didn't feel cool at all. Sluggish networks and clunky technology virtually snuffed out most consumers' enthusiasm for the wireless Internet. The industry leaders were concerned that users were fed up with overly hyped wireless Internet services that hadn't met expectations. But now that 2.5G networks were here to rescue the wireless industry, what next? As is the case for many exciting new technologies, it all began with a knee-jerk reaction.

"We still don't know how to use these network capabilities to generate new revenues," acknowledged a spokesman of SK Telecom, the first company in the world to launch commercial 2.5G service in October 2000. SK Telecom, which commanded half of South Korea's mobile subscribers, made the mistake of jumping onto the new-age wireless bandwagon without proper groundwork and was now scrambling to recover. For a country in the vanguard of hot wireless technology, nobody was able to figure out what's going to be in the 2.5G platform. A year since the commercial launch of its 2.5G service, to the frustration of SK Telecom managers, the subscribers were simply talking up against their ears in the traditional fashion.

The term 2.5G was a media phrase for the group of technologies that used packet switching to deliver data on demand at potential speeds of 144 Kbit/s to 384 Kbit/s. These bridging technologies, also available on other second-generation systems—CDMA and TDMA—were collectively known as 2.5G solutions since they were built on the existing networks using the same transmission frequencies and much of the same equipment. However, because this technology, synonymous with 2.5 generation networks, was essentially the second-generation network upgraded to handle data, it was slower than 3G. On the positive side, despite being a long-planned upgrade to the existing wireless phone system, the technology was promising to offer nearly all the benefits of 3G for a fraction of the cost.

During all the fuss over 3G networks, the importance of 2.5G had largely been underplayed. That's partly because original objective behind 2.5G initiative was about making airwaves wider. But it became evident in the course of time that the very being of 2.5G initiative was crucial for the realization of wireless data services. One of the biggest advantages of 2.5 networks was that data traffic could travel in a channel separate from voice calls. Still, moving from voice to higher speed data and multimedia services was in a stark contrast to the earlier transition carried out to move from analog to digital air interface. This time, the wireless industry was going to grasp a new service platform and a myriad of technology challenges that fell outside its traditional engineering strengths.

The voice-centric wireless communication world was now operating in a new territory—data—and it wasn't an easy turf. And there came the rub. Embracing 2.5G technology wasn't only a matter of economy or time; it was also indispensable from the learning curve standpoint. A great market experience, GPRS would help make the case for different types of business models and there was a lot to do in the applications space. The interim step allowed operators to experiment with pricing models and popular data services in anticipation of rolling out the higher-speed 3G networks at a later stage. A slew of new services also allowed wireless operators to start rehearsing 3G applications in an attempt to incite demand. Last but not least, the success with 2.5G provided a vital experience in delivery and billing of wireless data services, which would subsequently help mobile companies master to succeed in 3G business.

But although the 2.5G initiative looked financially feasible, it wasn't without struggles of its own. The 2.5G technology, because it was a simple network upgrade, was supposed to be quick-and-easy way to get consumers hooked onto the higher-speed data transmission: enabling them to utilize fashionable services such as downloading movie trailers when booking a ticket, or watching sports highlights. However, it was neither easy nor quick. For a start, a key difference between 2.5G and 3G was speed, which translated to the amount of data that could be moved over the spectrum per second. Theoretically, 2.5G should provide speeds of 144 Kbit/s, and 3G should go up to 2 Mbit/s, but these speeds assumed maximum performance that could only be achieved by a stationary user located in a specific coverage area of the network with very few users.

Now service providers began preparing the market for the inevitable disappointment: the bandwidth will be somewhere between 28 Kbit/s to 128 Kbit/s. And as with any shared solution, the more users there are, the less bandwidth they will get. In the final analysis, even 28 Kbit/s was looking awfully optimistic. But while video wasn't a sure bet anyway, wireless industry could pioneer ways to beam music and other audio services over 2.5G-enabled phones. After all, cell phone was an audio machine: there was a headphone, a microphone, and selection buttons so users didn't have to change the interface to turn it into, for instance, a music player. At the time when college kids in the United States were downloading music in MP3 format from their personal computers, mobile phone users in Europe and Japan were overjoyed with downloads of simple ringtone music features.

Then Japanese mobile subscribers, moving a step further, started to download songs directly onto their handsets, which came equipped with headphones. Multimedia messaging service, or MMS, was another bright spot. Its predecessor, SMS, had enjoyed an extraordinary boom by allowing users to send short, telegram-like messages of up to 160 characters from one mobile phone to another. MMS allowed people to send and receive photographs. The industry cheerleaders like Nokia promoted messaging services such as MMS as a critical bridge to 3G wireless technology. They argued that wireless operators should offer a steady stream of improvements to the basic text messaging service— such as the addition of graphics and animation—in order to hold user's interest and encourage them to migrate gradually to more complex multimedia services.

NEXT STOP 3G

Even though the fate of 2.5G was not clear in the early 2000s, the course of technology that was determined with the emergence of the mobile Internet and multimedia services like MMS had made the process irreversible. The whole idea of 2.5G as a stepping stone to 3G wireless was for real, but there was a dilemma hidden in the bigger scheme of things. If 2.5G didn't work very well, it could really hamper the prospects of 3G. But if it was too successful, it could steal thunder from 3G and forestall it indefinitely. Would that mean that 3G networks could be entirely unnecessary in that case? While many analysts claimed that 2.5G was good enough to satisfy everyone's need as far as speed was concerned, mobile phone companies insisted that 2.5G in no way replaced the need for 3G.

Mobile phone operators saw it from a vantage point, and they were right. They argued that 3G would eventually prevail because it was cheaper to operate and that it offered data speeds up to ten times faster than 2.5G, permitting zippy services such as videophones and digital music players. The so-called 2.5G networks provided decent-enough data rates for WAP-based services, but as cell phone browsers matured beyond simple, text-heavy displays, the need arose for a bona fide 3G network capable of handling high-speed data rates. Moreover, the 3G technology was so complex that it would likely arrive years later than promised. And by the time it finally arrived, the 2.5G networks would have become clogged with traffic, and users would be more than willing to switch to 3G for better performance.

Wireless operators had built a great intelligence on voice usage over the decades, but for data, the infrastructure and efforts were generally not on par. Initially, there was little understanding of what consumers were doing, and which applications and services they were tuned to at any given instant. The cost of supporting data services exceeded the cost of managing voice services and the revenues from data services only became prominent after the coming of the iPhone. With the iPhone and credible rivals like Android phones came a growing appetite for data—first in the browser space and later with mobile apps. Case in point: the iPhones were the primary driver behind the unprecedented 5,000 percent increase in data usage on AT&T's network from 2007 to 2010.

Wireless operators now started to pay much more attention to the specifics at a very granular level and formulated business models and pricing plans as per the trends and forecasts. For them, after having started with thin, crispy 2.5G networks, upgrading the capacity to 3G networks was now imperative. The advent of smartphones had underscored the need for a specialized network, and there was no looking back. Smartphones, more than any other device, were able to take advantage of these faster networks to empower users to do or access nearly anything. The complete Internet experience that they've had on PC was now available in the mobile space and that was nothing short of a revolution. The union of ubiquitous high-speed wireless networks with mobile browsers meant people had the power of the web in the palm of their hand.

By 2009, AT&T had been turning heads with the sheer speed of its 3G network. The correlation between the mobile Internet and the 3G networks was evident in the post-iPhone wireless world. The mobile phone industry now had a challenge of wirelessly connecting millions users to the Internet, and for that, they needed far more capacity than required in providing plain voice services. Smartphones were now into the mainstream, and for this new value proposition, the wireless establishment needed to have faster networks so that they could manage the onslaught of this new gadgetry. The next generation of wireless networks was on its way.

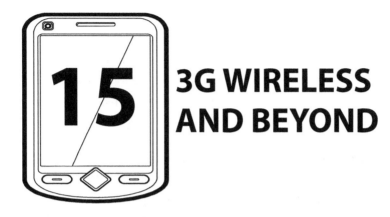

15 3G WIRELESS AND BEYOND

"They have created so much hype over 3G you would think it's the greatest thing since the rice cooker."
—Craig Ehrlich, chief of Hong Kong-based mobile operator Sunday Telecommunications in an interview in 2000

Although third-generation wireless or 3G had become a household name by 2010, it took a lot of stops and starts along the way. From its conception in the late 1990s to its worldwide popularity over the following decade, there were some serious hurdles that 3G had to overcome. Start with the early promises of 3G that created a marketing field of dreams in which video would vault mobile phones into an entirely new plane just like the addition of pictures to sound in the new medium of television had transformed broadcasting in the 1950s. The most compelling dimension of 3G, however, was the one that related to the build-out of the next-generation wireless infrastructure for the mobile Internet.

When European planners in Brussels were mapping out the timetable for 3G launch, across the Atlantic, the Internet was moving into the consumer realism and web startups like Netscape were stock market darlings. Understandably, while piecing together the 3G prototype,

Europeans became convinced that the rapid coalescence of mobility and the Internet would crystallize a new market with a powerful and timely convergence. A decade ago, similar central planning for GSM development had helped transform Europe into the world's richest wireless market, turning the regional wireless players, from Nokia to Vodafone, into global phenomenon. So while creating a new continental network—3G—high-speed Internet became Europe's audacious bid to lead the world in a crucial twenty-first-century technology.

For Europeans, it was now a megaproject, the equivalent in size, vision, and expense of America's Apollo space program in the 1960s. This was a unique blend of technological conquest and pervasive market drive borne out of Europe's GSM triumph. The pundits called 3G wireless the vision of the new century, a wireless nirvana where all dreams would come true. The so-called phone of the future riding on 3G platform would be powerful enough to provide high-speed Internet access, video-on-demand and countless whiz-bang features. According to early projections, 3G was to provide users with a whopping 2 Mbit/s of data, more than what most wired offices used to connect their users to the web at that time.

The predominantly European mobile phone establishment talked of such killer applications as videoconferencing and other pie-in-the-sky schemes that kept service providers and manufacturers busily plotting about how they would get into the act. But the 2 Mbit/s claim was not realistic in the first place because the key word "shared" was almost always omitted from the boastful projections. Worse still, the ugly truth was that many service providers couldn't even provide 2 Mbit/s solutions in the near term but instead would satisfy market needs with a mere 384 Kbit/s. Again, this bandwidth would be shared among all users within a particular cell.

The NTT DoCoMo and British Telecom (BT) rollouts in 2001, launched in a rush to become the world's first company to offer 3G services, showcased not only the shortcomings of the new technology but also the challenges that merely came with the first-mover status. Although both companies attributed the delays to software glitches, the Japanese and the Isle of Man rollouts offered the world a glimpse of the deployment issues as well as of the complexity and newness of 3G technology. Consumer acceptance also failed to match the launch-day hype. DoCoMo could sell only 42,000 3G handsets, far fewer than 150,000 phones target the company had modestly projected to achieve by March 2002. Third-generation handsets cost three to five times as much as conventional phones. They were clumsy and glitch-prone and had a relatively short battery life.

Spotty coverage, high prices, and shorter battery life of handsets continued to stifle the DoCoMo and BT services. Meanwhile, European operators either delayed or shelved plans to roll out 3G after spending about US$100 billion on government licenses for the services. Now that DoCoMo and BT had taken the plunge with 3G, Deutche Telekom's then-chairman Rolf Sommer said his company wouldn't launch commercial services until it was satisfied that the network was robust and working to the existing high-quality standards, and that suitable handsets and applications were ready. Similar views were echoed by the American wireless stalwart Motorola, who correctly figured out that the actual 3G rollout would kick off in 2004, followed by widescale deployment in 2005.

The original premise was rather too simple: beef up the wireless networks, and users will flock to them to browse the web, send e-mails with mile-long attachments, and even download video clips all on their phones. The execution proved to be a giant technical and financial challenge, however. Content and applications were clearly not ready yet, but every telecom outfit was standing in the line, saying that if 3G wireless takes off, it should not be the one left behind. The perception of data services had been driven more by the wireless industry's enthusiasm than by the market demand. Moreover, piping voice and data through a single converged handset required complex new partnerships among wireless, Internet, and a myriad of other service companies.

No one had actually figured out how to share the work let alone the revenue. There was no clear business plan; multinational giants jumped onto the 3G bandwagon with no more of a clue than barroom ideas of dotcom flameouts. Another important lesson from this mobile extravagance was that there is never enough bandwidth, so one must focus on technology that optimizes the use of bandwidth. The telecommunications industry has long been a victim of the myth that bandwidth was everything. But a mere progression in speed could be meaningless in many ways because what applications do at the desktop is substantially different from what they do on a cell phone. The i-mode example had affirmed that mobile phone companies needed to think about making attractive offers to end users instead of merely relying on fancy marketing campaigns.

Whether data services will generate sufficient revenue was also open to question in the early going. Would people really want to download high-bandwidth applications, such as video clips, to the tiny screens of mobile phones? No one was sure. And no one yet knew what the killer application might be to capture the imagination of people and hence turn 3G into a moneymaker. Still, with a spiffy new technology on the horizon, wireless companies felt they had no choice but to jump in.

TOO LITTLE TOO SOON

The 3G wireless story became a mystic tale of another technology debacle. What distinguished 3G mania from the dotcom bubble—which burst in the same time periphery—was the size of its victims and their ambitions: they were not little startups, but the world's major telecommunications operators and manufacturers. How did this happen? In the hindsight, it's easy to see why so many successful companies fell to the allure of 3G. An industry that grew fat on the booming cell phone business had to rush to master new specialties: Internet services and smartphones. While the cost and risk of 3G were frightening, the alternative for such high-flying firms as Ericsson and Orange of learning to live with near-saturated mobile-voice markets and declining user revenues would have been even worse.

Then the high-tech bubble burst. Carl Yankowski, a consumer electronics industry veteran and then-CEO of Palm Inc., compared 3G to the HDTV saga of the 1980s. Intel's Hans Geyer, while speaking at the 2001 GSM World Congress in Cannes, France went a step further when he told attendees that the wireless industry was heading for bankruptcy before even a single 3G call was made. The technology press quickly scaled down 3G to small talk, and the focus turned away from 3G toward tangible applications such as ringtones and camera phones. The mobile phone industry nevertheless maintained its relentless pursuit of advancements through user appetite for multimedia-rich features.

It's easy to overdo the pessimism. While majority of industry analysts decried the lethargy in 3G, a group of cheerleaders still believed it was the only way forward. Third-generation wireless was about a new paradigm, they said, which needed to adapt to a radically different world and it couldn't happen overnight. In the long term, 3G could prove a boon, but it would take time. These evangelists reminded doomsayers, on the other side of the fence, that GSM was not an instant success either; it too had its teething problems. Even text messaging found a lukewarm response in its early going. The apparent shift to multimedia services, they said, favored 3G, whose bigger transmission pipes required rich content to justify new applications.

Everybody knew wireless bandwidth was a hodgepodge, so problems were bound to erupt. To some industry observers, the rhetoric about 3G fallout was reminiscent of the early days of GSM when people said whatever digital offered could be done with analog. After all, voice is analog, these critics used to argue. A more granular view of this whole episode reveals that what really had bitten the dust was not 3G itself but over-hyped visions of its potential. One couldn't simply blame the 3G fiasco on the great rush for the riches of the mobile web because the idea of 3G services was presumed to be much more than just the wireless version of the Internet.

In retrospect, the optimists were right. For 3G, one of the biggest gambles in business history, payback came in rather strange ways. First, the transformation from 2G to 3G was evolutionary and would take several years. A full suite of voice and data offerings would eventually make the transition from 2.5G to 3G inevitable anyway. Second, multimedia-messaging services like MMS, which evolved from SMS, subsequently played a key role in driving the network rollout. Such services would allow downloads of ringtones for mobile phones in stereo or MIDI files along with video clips and animations. People could send and receive non-real-time transmission of text synchronized with audio and video images over cell phones. Third, despite this maze of sophisticated data services, voice remained the real breadwinner for mobile phone operators.

In an ironic twist, voice came forward to provide a safety net for the shaky 3G start. At one stage, it looked as if the focus of 3G would move away from commercially unproven multimedia services, such as videoconferencing, toward providing extra capacity for voice traffic. The existing cellular networks were already approaching full capacity as users continued exodus from fixed to mobile networks. In pure voice terms, 3G networks would offer at least twenty times the capacity of the existing GSM networks. In theory, an operator could go on creating extra capacity by splitting cells in the GSM system, but the wireless industry was reaching a point where it would be more costly to do this than build a new 3G network.

The 3G story started with an optimism that became the hallmark of the wireless communication's vision for the twenty-first century. Then pessimism followed, creating shock waves throughout the mobile phone world. But once the cycle of boom and gloom was over, the new buzzword was realism that said, "3G is here, taking one step forward on its course, and that it's based on some real business case." Now there was more awareness on difficulties ahead and on the need to create new business models and revenue streams.

Third-generation wireless was likely to cost an operator several billion dollars, and many years to roll out a complete network. Plus, 3G coverage could remain spotty and sporadic for years. But the wheels of history won't go backward. After the days of blind faith in this mobile enigma ended, 3G was still standing as the only viable solution for the longer term. Although the elusive quest for an unprecedented wireless glory was over, the future 3G landscape was starting to take shape nevertheless. After early alarm bells, by 2002, network construction was well under way in Europe, albeit at a slower pace than predicted earlier. Buying into 3G would mean buying a promise: wait a few more years, and buying the real thing would be more likely.

That real thing came by in the late 2000s when the iPhone and Android pushed the mobile Internet well into the mainstream with hundreds of millions of subscribers. The wireless industry—then focused on voice and messaging services—was caught unguarded by the explosive growth in mobile Internet traffic. For them, data were only useful to the mobile enterprise workers and early adapters who used smartphones with clunky browsers that made mobile web surfing far less appealing. But Apple's seductive phone with its powerful software and intuitive interface encouraged users to go online and stay online. The iPhone users consumed an average of up to ten times the bandwidth of mobile subscribers. They were playing games on their phones, sending video messages, or downloading music. A new network regime was taking shape in a wireless order created by smartphones like the iPhone.

3G: EVOLUTION, NOT REVOLUTION

History tells us that technology companies are far too eager to proclaim a revolution. The relics of telecommunications are a testament of the fact that whenever a technology was hyped, it only found difficulties and delays in development later. The scenes of the 3G carnage were yet again remindful of what the telecommunications world had witnessed in its ISDN conquest during the 1980s and the ATM euphoria of the 1990s. But while ISDN and ATM technologies had little leverage in their course of evolution, the 3G wireless archetype had something to fall back on. Once the smoke and mirrors were gone, what we were left with was a rather dull upgrade from the existing second-generation wireless networks under the 2.5G manifestation.

Despite setbacks on both technological and marketing fronts, wireless industry's commitment to deploy 3G was unshaken. The wireless companies picked up the pieces and began to work out the kinks for providing quality phones that would subsequently accompany this new technology. Once 3G began to take hold, cell phone companies worked tirelessly to find ways to utilize this new technology. Although mobile phones long had the ability to access data networks such as the Internet, it was not until the widespread availability of good-quality 3G coverage in the mid-2000s that specialized devices like the iPhone emerged to access the mobile Internet. As coffee-shop goers and technology experts alike attested, the speed was completely real.

The wireless industry didn't see meaningful deployment of 3G networks until 2003 in Japan, 2004 in Europe, and 2005 in the United States. In the start, while 3G was in its infancy, still-slow GSM networks evolved even further from GPRS to a zippier technology dubbed as Enhanced Data Rates for GSM Evolution, or EDGE for short. The preferred U.S. version read EDGE as Enhanced Data Rates for Global Evolution. The EDGE technology (GSM 384) was originally developed by Ericsson for operators with no Universal Mobile Telecommunications System (UMTS)-based 3G spectrum; EDGE enabled both GSM and TDMA operators to offer data services at speeds closer to those available at UMTS-based 3G networks.

In fact, in the early stages, the GSM-centric evolution path to 3G in the predominantly European UMTS technology had left the U.S. TDMA operators like AT&T Wireless in the cold. So they had little choice but to embrace EDGE before moving on to an entirely new 3G technology: UMTS based on wideband-CDMA air interface.

EDGE was an unsung hero of the 3G tale and its rise to the occasion was a testament to 3G's evolutionary journey. EDGE brought an immediate solution for higher-speed data services for TDMA/GSM operators and its close proximity with GPRS was surely helpful. It's worthwhile to note that despite EDGE's ability to support 384 Kbit/s—up from 144 Kbit/s via GPRS—and the fact that EDGE cost a fraction of a 3G network, the cash-starved wireless operators long ignored this technology. The interest in EDGE finally came from the United States which wasn't bitten by the 3G spectrum blues. The first EDGE network was rolled out by Cingular Wireless in 2003 with theoretical speeds reaching 236.8 Kbit/s, though in reality it reached nowhere near that.

The 3G affair, sparked by the promised merger of voice and data onto cellular handset, started reaching critical mass by the mid-2000s. First, wireless networks became more economical to deploy after some years of manufacture and costs dropped for both 3G base stations and handsets. At around the same time, the emergence of dual-core processors allowed handsets to offload much of the work to a separate application processor, which inevitably helped 3G take-off. The technology's reputation as a power hog also eased after handset designers adapted new power-management techniques to optimize battery life. Meanwhile, memory capacity kept increasing to support new features. During the mid-2000s, an average handset carried 4 MBytes of storage; within two to three years, memory capacities surpassed 64 MBytes mark.

The 3G standard was built around the Third Generation Partnership Project (3GPP) which structured specifications as Releases. Release 5, generally known as High-Speed Downlink Packet Access (HSDPA), was eventually upgraded to Release 7, High Speed Packet Access Plus (HSPA+), which employed a multiple-antenna technique for improved system performance, high-efficiency modulation technique for extra capacity and improved packet efficiency, and continuous packet connectivity (CPC) functionality for reducing power consumption of mobile devices. Subsequent standard or Release 8 was Long Term Evolution (LTE), which was considered to be much more evolutionary than its predecessors. LTE, based on orthogonal frequency-division multiplexing (OFDM) and multiple-input multiple-output (MIMO) techniques, was going to be adapted by major wireless operators like Vodafone, Verizon, and NTT DoCoMo.

By 2007, there were 190 3G-based networks operating in forty countries, aiding the path for mobile TV, video communication, and location-based services. Eventually, 3G was defined not by the underlying technology but by its speed: up to 2 Mbit/s. For 3G deployment, Verizon and Sprint opted for CDMA-based systems, also referred to as EVDO, while GSM followers AT&T and T-Mobile used an enhanced version of 3G technology: HSDPA. This common heritage became a binding force in 2011 in AT&T's bid to acquire T-Mobile, a Deutsche Telekom company, which would make AT&T the largest wireless operator in the United States. A speedier version—HSPA+ or Turbo 3G—boasted download speeds of up to 14 Mbit/s.

By 2010, with the availability of media savvy devices such as iPad hooked onto wireless networks, it had become clear that at some point 3G networks would be overwhelmed by the growth of bandwidth-intensive applications like media streaming. So the industry began looking for data-optimized technologies with the promise of speed improvements up to tenfold over the existing 3G technologies. The next step was inevitably 4G.

4G: THE EVOLUTION GOES ON

Internet downloads and GPS position finding could be accomplished with 3G speeds but *Dick Tracy*-style video calling could not, at least not without significant compression techniques. Then there were applications like music and games that continued to grow at a rapid pace, making it imperative for the wireless establishment to ensure that bandwidth stays ahead of customer demand, now and in the future. As the saying goes "out with the old, in with the new," such was the case with 3G networks gradually making way for mobile Internet-centric fourth-generation wireless systems commonly known as 4G, the network that promises to be faster, stronger, and overall better than its predecessor. Wireless companies hope that 4G will revolutionize the way people connect to the Internet. They anticipate that 4G will eliminate the need for smartphones to search for Wi-Fi hot spots in order to connect to high-speed Internet.

In retrospect, 3G became a phrase which was very much part of the aphorism celebrating the long-awaited arrival of the smartphone. And now a handy, easy-to-grasp way of referring to the next generation of mobile communications, 4G was another rich phrase regularly used by the wireless industry. A loose term for the fourth generation of wireless systems, 4G technologies promise speeds that are about ten times faster than they are on 3G networks. And like its 3G predecessor, the industry gurus started calling it a game-changer the sooner 4G made its name on the wireless paraphernalia. But a closer look at this generation game reveals that the labels like 2G, 3G, and 4G don't really matter much. Advances are being made and will continue to be made, providing us with better, faster mobile broadband, whatever generation it is, and that's what matters.

There are distinctions, however. One of the key ways in which 4G differed technologically from 3G was in its elimination of circuit switching technology which was common in traditional telecom settings—instead employing an all-IP network. The crucial significance of 4G was lying in the fact that it aimed to be a full-fledged wireless IP network and that it would potentially close the capability gap between wired and wireless worlds. Earlier, the ITU had standardized 3G networks for use with both circuit- and packet-based networks, meaning the data would be transferred both through telecom- and Internet-based networks. Conversely, 4G—being the first generation of network technology which is completely IP-based—is set up to use the packet technology only.

The general idea behind 4G is to provide a comprehensive and secure all-IP based solution that, in turn, would facilitate IP telephony, ultrabroadband Internet access, gaming services and streamed multimedia to mobile users. With an all-IP makeup, 4G would allow a treatment of voice calls just like any other type of streaming audio media, thus providing a key turning point for smartphones, which were now increasingly comparable to PCs. But while video chat and media consumption are seen to be two key-use cases on the upcoming smart devices, streaming a single video from the web to a mobile device takes as much bandwidth as ten phone calls. The 4G services claimed to be capable of download speeds in excess of 5 Mbit/s compared with 3G's real-world maximum of around 1.4 Mbit/s.

The generational change had arrived just in time or so said the marketers. They had grown ever bolder on the heels of the second coming of 3G and the meteoric rise of smartphones. The marketers, who didn't see any trouble in establishing credibility this time around, would declare any new network rollout to be 4G. Take the example of T-Mobile, which rebranded its HSPA+ network as 4G, but the rivals claimed that it was a 3G system that had evolved into 3.5G network over the years. Thanks to bold marketing campaign from T-Mobile, HSPA+, an enhanced version of a 3G technology, became the third flavor of 4G along with frontrunners LTE and WiMax. The marketers also frequently quoted Verizon who had been keen to get its 4G network in place before concluding any agreement with Apple to start selling the iPhone.

Once again, like the early go-go days of 3G, marketing departments had started pushing out the concept of 4G networks years before the engineers even decided what a 4G network was. And when the engineers finally got together to determine what makes a 4G network under the ITU umbrella in October 2009, they were about two years behind the market trajectory. The ITU set the main criteria that required speed boosts, but more importantly, 4G was required to make more efficient use of the spectrum. The standard body also required that equipment makers provide features that will help guarantee the quality-of-service on wireless networks. True 4G called for peak speeds of 100 Mbit/s for mobile applications such as driving a car down the road and 1 Gbit/s for fixed networks. That it's going to take four or five years before people start rolling out anything like the ITU's version of 4G was evident.

Much like the transition from 2G to 3G, several technologies were developed to provide performance boosts until a proper 4G network is christened. Just as was the case in the 3G saga, marked by the GSM versus CDMA split, 4G technologies separate into two broad camps: LTE and WiMax. Both technologies are designed to move data rather than voice and offer data speeds of up to 10 Mbit/s. WiMax hit the technology market a little earlier than LTE and charged a bit less as well. In the United States, Sprint initially threw its weight behind WiMax, while AT&T and Verizon chose LTE. There's been quite a bit of debate on whether LTE and WiMax meet all the technical requirements to be classified as 4G technologies.

Long Term Evolution or LTE, as the name suggests, is actually an evolution of the existing 3G standard, not a new standard in its own right. LTE is simply an advanced form of 3G, also called 3.9G by the ITU. The concept of the Long Term Evolution of 3G network based on the European wideband-CDMA technology was first outlined during a Radio Access Network workshop held in Toronto in 2004. The official name designated for the technology was "3G Long-Term Evolution."

LTE offered a theoretical capacity of up to 100 Mbit/s in the downlink and 50 Mbit/s in the uplink, and more if MIMO technique for antenna arrays is used. Although it technically doesn't comply with all of the 4G specs, mostly in terms of speed, LTE is where most wireless networks seem to be headed. The technology promises to bring speed boost as well as latency improvements. Moreover, the LTE architecture is the first to allow non-hierarchical IP traffic. All the other network schemes specify that all traffic is backhauled to a centralized control point and sent back out, and this could reduce by an order of magnitude the capacity that has to be built into networks. LTE, on the other hand, lets traffic route between base stations.

On December 14, 2009, the first commercial LTE network was deployed in the Scandinavian capitals of Stockholm and Oslo by the Swedish-Finnish network operator TeliaSonera and its Norwegian brand NetCom. TeliaSonera branded the network as 4G. The user terminals and modem devices on offer were manufactured by Samsung, and the network infrastructure was provided by Huawei in Oslo and Ericsson in Stockholm, respectively. The early LTE handsets started to become available by 2011 and they demonstrated how superb LTE was in transporting video. Still, industry watchers wouldn't expect the widescale deployment of LTE in handsets before 2015. Evolution was the essence here, after all.

ONE NETWORK AT A TIME

In 2010, when Sprint pushed the rollout of its WiMax network in the United States, the wireless operator named it "4G" to appeal consumers now comfortable with 3G lingua franca. The higher speeds of 4G coupled with Sprint's aggressive pricing began to expose consumers to connectivity that was similar to what they've had at home, but in this case, it always lived inside their mobile devices. The world's first commercial mobile WiMax service, however, was launched by Korea Telecom in Seoul on June 30, 2006. The mobile WiMax standard was also an evolution of the original fixed-wireless WiMax specification and aimed to fulfill the criteria of 1 Gbit/s for stationary reception and 100 Mbit/s for mobile reception.

The crucial benefit that WiMax had was time-to-market. WiMax, a sort of faster and longer-range version of Wi-Fi, was available by the end of 2000s whereas LTE was expected to take a while to reach its full potential. WiMax also gave the world the first 4G smartphone, the HTC Evo, a gorgeous device with a 4.3-inch touchscreen which ran Google's Android operating system and featured two cameras, GPS navigation, HDMI output, and mobile hot-spot capability. The phone cost US$200 with a two-year contract. Sprint also charged an extra US$10 a month, in addition to its standard data plan, as a service fee to access the 4G network. Ironically, just as Sprint was introducing its 4G network, Apple was unveiling its iPhone 4, which, name aside, ran only on AT&T's 3G network.

LTE was still at the standards stage when WiMax became a commercial reality. Still, LTE was winning the game when it came to mass deployment. As more and more wireless operators embraced LTE, the technology started gaining momentum among handset and infrastructure makers. Two large wireless equipment makers Ericsson and Nokia were not making WiMax gear. Nokia distanced itself from WiMax because the technology had no direct links with the existing 3G approach. Sweden's Ericsson, the world's number one telecom-equipment maker, argued that WiMax is not optimized for voice calls on the move. Even Verizon, a longtime loyalist of Qualcomm and its CDMA technology, started considering LTE, thus joining the camp vying for the evolution to the GSM standards path.

If Verizon, the largest wireless service provider in the United States, threw its weight behind LTE, the long reign of Qualcomm's CDMA products could finally come to an end. Indeed, the era of ideological standard wars of the 1990s was over. It was now increasingly apparent that the 4G technology of choice would be LTE and that interest in WiMax would decline correspondingly. WiMax though could have a niche role to play, predominantly providing DSL replacement services in emerging markets. The fact that GSM/wideband-CDMA/LTE was going to be the winning approach had become palpable because of a peculiar development in summer 2010. When Intel acquired the communication business of German chipmaker Infineon Technologies, which had made some good progress on LTE products, it became obvious that WiMax now was on an uneven footing.

Intel was "WiMax, WiMax, WiMax" at one point. It sank at least US$1.2 billion into supporting WiMax operators like Clearwire. But this early backer of WiMax subsequently shifted gears in favor of LTE, which was clearly gaining favor with the handset makers after winning over wireless operators around the world. Despite the fact that WiMax came from the data networking world, and it was a crucial merit, LTE's history took it back to the GSM and its offshoot GPRS. So it carried a much more effective upgrade path for GSM- and wideband CDMA-focused 3G operators. As a result, LTE offered huge benefits of scale which could eventually drive the price down to such a point that WiMax wouldn't be competitive anymore.

If displays and processors were the main focus for smartphones in 2010, mounting data traffic made 4G the keyword for 2011. The overwhelmed AT&T network had shown the wireless world that smartphones weren't the only value proposition; they also needed faster networks. The demand for wireless data around the globe is expected to double each year through 2014 as the population turns to smartphones and similar data-centric devices for instant information everywhere. By 2015 or so, a predominant majority of phones could have 4G capabilities, but they are likely to complement it with 3G. Rather than a sudden revolution, consumers are more likely to go through a gradual transition to the new technology offering increasing speeds.

THE REAL WIRELESS INTERNET

The book has shed light on how mobile handsets are primed for video and how exciting moment this is in the mobile technology evolution. That, however, also leads to the fears that casual scans of YouTube clips and streams of Netflix movie downloads could crush cellular networks. Moreover, gulping gigabytes of bandwidth for high-definition movie streams could also eat through consumers' wallets and disrupt wireless operators' data plans. Here, Wi-Fi came to rescue the dream of adaption of rich content on smartphones through intelligent and predictive models, which call for automatic detection when a mobile device hits a Wi-Fi network and starts downloading content such as movies and e-books. The fact that Wi-Fi is cheaper and faster than cellular networks makes it a crucial deterrent against the strains of video communications.

Most smartphones offered by AT&T have built-in Wi-Fi, which lets the handset automatically switch from the wireless network to a Wi-Fi hotspot without a need for prompting. Customers could use their devices on Wi-Fi networks, including those at more than twenty thousand AT&T Wi-Fi Hot Spots in the United States, without counting against their monthly data allotments. The hybrid phones let people make connections using a local Wi-Fi access point and seamlessly switch over to a cell phone network once outside the hot spot. The net result is greater flexibility in mobile communications as well as potential cost savings gained by shifting call minutes from cell phone plan onto the Wi-Fi-based local wireless network.

The early 2000s saw an unstoppable grassroots movement to deploy Wi-Fi in business and public access points, attempting to recreate the Internet in the wireless space all over again. Folks began calling it the real wireless Internet. The office version of Wi-Fi was widely adapted in the U.S. market, while the rest of the world caught up with that over time. Public Wi-Fi, on the other hand, targeted small user segments while offering wireless Internet service to indoors such as at cafés, air-ports, and homes. Wi-Fi marked a promising inflection point at a time when 3G had stumbled, inspiring a mania that was unseen since the great Internet boom of 1990s. No wonder, Wi-Fi turned into a poster child for pervasive computing—another manifestation of always-on computing—and became the undoing of a tradition of controlled use of bandwidth in the wireless world. Wi-Fi eventually became wireless data's second silver bullet after SMS.

During this time, some industry watchers began asking who needed 3G anyway. In the hindsight, however, Wi-Fi wasn't a 3G replacement, as some people had dreamed, but was an appendage to 3G networks. But wireless operators began to treat Wi-Fi as an extension of their cellular networks only after the initial knee-jerk reaction. Cellular carriers, who spent billions of dollars to upgrade their systems for high-speed 3G net-works, were initially skeptical to integrate Wi-Fi and cellular into a single package. Early versions of Wi-Fi cell phones failed miserably because of the enormous drain on the batteries and because users were forced to manually switch between networks. Wireless operator's fetish with accounting and billing systems was another major hurdle in the way of widescale Wi-Fi acceptance. But then Motorola solved the automatic transfer problem by allowing the cell phone sense when it's about to reach the end of Wi-Fi coverage. While the call still traveled through the Wi-Fi network, the phone would register the call onto cellular network, in essence creating a second phone line for the call.

Despite the fact that Wi-Fi was touted as a 3G killer in the early phase of its success, after Apple's iPhone launch, it became an important com-plement to 3G and accomplished broad industry recognition. While 3G is mobile, Wi-Fi is nomadic. What's critical was to make Wi-Fi seamless with cellular networks. One of Apple's radical efforts in the smartphone arena is the effective integration of Wi-Fi into the cellular mainstream. The iPhone really reinvigorated the ability to use both kinds of networks on a single device and allowed consumers to take advantage of the best each has to offer.

The iPhone liaison with AT&T had led to the first meaningful Wi-Fi integration into the mobile phone world. Then in 2010, at the launch of the iPhone 4, Apple raised Wi-Fi's strategic value by moving FaceTime-based video calling onto Wi-Fi networks exclusively. Now, after Apple showed them the way, the mobile phone establishment is finally angling to join the Wi-Fi movement. The hot-spot companies have started signing roaming agreements with cell phone service providers in anticipation of more widespread use of Wi-Fi in cell phones.

ON THE WAY TO 5G

There were several keys to the fulfillment of the mobile Internet paragon: the evolution of user terminals being the prime one. The next biggest barrier was speed. Sending data to and from mobile devices was initially a tedious process. Speed peaked at around 14.4 Kbit/s, a rate that didn't allow much beyond transmitting e-mail and checking websites for small amounts of information. A lot was at stake for faster wireless data networks. For instance, the future of m-commerce was intertwined with the realization of fatter network pipes. Mobile commerce could become a substantial source of revenue for wireless companies once speedier networks were built out. The capabilities of 3G and 4G technologies would also make applications such as messaging and web browsing more appealing to users and enable new ones such as MP3 downloads, streaming audio, and mobile gaming. Although these activities would chiefly be used for non-commerce applications, they would give m-commerce a boost, nonetheless.

Wireless operators were not fighting to become dumb pipes after all. But the flip side for the network operators like AT&T was that they had a lot more at stake than did the newcomers like Apple and Google. For instance, the more iPhones are sold, the more unhappy users could be because of the network congestion. The iPhone had created the consumerization of wireless data services, but on platforms like the mobile Internet, consumerization was generally accompanied by a catastrophic decline in per bit revenue for the service provider. The world had witnessed how Apple's iPhone crippled swaths of AT&T's network as users complained about dropped calls and network delays. The network that pioneered the smartphone in collaboration with the iPhone also topped in consumer dissatisfaction.

The post-iPhone smartphone enterprise had brought down the traditional "one-size-fits-all" pricing model that originally made the mobile Internet more affordable to a greater number of people. Consumers naturally preferred unlimited plans, but carriers couldn't provide them due to limited infrastructure, so they dropped unlimited plans in favor of tiered data buckets. AT&T management also claimed to have adapted the lower pricing tiers so it could evenly distribute the riches of smartphones to a broader range of customers. Wireless data are a finite resource, so mobile phone carriers, responding to the bandwidth challenge, chose not to treat data like a commodity. They carved off 5 percent of their heaviest users and stigmatized them.

Now the phrases like "data hogs" and "bandwidth hogs" began to circulate in trade media for power smartphone users. The era of unlimited plans had come to an end within short three years after the iPhone had provided the first viable mobile Internet experience over handsets. The wireless operators are not only thinking ahead in terms of increasing bandwidth demand to bring people the joys of downloads, video chats and digital media on their palmtop, they also find it vital to predict usage and work out plans for different classes of users. Wireless data move from a tower to a mobile phone over air waves, which carriers buy at auction from the government. Each wireless carrier has a limited amount of spectrum, yet that limited amount renews itself, moment after moment.

How consumers respond to these tiered wireless data packages would define the course of action for wireless operators, but one thing is certain: the relentless march of bandwidth had started with ascent of smartphones, and it would only go farther. People had started streaming shows and movies from the web to their smartphones, iPads, and other mobile gadgets, and as they did so, the demand for mobile bandwidth grew faster than anyone ever imagined. Moreover, the faster the network, the more that people will use it.

The basic vision of the mobile Internet is now in place. With a steady rise in network speed, it is hoped that the dream of a pervasive wireless network would turn into a reality with smartphones making wireless data services as easy to use as the ubiquitous telephone system.

Mark Weiser's work on "ubiquitous computing" at Xerox PARC back in 1988 had conceived of wireless data as the underlying technology for a computing environment that enabled personal mobility of computer users. Now some people in the wireless industry believed that the next generation of wireless networks—4G—could embody that vision of "ubiquitous computing" by offering the ability to access the applications we want from any platform, anywhere, any time. To create such an environment, one needs to integrate various applications working in harmony with intelligent sensor networks. For instance, users' cars would send SMS to their mobile phones, if someone tried to open the door, while they were away from their cars. Or a user's home security camera is hooked to the Internet, so that he or she can view the sitting room on his or her mobile phone screen.

16 SMARTPHONE ON STEROIDS

"It's a foregone conclusion that the personal computer of the future is this size. You could add wireless HDMI to it someday, and it could also be your set-top box."
—Nvidia CEO Jen-Hsun Huang holding up his smartphone while speaking at the company's annual conference in 2010

The personal computer had been the primary driver for the electronics industry for almost three decades, leading to the U.S. technology leadership in chips, software and networking arenas. After its inception in the late 1970s, over the course of next thirty years, the PC's role changed dramatically, from enterprise productivity to communications to entertainment. During this period, Intel and Microsoft collaborated and built the computing platform that singlehandedly transformed corporate IT, opened computing to the masses—both inside and outside of corporations—and unseated such giants as IBM, HP, Digital Equipment and even Apple. The two companies have been collaborating since before IBM introduced its first PC in 1981, a machine that used Microsoft's DOS operating system and the Intel chip design known by the designation x86. The relationship became particularly lucrative after Microsoft's 1985 launch of icon-rich and easier-to-use Windows software helped make PCs a mainstay in homes and offices.

The collaboration between Intel and Microsoft is legendary in the annals of the technology history. Intel and Microsoft controlled the two most important standards in computing: Windows operating system for PCs and the x86 processors-based Intel Architecture, the set of rules governing how software interacts with the processor it runs on. Microsoft's Windows OS software is written to Intel's x86 architecture, and that insures that the Wintel pact remains in force for the foreseeable future. The marriage of Intel chip and Windows operating system shaped the PC business since the early 1980s by defining the standard for which software developers created applications. However, the PC's reign at the top of the electronics food chain was falling apart by the late 2000s when new technology forces started to push computing into data centers and onto mobile phones, businesses that Intel and Microsoft didn't dominate.

The world's two most powerful technology companies who had built a PC empire over a period of two decades were seeing a steady end of their market hegemony. Especially, in the mobile devices domain, all they could show was a troubled track record. The PC kingpins had been playing catch-up in the mobile market for about a decade. Despite cash, power, research and market might they had been left with nothing in the phone arena. Amid PC's shrinking in relevance industry observers began pinning the question if Intel and Microsoft would thrive or even survive in the post-PC era? Either alone or in partnership, they hadn't been particularly successful in extending into the cell phone market where sales were now hovering around 1 billion units per year by 2010.

That meant a lot of chips and software, and for both Intel and Microsoft, they were not used to leaving that much money on the table. Although the PC business was still a cash cow, it now came with a sinking feeling as they could see the future of computing moving to a new turf:

portable mobile devices. Further into this turmoil, the PDA episode was a sucker punch for Intel and Microsoft alike, as they both saw this class of gadgets as promising and they had heavily invested in it. Compaq's iPaq PDA—which initially won over business buyers with feature-packed devices—was based on Intel's StrongARM processor and Microsoft's PocketPC operating system.

In fact, the Compaq eye-catching iPaq handheld computer, introduced in summer 2000, was hailed as having rescued Microsoft's mobile strategy for a little while. Until Compaq began shipping the hot-selling iPaq with a high-resolution color screen, improved interface, and multimedia features, receiving favorable attention from different corners of the industry, Microsoft's mobile initiatives were on the brink of disaster. Compaq pulled them back from precipice. But, in the end, PDAs were dwarfed by smartphones, and folks began drawing scenarios of how smartphones eventually would replace personal computers. The IT's most-hated power couple was in spotlight, and its technological prowess was in question, especially Microsoft, which had yet to deliver a competitive version of Windows for smartphones and other molds of portable computing.

The Wintel couple—short for Windows and Intel—was now increasingly seen as the past icon trying to hold on to its franchise. The forces of disruption had reached at their doorstep, but both were battle hardened and wouldn't go without a fight. Just when the technology industry's most lucrative partnership was coming to a day of reckoning, there came a twist: a new class of portable and mobile devices being developed at the fringes of the smartphone archetype. The footprint of Wintel was all over these undertakings. Smoke-and-mirrors departments in Redmond, Washington, and Santa Clara, California, were clearly up for something new. This time around, both Intel and Microsoft were taking a more holistic approach to steal the show from then up-and-coming smartphone business. They probably knew that they were going against the wind; still, it was a calculated risk they seemed to think was worth taking.

UMPC: THE COMPUTER ON THE GO

In 2006, Microsoft launched a project codenamed Origami in collaboration with Intel, Samsung and a few others. The new venture aimed to seek alternative form factors and input methods for mobile computing devices and to market them to consumers. The underlying idea was to build ultra-mobile computers running off-the-shelf PC components with desktop operating system and a flexible applications suite at a consumer-level cost. When Microsoft announced the project at CeBIT show in March 2006, the initiative received a healthy boost from OEMs who vowed to produce devices in line with the Microsoft Origami concept. The ad on the Microsoft website boasted, "I am your go everywhere do everything device."

In May 2007, Microsoft loosened the design description to "any portable computer running full Windows with a screen size of 7 inches or smaller." As per technical specs, the touchscreen, incorporated as an onscreen keyboard, would offer a minimum resolution of 800 × 480 pixels. The Origami-based devices were to be lighter than most tablet-type PCs, weighing about 2 pounds with a 7-inch touchscreen. They promised a new class of mobile gizmo that would use an Intel processor and a modified version of Microsoft's Windows XP Tablet PC software. Apart from running a full version of Windows XP operating system software, the devices would come equipped with popular Microsoft applications such as Outlook and Office, which includes Word, Excel, and PowerPoint.

Microsoft's Origami devices were dubbed by Intel as ultra-mobile PCs (UMPCs). Intel's first introduction of the concept was referred to as "handtop" by CEO Paul Otellini, but was later dropped in favor of UMPC. Otellini at his CES keynote speech touted these ultra-low-power mobile Internet devices as "the next big thing in computing." Likewise, Microsoft described the UMPC as "a device-like computer that is small, mobile, and runs the full Windows operating system. The UMPC goes anywhere and does anything that your current computer can do." It's worthwhile to note that, early in this dash, Microsoft's Wintel partner had announced its own Intel Ultra Mobile Platform 2007 based on low-power processor designed to drive both UMPCs and yet another category, mobile Internet devices (MIDs). Intel MIDs wouldn't necessarily have to run standard Windows software.

The term *UMPC* was first used for mobile devices with 7-inch diagonal screen marketed in 2006 by Samsung, Asus, Founder, and TabletKiosk. The first UMPCs on the market were AMtek's T700 and Samsung's Q1. These devices were tablets with touch capability running Windows Tablet PC Edition or Windows XP with Touch Pack. UMPC was not a sit-on-your-desk computer, nor was it even a sit-on-your-lap computer; rather, it was a hold-in-your-hands computer. UMPC, which offered all kinds of mobility and connectivity features, would be so small and light-weight that a user could hold it in his or her hands, much like one would hold a book. Simply put, UMPC was a notebook computer in extremely small, handheld proportions while still offering elements of productiv-ity, connectivity, and entertainment.

Internet access was, it was claimed, made easy; running Internet Explorer or other web browsers, users could access Wi-Fi hotspots or even use a Bluetooth-enabled mobile phone to surf the Internet. Another help-ful piece of functionality came through inclusion of GPS-based loca-tion technology. Whether using the Microsoft version or software from another provider, a user could use the UMPC to get directions from place to place while tracking his or her progress on the computer in hand. The navigation could take place both via satellite or Wi-Fi network chennels. Users could also sync the UMPC with their home or work computer and take their documents, spreadsheets, audio and video files, e-mail, calen-dar, and contacts with them while on the move.

Intel and Microsoft created a massive marketing blaze for the so-called UMPC with Origami Experience. Intel, especially, made a huge bet on this new product category. The top chipmaker had made a similar kind of bet on Wi-Fi technology in the early 2000s: within four years, 96 percent of laptops were Wi-Fi-enabled, mostly powered by its Centrino chips. Now the world's largest silicon producer was envisioning a computing environment populated with devices like UMPCs and MIDs, which, in turn, could provide a chance to re-create a new genre of mobile computing devices. In parallel, Intel was sewing up a new chip architecture to enable Internet-focused, multimedia-enhanced handheld gadgets powered by light, cool, energy-stingy Atom chips. It was Intel's biggest repositioning since its drive into notebook-specific processors in 2000, which led to the launch of the Centrino chip.

But this time around, the risk was that this isn't so much a category that people really need, but an in-betweener that nobody needs. Some people in the industry were calling UMPC and MID a product category that would sit in the gap between the smartphone and the laptop computer. In the end, as many industry watchers had feared, these novel computing machines just proved to be smartphones on steroids. Their design cycle was relatively long compared to the market's wheeling and dealing, so the Wintel had to have this magic ability to predict what people would like to have in three years from 2007. To get to the kind of market acceptance with the unknown and untried UMPC, Intel and Microsoft were going to need their past magic more than ever. Just a few years ago, the PDA camp, which included the Wintel in earnest, had touted that people could leave the PC at home and office, and that it'd be fine if they just bring the PDA.

Whether it was bad marketing, bad design, or bad timing, UMPC devices never reached mass-market acceptance and remained, as had previous mobile PCs, in the geek, pro-mobile markets. The lack of battery technology advances, poor transfer of the desktop interface to touch- and keyboard-driven interfaces, and the introduction of a new, more heavyweight operating system in Windows Vista also prevented OEMs from producing successful, low-cost alternatives. The UMPC was initially conceived as a small, inexpensive slate with a touchscreen, but the concept didn't catch on due to low performance and higher-than-expected pricing. At a time when processor speed, memory size, and hard drive capacity were ballooning, UMPC price points were gravitating toward the high-end target price range of US$500 to US$1,000. With prices like these, it's hardly a surprise that the UMPC couldn't take the consumer market by storm.

Intel had also made a case for the UMPC hoping it would go into non-consumer, vertical applications where it would provide access to some specific Windows apps for an on-the-go workforce, like FedEx, or for people in the hospitality industry. The idea was that there are large companies out there who have specific Windows applications that people need access to in non-desktop situations, so they're supposed to go out and buy UMPCs to run those Windows-based applications. However, the form factor simply didn't work for the desktop-like Windows software applications. Moreover, something as experimental as deploying a fleet of brand-new mobile computing devices just to run some legacy in-house Windows applications seemed unsettling for IT bosses.

Eventually, UMPC ended up as a product splash that appears to have collapsed under the weight of unbridled feature creep. All who invested in it were assuming that price was the barrier, but when Moore's Law eventually brought the prices down, it became apparent that the UMPC was simply an awkward, neither-fish-nor-fowl solution looking for a problem. Intel, meanwhile, quietly dropped UMPCs from its marketing agenda, choosing instead to focus on two new categories of ultramobile devices as target market for its low-power Atom processors and associated chipsets: MIDs and netbooks. Intel began to deemphasize the term "UMPC" in the spring of 2008 while it started to talk about MID. The trouble was that to some industry watchers, MIDs were merely a recycled form of the original UMPC concept.

In a way, MIDs evolved out of the UMPC category, replacing pocket computers' exorbitant costs and bloat with more streamlined features designed for consumers. The entertainment focus was evident in early UMPC campaigns as the product was originally marketed as lifestyle gizmo. But few people were ever going to make a relatively bulky 7-inch display device with a two-hour battery life part of their lifestyle. On the other hand, in a smaller form factor—say a 5-inch display, and a battery life of at least five or six hours—MID could fit the bill. As long as it's small enough to carry around in a purse or jacket pocket, and cheap enough to be written off as an entertainment expense like a Netflix subscription or a new TV, Intel managers believed MID had a good stab at becoming a lifestyle product.

MID: IN SEARCH OF ITSELF

In June of 2006, Intel had sold its mobile phone–oriented XScale processor business to Marvel Technology. Now when one is Intel and has dominated the desktop and laptop PC market, which account for annual shipments of about 300 million processors, but has virtually no presence in the mobile phone market, which now consumes over a billion processors annually, what does a company do? You invent a new device category that targets the gap between mobile phones and notebook PCs, and use Intel architecture (a.k.a. x86) processors to fill that gap. So, early in 2007, Intel launched a new category of devices dubbed mobile Internet device or MID, aimed at letting connected consumers take a "full Internet experience" on-the-go. This was in a stark contrast with the limited web browsing capabilities of mobile handsets in the pre-iPhone era.

When Intel introduced a prototype MID in 2007, the concept of a multimedia-capable, always-on handheld with wireless Internet access captivated many in the consumer electronics world, including Intel's competitors. The formation of MID as a new class of products would serve to fill the gap between smartphones and UMPCs; an MID would be larger than a smartphone but smaller than a UMPC. The common perception was that MIDs wouldn't be designed to replace mobile phones or smartphones but to be used as companion devices. For a handheld device, MIDs would have an unprecedented level of multimedia capabilities and typically would come in a tablet-like form factor. These novel units were generally being seen as designed to provide entertainment, information, and location-based services for personal use rather than for corporate use.

In spring 2007, Intel showcased a prototype MID at the Intel Developer Forum in Beijing. An MID development kit by Sophia Systems using Intel's Atom processor was later released in April 2008. Then in January 2009, at the CES, Intel showed off several Atom-powered MID prototypes at its sprawling display booths. During his CES keynote, Intel chairman Craig Barrett demoed the OQO device, an MID that was to be made available in the first half of 2009 for US$999. Sporting a 5-inch touchscreen with a slide-out 58-key physical keyboard, it would run an Intel Atom processor, have up to 60 GBytes of storage, and run Windows Vista or XP. But it never happened. OQO Inc., founded in San Francisco earlier in 2000, closed its doors without shipping a single MID. Industry watchers said that OQO devices came with physical keyboards that were too small for users wanting to type long documents.

Although a touchscreen was standard, some devices relied solely on the screen for data entry, while others also had a physical keyboard that folded or slid out. MIDs were generally designed to fit in the pocket: their screen sizes ranged from 4 inches to 7 inches, making them a little larger than a smartphone, but smaller than most tablets and netbooks. The device was marketed as pocketable gizmo, but at 6.5 inches on its longest side, it was too big to slide into most pockets. Moreover, some MIDs supported voice while others just offered data-centric features. Finally, the fact that full-fledged Windows software ran on these small displays also didn't speak well for the MID user interface.

In all fairness to Intel, MIDs brought some impressive hardware bytes. Weighing as little as one pound and often measuring less than half an inch thick, MIDs made carrying the Internet everywhere relatively easy. As a web browsing device, MIDs relied on Wi-Fi connectivity, though some MIDs also offered 3G capabilities for connecting to a wireless provider for data access. Bluetooth, TV outputs, and other connections could also be included in some MID designs. MIDs could run a variety of operating systems, with some of the MIDs based on Google's Android platform. And since they didn't require the full processing power of a notebook or even larger UMPCs, MIDs could become relatively affordable.

What could potentially help MIDs were lower prices, more rugged designs, and some MID-optimized software. Probably the most critical issue was that what form factor was going to look like because people designing MIDs were coming both from the PC side of the house and from the mobile side. Not surprisingly, therefore, two different design paradigms began to take shape: MIDs, which were more or less a miniature version of a laptop, and mobile-optimized devices, which looked more like smartphones. That's exactly where MID opponents like Intel's alter ego Advanced Micro Devices (AMD) would hit the promise of this device. They said that MIDs would only have some play in the short term, calling them a tweener between a cell phone and a notebook. The concept of an MID—a device midway in size and capability between smartphones and the smallest notebooks—was under tremendous pressure from both sides.

Amid all this uncertainty, the Intel guys stepped in and gave some hints by elaborating that MIDs would sit below the UMPC but above the smartphone, but they would look a lot like a smartphone. They said the Internet experience on the handheld is suboptimal and that Intel could close the gap using its IT technology prowess. But then some people defined this new product category as a grown-up smartphone. Further muddying the waters were industry terms such as "smart book," "Internet tablet" (IT), "Internet media tablet" (IMT), "pocket PC," and "UMPC," which were sometimes used interchangeably with MID. Now people began to wonder what was an MID, exactly? Was it just a marketing term?

The name MID had been around since 2005. Many in the industry believed that MID was a marketing gimmick straight out from Intel's PR machine and that it was designed to make the product sound new and different, when MIDs were primarily a variation of the existing mobile devices. Analysts and mobile device manufacturers generally detested the term, partly because there were so many different variations within the product category. There was also a question if there was a real market for the MID. Consumers had so many choices. Would they buy another electronic device that's larger than a cellular phone, but significantly smaller than a notebook computer? One of the world's smartest and biggest technology companies was betting they would.

However, Intel's premise that MID would serve well to lifestyle boomers, generation Y social networkers, young gamers, frugal generalists, and multimedia enthusiasts proved yet another paradox. Early MIDs—including devices from Compal and Casio demonstrated in 2008—had flopped. They were power hungry and were built in a clunky form factor. Some even had a fan inside. So why didn't MID take off? Was Intel's vision of MIDs flawed? Analysts and industry insiders said that using a term like MID was just confusing to buyers and end users. Who needs an MID when one is already toting around a laptop, a mobile phone, and an iPod? Consumers who wanted on-the-go Internet already carried a smartphone and perhaps an iPad. They simply didn't want to lug around yet another device.

For some observers, what Intel was talking about was already here—and powered by ARM-based processors. One of the first true MIDs to reach market appears to be Nokia's N Series Internet tablets, and these devices, which debuted some two years before Intel's invention of the MID, had only met with limited success. Consequently, few portable device makers followed Nokia over the MID cliff. Then, in June of 2007, just a few short months after Intel officially launched the MID device category, Apple began shipping iPhones and the air was let out of the MID balloon. Some consolation came when market research firm Gartner quoted iPod Touch as an example of MIDs; Gartner also considered e-readers like the Amazon Kindle that were connected to the Internet to be MIDs.

INTEL AT ARM'S LENGTH

The lines between smartphones and ordinary mobile phones are continuously blurring. Smartphone could thus replace millions of mobile phones at the lower end of sophistication scale and laptop computers at the higher end. Suddenly, the cookie-cutter, web-enabled smartphones had attained a make-or-break status that the world's largest chipmaker simply couldn't afford to overlook. Intel chief Paul Otellini told his audience at the Intel Development Forum in September 2010 that "there are 2.8 billion smart devices out there right now and that number will double by 2014." What seemed to worry Intel the most was the prognostication that ARM system architecture will triumph over Intel as smartphones and tablets disrupt the x86-centric PC industry. By 2010, it was clear that mobile computing would bring disruption to all of kinds computing devices. That was also about the time when ARM had become the fastest-growing processor and the instruction set architecture of choice of mobile computing.

Intel continued to dominate in the enterprise IT world, and while it tried, the silicon industry bellwether didn't succeed to grab land in the smartphone bonanza. In the smartphone world, the predominant system platform is shaped by ARM Holdings, a U.K.–based maker of chip technology that largely goes unnoticed outside the semiconductor industry. Another company called MIPS Technologies Inc. does roughly the same thing on a smaller scale and for less portable devices like set-top boxes and servers. ARM sells designs for energy-efficient chips licensed by processor chipmakers like Texas Instruments, Qualcomm, and STMicroelectronics. These companies already have a hold on the mobile phone market, one area Intel has failed to penetrate.

ARM and its licensees have pretty much captured the smartphone market that Intel desired, and now they have momentum in tablets, thanks to Apple's iPad. Apple utilized ARM-based architecture for its mobile devices, including iPhones and iPads. The ARM-based architecture has lower power consumption and offers the ability to easily integrate audio and video functions onto a single chip. But ever since Intel said that it wanted to push into the mobile Internet device market, there has been a sense of anticipation of a pitched battle between ARM and Intel over the control of smartphone hardware. From the day Intel introduced the Atom processor, it has been working feverishly to lower the power requirements, but while it does that, ARM has been working relentlessly to boost the performance of its own processors. So will future mobile handhelds run on an x86 or ARM instruction set architecture? According to market watchers, it's still too early to make a call.

Intel and ARM are not directly comparable. One is a high-volume chip company that sets standards; the other licenses intellectual property (IP) and partners with multiple silicon companies. ARM has one hundred to two hundred partners in its ecosystem, and it allows them to build processors with its IP software. One is king of the PC and notebook computer, and the other has become ubiquitous in the mobile phone by trying not to impose itself too much. Intel led in chip manufacturing and ARM led in mobile architecture. While Intel couldn't be discounted given its resources, innovation history, and ability to displace incumbents, its computing architecture, predominant in PCs and enterprise environments, requires a lot more power. There has been even a talk for Intel to abandon its x86 architecture for portable devices and embrace ARM-like system-on-chip designs to be more competitive in the smartphone realm.

Intel has been trying very hard to lower the power consumption of its chips, and has made progress in the area. Still, Intel's chips haven't been able to match the low-power consumption of chips based on designs licensed from ARM. An ARM processor is also much more accepted in price-sensitive markets such as mobile phones. Moreover, thousands of applications can already run on an ARM processor, and there's a huge ecosystem of tools made available for developing these applications. About 98 percent of the more than 1 billion mobile phones sold each year use at least one ARM processor. ARM processors are also used extensively in consumer electronics products like digital media and music players, handheld game consoles, calculators, and computer peripherals such as hard drives and routers.

The story of ARM's life starts with the original development of a processor and ends with the establishment of ARM Ltd as a global force in the microprocessor industry. The history of the ARM processor family is also closely intertwined with that of the British personal computer industry. The first ARM chip—the Acorn RISC Machine (ARM)—was developed between 1983 and 1985 by a team at Acorn Computers Ltd, a pioneering developer of personal computers in Britain. Acorn was one of the leading names in the British PC market at that time. The release of the BBC Micro in 1982 caught the crest of the home computer wave in Britain, and the BBC name gave Acorn's design an added credibility compared with competing machines from a plethora of developers in the market. Acorn had also been renowned for the caliber of its research and development staff. It was able to pick the cream of graduates from Cambridge University, home of a highly regarded computer science faculty.

ARM was originally conceived as a processor for desktop PCs produced by Acorn Computers, a market that was later to be dominated by the x86 family used by IBM compatible PCs. Inevitably, after achieving some success with the BBC Micro computer, Acorn Computers considered how to move on from the relatively simple processor to address business markets outside the pedigree of the IBM PC launched earlier in 1981. The official Acorn RISC Machine project started in October

1983, boasting Steve Furber, Robert Heaton, and Roger Wilson as key designers. VLSI Technology Inc. was chosen as the silicon partner since it already supplied Acorn with chips.

VLSI produced the first ARM silicon device on April 26, 1985—it worked first time and came to be known as ARM 1. The first "real" production systems named ARM 2 were made available the following year. Such was the secrecy surrounding the ARM processor project that when Olivetti was negotiating to take a controlling share of Acorn in 1985, it was not told about the ARM development team until after the negotiations had been finalized. Another interesting aspect of this project was the relative simplicity of ARM processors that later made them suitable for the low-power mobile phone environment. One of the reasons the ARM was designed as a small-scale processor was that the available resources to design it were not sufficient to allow the creation of a large and complex processor. While now presented as a technical plus for the ARM processor, it began as a necessity for a processor designed by a team of talented but inexperienced designers.

In the late 1980s, Apple Computer and VLSI Technology started working with Acorn on newer versions of the ARM processor. The work got such a momentum that in 1990, Acorn spun off the design team into a new company called Advanced RISC Machines Ltd. For this reason, the ARM acronym is sometimes expanded as Advanced RISC Machine instead of Acorn RISC Machine. Advanced RISC Machines became ARM Ltd when its parent company, ARM Holdings PLC, floated on the London Stock Exchange and NASDAQ in 1998. Apple had used the ARM 6-based processor as the basis for its Newton PDA. The origin of ARM's success in mobile phone space is largely traced to Symbian's decision to exclusively support ARM instruction set architecture (ISA). This, in turn, was the consequence of a mid-1990s decision by Texas Instruments to use ARM in its mobile phone application specific integrated circuits (ASICs) for Nokia, the driving force behind the inception of Symbian smartphone project.

A decade after the Symbian breakthrough, ARM is squarely positioned in the smartphone game and challenger to Intel with a compelling business model. Intel and ARM are now holding the two bookends of large desktop market and growing mobile device market as their respective strongholds, and with a steady progression of convergence between the two, they could be heading for a collision course for the hardware riches of smartphone nirvana. Here, if history is any guide, the rising tide seems to be on ARM's side. Start with the computing strength that Intel so keenly claims as its key leverage. Industry chronicles are a testament to the fact that, in mobile settings, this leverage proved to be of little use for its Wintel associate, Microsoft. The far greater harbinger is the fact that software developers are increasingly focusing on ARM platforms. As the number of applications grows on ARM platforms it makes it more difficult for Intel to turn the market in its direction. This could well turn into a PDA déjà vu.

INTEL'S ANSWER: ATOM

In 2010, Intel wanted to be inside everything—again—and this time, it was counting on its Atom processors to help break its dependence on the slowing PC market. Intel wanted to equip the x86-based hardware to invade ARM's traditional domain—low-power handhelds—and for that Intel had to significantly improve performance while reducing the chip footprint and power consumption. Intel managers are quick to point out that the software community is already hooked on to x86 platform. But that could be boon as well as bane. For a start, in Intel's case, the problem is largely the one of power consumption. To maintain backward compatibility with the x86 processors meant that all applications that run on a PC will also run on the scaled-down Atom chip and that required a fair amount of processing power. Getting performance that most people consider acceptable also depletes the battery quickly. So while the Atom chip would power a netbook for eight hours, it won't power a smartphone for very long because it would be running on a battery that's light enough to carry comfortably in a pocket.

Then there was this claim the company managers made that by letting Intel define the platform architecture, handset OEMs get assurances about software compatibility, power consumption, and faster time-to-market. But that too could go to Intel's disadvantage as it sounded like a threatening loss of control to the likes of Nokia and Motorola. Intel's heavy-handed dominance of the PC makers in the 1980s and 1990s had earlier led communications OEMs to engage Intel with caution. In fact, this was Intel's second major tilt at the wireless communications market, and again, there was no clear indication as to why the company should be any more successful this time around. Intel had launched itself in the handset market in the early 2000s with the XScale wireless processor based on an ARM architectural license. The PC titan found getting traction with non-U.S. customers difficult and ended up selling the business to Marvell Technology in 2006.

Acquisitions have been the cornerstone of Intel's foray into consumer electronics and communications markets. This time, however, it's different in a sense that its homegrown Atom processor architecture has taken the center-stage, while Intel is making carefully thought acquisitions to complement its broad Atom strategy. Take the case of Infineon whose wireless business unit was taken over by Intel for about US$1.4 billion in cash in August 2010. Among other objectives, it was aimed to fill a gap in Intel's wireless communication roadmap: the Atom processor didn't support mobile telephony intrinsically, except as Voice-over-Internet protocol (VoIP) feature. The transaction also gave Intel a sizable presence in the cellular silicon market with top OEM customers including Apple, Nokia, Samsung, LG, and others. Furthermore, the deal provided Intel with a ticket to bet on the ongoing convergence of communications and computing.

At the same time, however, Intel's move looked like an admission that by turning its back on wireless communications earlier in 2006, the company got it wrong. Moreover, Infineon's wireless chips were based on ARM architecture, so Atom- and ARM-based chips going hand-in-hand could be confusing to customers. In a somewhat similar move, back in 2009, Intel had baffled the market when it paid US$884 million to acquire Wind River Systems Inc., which designed operating systems software for cars, mobile phones, and industrial machinery. Wind River, a major supporter of the open-source initiatives, was the Mobile Linux Group's chosen system integrator. Among its customers were the biggest OEMs in the wireless market, including Ericsson and Motorola.

Intel's acquisition of Wind River was yet another indicator of the sea change occurring inside the semiconductor world's perennial leader. It was now inevitable that the x86 giant expands beyond desktop PC market and into designs targeting a wide range of mobile devices, and for that, Wind River could help tailor commonly used operating systems such as Linux and Android to work on Intel's chips. Operating systems are the basic set of programs that control any device with a chip. If Wind River didn't customize that software, the manufacturers would have to do so themselves at considerable effort and expense.

While industry observers were still busy in reading through the lines, Intel threw out yet another stunner in mid-August 2010 when, in its largest acquisition ever, it purchased security software maker McAfee for US$7.68 billion. Now there were clear parallels between the McAfee deal and Intel's 2009 acquisition of embedded software vendor Wind River. The rationale: with all these Atom-powered gizmos connected to the Internet, they're going to come under the same kind of virus and malware attacks that plague PCs. That will require protection, which works better if the software is built into the chips.

The rise of tablet and mobile phone-based computing platforms brought with it increased demand for security intelligence that is embedded at the hardware level. Otherwise, Intel's acquisition of McAfee didn't seem to have a profound impact in the PC market; it only allowed Intel to benefit significantly from rising demand for embedded security in smart mobile devices. For smartphones, highly popular among many corporate executives, who often use their phones to transmit confidential information, perhaps one of the most challenging considerations for the future could be security. One downside to the openness and configurability of smartphones was that it also made them susceptible to viruses. Hackers had already written viruses that attacked Symbian-based smartphones.

Now with the acquisition of McAfee and Wind River—a Microsoft rival in the embedded OS space—it was evident that Intel was building up the software side of the smartphone business. Ironically, its Wintel associate on the software side, Microsoft, had a surprise in store as well. In late 2010, Microsoft announced that the next version of its Windows Phone 7 operating system would support ARM-based chips, confirming months of speculation that the software giant would broaden support for Windows beyond x86 platforms. Intel countered by optimizing the x86 platform for Android software.

These developments not only showed cracks in the Wintel duopoly, they also mirrored the clearest sign that interests of Intel and Microsoft were diverging in the epic battle of smartphones. It also demonstrated how important the new architecture Atom was to Intel and that the Santa Clara–based company was behind it for the long-haul. It's a new architecture that offers Pentium M-class performance within a 2-watt power envelope and is packed into a 25-mm^2 die size. These numbers are quite small for an x86 processor, but they allow Atom to target embedded applications like mobile Internet devices where the x86 has not been used before. Then there is the argument that is harder to resist: Intel's powerful manufacturing machine. At the same time, however, just don't expect Atom to dethrone ARM as king of the mobile world any time soon. The processor-centric hardware fight between ARM and Intel would probably see a lot more action in the coming years.

THE END OF WINTEL

In the early 2000s, the Wintel pair joined the PDA bandwagon to emulate the PC story on a smaller footprint, and after failing on that front, they decided a solo flight to re-create a new mobile computing experience not on a single, but a dual platform. UMPC carried more of a PC genesis while MID tilted somewhat toward the phone side of the house. But they both failed to make any impact at a time when the smartphone market was roaring ahead. Smartphones were exploding in an unprecedented way and anything remotely similar in agenda was just washed away.

The backroom power brokers of the computer industry—Intel and Microsoft—were looking for new things that they could sell amid the maturing PC market and its shrinking profit margins. But UMPC and MID, loaded with expensive Windows software, operated sluggishly and had shorter battery lives. That was because the huge operating system was designed to run desktop computers, not small portables. The large PC makers like Dell, HP, and Lenovo cautiously assessed the so-called tweener opportunity from the fringes while evaluating a meaningful balance between weight, screen size, battery life, and price. By 2010, the Wintel's gadget galore looked more like an evolutionary blip on a playbook that had the smartphone gospel written all over it.

The horizontal business model that Microsoft and Intel pioneered in the PC era now clearly seemed out of place. The success of Apple's iPhone and iPad had driven a deeper wedge between Microsoft and Intel. Technology industry veterans saw the failure of UMPC and MID as yet another signpost in the slow decay of the Wintel alliance. According to Steve Perlman, chief executive of online game startup OnLive Inc. and veteran of Microsoft and Apple, Wintel technologies were so deeply embedded in the information-era economy that they will exist for a long time, but the legacy products from these two companies will miss out on big growth opportunities in the mobile business. For Perlman, the Wintel system was too heavy to move into these upscale new markets.

In a nutshell, what people are seeing in the smartphone coming-of-age story is bigger than anything that has come before it; much bigger than the PC. By 2010, with the explosion of iOS and Android devices, and the relative stagnation of the desktop market, mobile OS shipments had surpassed desktop OS shipments. Consequently, Apple and Google, not Intel and Microsoft, are calling the shots in the new gadget order. Google CEO Eric Schmidt claimed that in the second quarter of 2010, the company was activating 200,000 units a day or over 18 million a quarter. The smartphone was the way to go.

In fact, the game was moving far beyond smartphones and into the realm of portable computing gadgets like tablets. According to a study from Goldman Sachs released in late 2010, one out of three tablets that's bought will replace a PC, and that's where both Intel and Microsoft came across as most vulnerable. Microsoft's tablet response especially hadn't been forthcoming: the company vowed to have Windows 7-based slates in stores by Christmas 2010, but that failed to materialize. Another rude awakening came in early 2011 when Hewlett-Packard, while announcing the series of mobile products, both smartphones and tablets, based on its newly acquired webOS operating system, rocked the industry by telling the audience that some of its notebooks will run on its in-house webOS software. The world's largest PC maker was making it known that it will put a non-Windows operating system on its notebook PCs.

However one put it, things were changing. The failure of Intel and Microsoft to effectively respond to the grave threat to their core franchises posed by new computing form factors like tablets and smartphones could result in vastly downsized companies struggling for relevance in tech's most important markets. Perhaps the biggest threat to once-invincible Microsoft came from Google, which aimed twin daggers at Microsoft's heart. While Android was all about limiting Microsoft's future in the smartphone and tablet domains, the Chrome operating system was taking a dead aim at its great twin cash cows: the Windows and Microsoft Office franchises. There had been a lot of ambiguity in trade press about Chrome OS and what was it actually aiming for in the presence of the unstoppable Android. Google chief Eric Schmidt differentiated Chrome OS from Android software in a way that finally made sense: Chrome OS is for devices with keyboards, he explained, and Android is for devices that respond to touch.

The cat was out of the bag: Google was positioning Chrome OS against Microsoft as a lightweight operating system that would enable portable devices like netbooks to handle everyday computing with web-based applications without need of any native software. With Chrome OS, Google was waiting in the wings to seize the moment, hoping that disruption may eventually provide it a window of opportunity to dismantle the Windows stronghold. The next two chapters dig deeper into the anatomy of new form factors like netbooks and tablets and establish how these new gadgets would relate in the ongoing smartphone revolution.

THE NEW COMPUTING ORDER

"Desktops PCs are effectively a flatlining commodity and in five years' time you'll wonder why you need a PC at all. The idea of sitting at a desk to view a web page is inherently annoying; phone screens are small but the size of the display relative to the phone size is growing and the resolution of screens is growing very rapidly."
—Nigel Clifford, CEO of Symbian, while speaking at the Symbian Smartphone Show in London in October 2006

Within a month after the launch of the Windows 95 on August 24, 1995, Microsoft's new operating system had sold more than 250 million copies. The Windows 95 craze exploded as the glowing press coverage rocked the PC industry as well as Microsoft's competition. The writing on the wall was very clear: PCs—running Microsoft software—would be the single most important device in our lives. At the height of this euphoria, on September 4, 1995, Larry Ellison, chairman and CEO of Oracle Corp., the world's largest database company, called the PC a ridiculous device while speaking at the International Data Corp.'s European IT Forum in Paris: "What the world really wants is to plug into a wall to get electronic power, and plug in to get data." In his article "Time Your Attack: Oracle's Lost Revolution," published in *Wired* magazine on

December 21, 2009, Daniel Roth describes how Ellison argued that ordinary desktop PCs were too expensive and too complicated for most people. They needed lots of memory and a fast processor in order to run Windows, and they needed hard disks and CD-ROM drives to store additional programs such as word processors and spreadsheets.

The centerpiece of Allison's yearnings for a world without PC hegemony was based on a concept called network computer or NC. It worked like this: a simple machine with a small hard drive would access computing applications online while all the data—videos, documents, pictures—would be stored in databases instead of on the computer itself. At about the time when commercial Internet was just taking off, it was indeed a powerful idea that enchanted companies and analysts throughout the IT industry. IBM saw the light quickly and immediately established the Network Computer (NC) division to build the bare-bones machines. Netscape co-founder Marc Andreessen declared the NC a pretty major new business opportunity; in place of an operating system, this machine would work with programs and files through browsers like Netscape Navigator. But Sun Microsystems and its then-CTO Eric Schmidt were probably most excited as Sun built an NC prototype within months and began developing a lean operating system to run on it. The server company spearheading the development of the Java programming language, which powered most of the web applications, hoped its work would gain new relevance for the Java juggernaut.

The term *network computer* was used somewhat interchangeably to describe a diskless desktop computer or a thin client. "The NC story just exploded beyond anything I imagined," Ellison said in the heydays of NC marketing blaze. "It took on a life of its own." He also made it clear that NC was part of his strategy to dethrone Microsoft. "IBM is the past. Microsoft is the present. Oracle is the future." By 1999, however, IBM had sold only 10,000 units and eventually shuttered its NC division. Netscape was crushed by Microsoft's Internet Explorer. Oracle, after spending four years and losing nearly US$175 million, pulled the plug on its NC business. The enterprise software titan changed the name of its network computer spinoff to Liberate Technologies and began shifting its business focus toward set-top box software for interactive television.

These early attempts to establish legacy mini-computers as a new class of mainstream personal computing devices largely failed because they were built around comparatively expensive platforms requiring proprietary software applications and they imposed severe usability limitations. But while Ellison's network computer venture failed as a product and as a business, it seeded an idea—and a group of technologies—that would go on to remake the computing world in the coming years. Subsequently, a decade after the NC foray, a wave of cheap, underpowered computers with small or no hard drives and omnipresent Internet connections did challenge the traditional PC. Ellison's vision of widget-based online software had become a reality with the advent of smartphones, tablets, and cloud computing.

Also, about a decade later, Ellison's bold prognostication was echoed by his friend Steve Jobs, who during his speech at the 2010 D8 Conference, pronounced the end of the PC era and few would question his verdict. He said, "When we were an agrarian nation, all cars were trucks. But as people moved more toward urban centers, people started to get into cars. I think PCs are going to be like trucks. Less people will need them. And this is going to make some people uneasy." Jobs's unflattering comparison between PCs and trucks brought the debate about the future of PC to a logical conclusion because what he said had already been overwhelmingly endorsed by other industry luminaries. Many in the industry shared this notion that with the launch of the iPhone, it is game over for PCs.

"The PC as we know it is slowly dying due to increased desire of the marketplace to be mobile," remarked Tom Golway, global technology director at Thomson Reuters. "Lightweight platforms, both physically and software-wise, are not just a requirement; they will be the expectation of the next generation of users." Google's European head John Herlihy joined the chorus while speaking at the 2010 Digital Landscapes conference in Ireland, "In three years time, desktops will be irrelevant. Instead, mobile devices will be the main way people interact with the Internet." The smartphone-led shift had ended the reign of pagers, some standalone music players, and the PDA. A new era was beginning and the PC could become the next casualty. Could smartphones do to the PC, at least partially, what they had done to PDAs? There are people who reckon that PC and its operating system will be quaint metaphors the same way we look back at eight-track tapes, rotary phones, and typewriters.

With gigahertz processors, the divide between the smartphone and PC has narrowed, but smartphones are not likely to overcome usability limitations anytime soon. The meteoric rise of smartphones and the rhetoric about PC being a passé in fact symbolized the ongoing evolution in the computing experience. Otherwise, smartphones aren't going to literally replace PCs in the near future for several key reasons. First and foremost, the screen is not large enough despite smartphones doing some amazing stuff with great resolution and quality. The size of the screen limits the smartphone's usability even if it presents resolution between standard TV and high-definition TV (HDTV) with a potential to reach HDTV resolution soon. Second, the input mechanism is much less convenient than normal keyboard and mouse, again due to the small form factor. Third, the battery doesn't provide enough working time, so using the smartphone for applications rather than just voice calls reduces the time between charges.

TRANSITION TO POST-PC WORLD

By late 2000s, however, computer users were warming up to PC alternatives like tablets and smartphones, and all these changes bolstered the idea of the post-PC era. It was a foregone conclusion that once these faster, more location-aware, service-oriented devices become more pervasive, the PC would quickly become a secondary object of desire. In a few years time, the smartphone wouldn't just be an essential traveling companion—it could become users' one-stop connection to the computing world at large. The PC wasn't going away altogether, but it wouldn't be the center of IT universe anymore. The chickens were coming to roost for PC makers who failed to anticipate the popularity of mobile Internet devices like the iPhone and the iPad. But was the computing industry ready for the post-PC era? Was another technological disruption at the doorstep?

On the other side of the fence, PC proponents argued that each new, disruptive innovation was actually a sign of health and of sustained public desire for the life-enhancing benefits of personal computing and being connected online. "We've only just begun to experience what a truly *personal* computing experience can look and feel like," said Dirk Meyer, president and CEO of AMD. Smartphones, netbooks, tablets, cloud computing, mobility, new operating systems—the introduction of each has sparked and stoked this "death of the PC" debate. Yet these are all personal computing devices, Meyer asserted. In fact, if one followed this viewpoint, the general-purpose PC was steadily morphing into a range of new form factors, which in turn, was leading to the creation of a new genre of powerful portable computing devices. This part of the chapter highlights a few prominent antecedents that chronicle the journey of personal computing toward a more portable space.

When Goldman Sachs employees were given BlackBerry phones in the early 2000s, their use of laptops fell by 45 percent. A fifth of BlackBerry users stopped using their laptops altogether. People loved their laptops because they were mobile, but when mobile phones started hosting similar content and applications, it was far more simple and cool to have it all on a smartphone. Clearly, to some people, laptops started looking like an old and bulky cousin of mobile phone. Conversely, this was yet another sign that smartphones would start to behave more and more like notebook computers. With the fast processors, smartphones could finally run full-blown apps such as Adobe Photoshop—and not just with the limited features as offered in the early Photoshop apps. There were other forthcoming power apps on the horizon that could handle photographic effects and process large, high-resolution images and videos.

After BlackBerry created the unwired enterprise by making its way into corporate sector, the round two of smartphone pioneer era came in 2007 when the iPhone put the web in people's hands. By 2010, smart-phones had plunged into the mainstream, giving millions of people the ability to browse the Internet, watch movies and stream music any-where they could maintain a cellular or Wi-Fi connection—and without having to find a place to sit down and boot up a laptop. These über-phones featured larger displays and used touch input, but the back-ground processing and abundant memory played a crucial role in ena-bling the full PC-like capabilities. Devices like smartphones and tablets, which facilitated better on-the-go content consumption, had inevitably had an impact on the growth of the PC market.

The next anecdote relates to the iPhone adaption in the enterprise world and is no less intriguing. Many companies initially chafed at letting employees use iPhones amid concerns that they could poten-tially pose security risks to corporate data. Even the iPhone's lack of a physical keyboard was initially used as a pretext as to why the iPhone shouldn't be adapted in the business world. But Apple knew the stakes on the enterprise side of the business and methodically incorporated the enterprise "must-haves" with hardware and software updates after the iPhone's release in 2007. However, it was really the iPhone's extend-ibility through third-party applications where things started to get interesting. Companies like Oracle and Salesforce.com found a plenty of applications in the App Store to empower their mobile workers. For example, applications like Quickoffice and Documents to Go addressed important corporate needs by providing the ability to read, edit, and create Microsoft Office documents.

In retrospect, the June 2008 launch of the iPhone 3GS was a turning point in the enterprise adaption of this stellar phone. The iPhone 3GS was far more business-friendly than was its predecessor; the resistance ebbed after Apple released a version of iPhone software with beefed-up security and better support for corporate e-mail. The focus of the iPhone's enterprise chops had been predominantly about e-mail, calendaring, contacts, and security—which were all keys to the success for any smartphone. With the release of the 3.0 iOS upgrade came the support for Microsoft Exchange and beefed-up enterprise security and deployment features, and from here on, there was this general agreement that the iPhone might "just" be ready for the corporate use.

Motorola's Atrix, a souped-up 4G smartphone that doubled as the engine for a notebook and netbook running on Android, was another testimony of the blurring lines between a phone and a notebook computer. A user could slap it into a stand-alone dock connected to a screen or slip it into the laptop dock, and the phone would become the conduit for an Android notebook. Combine Nvidia's dual-core processor with 1 GByte of notebook-grade RAM and 16 GBytes of built-in memory, and what the user had in hand was truly a pocket-size computer. The phone would retain all the user's apps, and would potentially virtualized applications in the cloud, while the user could go from dock to dock, bringing what Motorola called one's "webtop" with him or her. All the processing, storage, and important stuff stayed on the phone. That allowed the laptop dock—essentially a screen and a battery—to be very thin, light, and light on battery consumption.

Intel, which practically owned the notebook hardware segment, clearly felt the pressure. The world's largest chipmaker now led the makeover effort on behalf of conventional notebook makers, and the goal was to deliver thinner and lighter notebooks. Intel called this category "Ultrabooks."

The final testament of blurring lines between the PC and the mobile once again takes us to the doorstep of Apple, where its three iconic products—the iPod Touch, the iPhone, and the iPad—were all running on a single platform: iOS. Apple wanted to take this synergy between the PC and mobile platforms a step further by blending traits of iOS into the next version of Mac operating system, OS X. Ahead of its OS X release, Apple introduced an "app store" for the Mac, which would make it simpler for users to find and download programs and to be notified of updates. The new version of Mac operating system, called Lion, allowed users to display icons for their applications in a way similar to the iPhone, the iPad, and the iPod Touch and manipulate them with a swipe of their fingers.

This also showed how Apple wanted to make its computers behave more like the iPad and the iPhone without losing their greater power; more traditional keyboards, touchpads, and mice; and the ability to run conventional programs. Case in point: MacBook Air notebook computers felt more like iPads and iPhones without sacrificing their ability to work like regular computers. Jobs described the newly renovated machine a result of combining features from the iPad with those of the Mac. "We think all notebooks will be like this one day," he asserted. Earlier in October 2010, Apple had infused two of its new MacBook Air notebook computers with popular features from the iPhone and iPad, while the company promised to update software for its Mac computers so that it could effectively borrow features from the iPhone and iPad. Like the iPhone, a restyled version of the ultrathin MacBook Air notebook used flash memory rather than a disk drive for data storage, contributing to longer battery life and smaller size and weight.

The fact that Macs were marginalized due to a relatively small number of titles from independent software developers—who focused their efforts on the bigger market for PCs that ran Microsoft Windows operating system—made a fascinating chapter in computer history books. But now in a dramatic reversal of fortunes, Apple was leveraging its iPhone and iPad platforms, a focal point for thousands of developers, to drive more user adaption to its notebook and desktop computer lineups.

Smart mobile devices had now become the dominant computing platform for humanity and were supplanting the IBM PC, which had stolen the market reign from Apple soon after it ignited the personal computer revolution in the early 1980s. Now, the old Mac platform of the 1980s was aiming to look less like yesterday's fussy personal computer and more like tomorrow's worry-free digital appliance.

BATTLE OF ECOSYSTEMS

Soon after Hewlett-Packard bought Palm in April 2010, trade media instantly started mulling over the outcome of this deal. The world's largest technology company could definitely save the Palm, market analysts said, but could Palm put HP in the smartphone spotlight? Many in the industry thought that the season for starting over in the smartphone world was nearly over. HP was likely to take a long time to evaluate its acquisition, they reckoned, but this market was moving very fast. When the trade press was finally done with all that pessimism, it was pointed out that HP has webOS, and that made it the only company besides Apple that had its own PCs, its own smartphones, and its own operating system. Next up, trade media circles began creating a series of scenarios for Palm's revival, and then came the bombshell in June 2008 from then-HP CEO Mark Hurd: the acquisition hasn't much to do with smartphones.

Palm, after taking a stab at smartphones, had fallen into HP's lap and it happened just after Apple haters had started seeing Palm as a savior and its latest phone Palm Pre as the iPhone killer. At a Merrill Lynch technology symposium, Hurd laid out his reasoning behind the purchase of Palm, saying that webOS is a ground-up piece of software that has been built as a web operating environment. "We have tens of millions of HP small form factor web-connected devices," he said. "Now imagine that being a web-connected environment where now you can get a common look and feel and a common set of services laid against that environment." HP had just witnessed the success of Apple's iPad, and it knew that this device category could open up in a big way and could pose a threat to not just netbooks but to laptops and desktops as well. In fact, a closer look revealed that HP was emulating Apple's verticalized mobile strategy by cultivating an in-house mobile ecosystem rather than relying on Microsoft, whose Windows Mobile software was an also-ran in smartphone space and lacked a credible tablet strategy.

Just a few months before HP's Palm affair, at the Mobile World Congress held in Barcelona in February 2010, Nokia had startled the technology world by announcing the launch of a Linux-based open-source operating system in collaboration with Intel. In the midst of a highly draining exercise of first taking control of Symbian and then turning it into an open-source undertaking, Nokia had decided to also adapt a separate operating system named as MeeGo. The world's biggest chipmaker and the world's largest mobile handset manufacturer were joining forces around a Linux platform to create an über-platform for the next generation of computing devices: tablets, pocket computers, netbooks, and more.

Nokia executives told a stunned crowd that through open innovation, MeeGo would create an ecosystem that would be second to none, drawing in players from different industries. Due to the spread of cloud computing and new advances in electronics and network technology, mobile devices would increasingly move beyond smartphones to include other computer-like gadgets such as tablets, and here the MeeGo platform would be an important asset for Nokia, they asserted. They also affirmed Nokia's commitment to the Symbian platform at that time, saying that MeeGo is conceived as the base for a wide array of new computing-centric devices given that market for handheld gadgets is deftly diversifying. MeeGo was a Linux-based software platform designed to work across a range of hardware architectures and devices including mobile computers, netbooks, tablets, media phones, connected TVs, and in-vehicle infotainment systems.

In 2009, Intel had started a project called Moblin for creating a Linux-based operating system designed specifically for netbooks. Separately, for smartphones and tablets, Nokia had been working on a Linux-centered software platform called Maemo. Both companies desperately wanted to claim a stake in the next-generation mobile OS market. Nokia had heavily relied on Symbian, which enjoyed popularity in terms of market share but was saddled with an archaic and needlessly complicated interface that hadn't adapted well to the world of touchscreen-based next-generation smartphones. Intel, while being successful in supplying its Atom chips to the netbook market, hadn't made significant inroads into the smartphone domain. The silicon bellwether was hoping that an OS might help it leverage its chip business into a new market. MeeGo wasn't completely out of the smartphone game.

There have been a number of theories as to why Intel and Nokia decided to unite their Moblin and Maemo mobile software platforms. But more than anything else, this new marriage underpinned the fact that both companies were serious about being leaders—and survivors—in a mostly open, truly flexible computing ecosystem. Two trends especially shaped the need for MeeGo. First, the ubiquitous Internet offering constant connection regardless of location and, second, the ability to access that connection through a variety of devices: in one's pocket, in the car or kitchen, in the living room, through the television, and so on. No single device will fit every need, but they will all be connected in some way. Intel and Nokia wanted to work together to make the best mobile computers in the world. Nokia, specifically, was looking to boost the competitiveness of its smartphones as well as widen its presence in other device areas through the MeeGo software.

During the late 2000s, the rise of smartphones and the growing popularity of tablets and streaming media players had opened the doors for the new operating systems that promised a better user experience. For instance, Android, launched in 2008 for smartphones, later spread to tablets and even birthed Google TV, a platform that combines cable TV programming with sites from the Internet. But why did Nokia pick the computing powerhouse Intel as its new partner? The reasoning probably lay in the fact that, while geared for mobile devices, Android and iOS, two of the most successful operating systems, stemmed more from a personal computer lineage compared with Symbian's mobile telecom roots. The dramatic swing on Nokia's part befitted the new era of mobile devices, which resemble small general-purpose computers more than single-purpose phones.

Customers now used the phone not just for calls and text messaging but also for e-mail, web browsing, and games. They added new software picked from application stores rather than using a limited list of applications supplied with the phone. Nokia was essentially positioning MeeGo as its go-to platform for its top-tier devices while Symbian would be used for its feature phone-like low- and mid-tier smartphones. Symbian wasn't supposed to compete with iOS and Android; that would be MeeGo's job. Symbian would instead firmly aim at the mid-range market. The MeeGo story came to a crossroads in February 2011 when Nokia's newly arrived CEO, Stephen Elops, sent shockwaves across the industry by going back to his former employer, Microsoft, in a bid to form a combined front against the iPhone and Android.

Although Nokia pledged its commitment with MeeGo and continued spending on its development along with Intel, developers would most likely abandon MeeGo just like Symbian. It was already an uphill battle for MeeGo because it was no more about developing OS software; the fight for the soul of the smartphone had turned into a battle of ecosystems. The collaboration between Intel and Nokia had failed to produce a phone in its first year. Now Nokia's selection of Windows Phone 7 practically marked the end of road for MeeGo software. Intel, apparently unhappy with this about-face, vowed to find new partners and carry on. However, given the transformation of mobile OS software into a vibrant ecosystem and the consolidation that would come as a consequence, such efforts might not stand much of a chance, even if an industry giant is behind them.

In parallel to this convoluted account of Nokia's new OS undertaking, a similarly intriguing tale was unraveling at the Google camp. The Chrome software, which initially embodied Google's youthful ambition in web browsing space, eventually reshaped into an operating system in 2009. When Google said it would be available for consumers in the second half of 2010, Chrome software was aimed at netbooks. The announcement ended months of speculation on whether Google might create for netbooks a variant of its smartphone operating system, Android. The pressing issue for the search giant was how netbooks would evolve, so when netbooks ended up as a product in transition, Google opted to create a lightweight OS for it under its Chrome browser, which it had released earlier in September 2008. Originally, according to press reports, Android was designed from the beginning to scale downward to feature phones and upward to MIDs and netbook-style products.

For the long term, the Android-Chrome OS split probably made sense for Google, who kept separate teams focused on smartphones and larger computer-like devices. By rolling out a separate software platform for netbooks, notebooks and desktops, Google also reduced the risk that its Android effort will become more fragmented, something that has happened to other mobile Linux variants. Apart from the early netbook craze, designers had also been eying Android for a wide range of mobile embedded systems. Google seems to have purposefully followed a dual strategy so that the company may adjust down the road, specifically, if there is a way for the two technologies to converge.

Part of this creative chaos in Google's game plan was due to the netbook phenomenon—the fact that netbooks took off. Part of what spurred netbooks was the fact that people got hold of these really tiny portables which were perceived from the beginning as being about the Internet and about the web. Google, a company unafraid to experiment, set the renewed laptop foray as the Chromebook in May 2011, two years after it laid out the Chrome vision that tended to bypass the desktop by employing the power of the Internet. Chromebooks, initially hooked on to the Verizon network for wireless Internet connectivity on a pilot basis, also offered limited capability for offline users: they would be able to work on Google Docs as well as open and save music, photos, and other files using SD cards and USB sticks.

The Chrome OS—embodying "go the web way, all the way" mantra—represented a paradigm shift to make the web anonymous with the computer. But there were other interesting developments as well. For instance, processors had gotten lower-power driven by the x86 architecture as well as the ARM architecture. The fact that the ARM stuff happened so fast, especially, had led to the prospect of really low-power computers, one thing that could make consumers really excited about the upcoming netbooks. Another critical aspect is the integration of the whole media player into Chrome OS software. Software programs like Chrome are built for the powerful portable computers from the ground-up. Instead of a mobile phone system trying to work in netbooks, or a desktop system trying to work on phones, Chrome aims to incorporate powerful computing in its DNA as it aspires to take advantage of new hardware form factors the industry hasn't even dreamed up.

The traditional makers of desktop and notebook computers were now increasingly willing to consider the alternatives that spanned from thin clients to tablets to netbooks. Since these devices comprised of fewer components, drew less power, and tended to last longer, these gadget wannabes also contributed to some kind of cost savings. Netbooks, just like smartphones, offered web-based services and some interesting new platforms, and aspired to take the place of the PC as the standard interface to the Internet. By 2008, the netbook had joined smartphones as a prime candidate to displace traditional PCs as many in the industry sided with the netbook, saying it had the advantage of size and that many people couldn't go by the smartphones' smaller screens.

RISE OF THE NETBOOK

Mary Lou Jepsen didn't set out to invent the netbook and turn the computer industry upside down when tapped to lead the development of the machine that would bring the One Laptop per Child (OLPC) project to life. In 2005, Nicholas Negroponte, the longtime MIT Media Lab head and multimedia visionary, launched this stellar goal hoping to create an inexpensive computer for children in developing countries. Jepsen, a pioneering LCD screen designer, was tasked to create a supercheap laptop that would have Wi-Fi, a color screen, and a full keyboard—and sell for about US$100. At that price, third-world governments could buy millions of units and hand them out freely in rural villages. Plus, it had to be small, incredibly rugged, and able to run on minimal power, as half the world's children have no regular access to electricity.

Given the resource constraints, Jepsen chose flash memory instead of using a spinning hard drive because flash drew very little energy. For software, she picked Linux and other free, open-source packages instead of paying for Microsoft's wares. Jepsen used an AMD Geode processor, which wasn't very fast and required less than a watt of power. And for screen, she devised an LCD panel that detected whether on-screen images are static and told the main processor to shut down, thus saving precious electricity resources. Taiwan's Quanta was the manufacturing partner in the OLPC project for creating subnotebooks intended to be marketed in developing countries, and here, the story took an intriguing twist. Quanta's archrival in Taiwan, Asustek, pronounced "a-soos-tech," somehow got wind of the project and the world's seventh-largest notebook maker began crafting its own inexpensive, low-performance computer for the retail market.

It was just about the time when Intel had started to deemphasize the UMPC mantra. When Asus—who was among the notebook makers that had jumped the UMPC bandwagon with zeal—approached Intel for its upcoming clamshell-like portable computing device, Intel saw in it the incarnation of UMPC. The timing of this call was also immaculate in a sense that Intel was in the midst of developing its low-cost, low-power Atom chip, which would help Asustek accomplish its goal. For three months, CEO Jonney Shih and the head of Asus's motherboard business, Jerry Shen personally worked out the basic concepts: what features to include—Wi-Fi, a touchpad, and a flash memory drive—and what to throw out—Microsoft Windows, initially, and a full-size keyboard. Then they brought in a team of engineers to make their ideas real. At one point, as they struggled over the machine's software interface, Shen had to lock the team members in a Taipei hot-springs hotel for two days so they could bring it to a close.

They finally emerged with the answers. Asus, following in the footsteps of OLPC and Intel, developed the tiny Eee PC, an ultra mobile Internet device targeted at the general public. The slender 7-inch screen device had no DVD drive and wasn't potent enough to run programs like Photoshop. Indeed, Asustek intended it mainly just for checking e-mail and surfing the web. Their customers, they figured, would be children, seniors, and the emerging middle class in India and China who can't afford a full US$1,000 laptop. The Eee PC carrying a price tag of US$399 was derided by rivals as a low-power plaything when it hit stores in October 2007. But when the first few thousand Eee PC units went on sale in Taiwan, they sold out in thirty minutes. As it turned out, Eee PCs weren't bought by people in poor countries but by middle-class consumers in Western Europe and the United States, people who wanted a second laptop to carry in a handbag for peeking at YouTube or Facebook wherever they wanted.

The Asus Eee PC was a 2-pound laptop with a 7-inch screen, standard ports, built-in Wi-Fi, and a webcam, all at a starting price below US$300. It was tight, compact, easy to use, and inexpensive. Intel quickly coined the term "netbook" for the new category of companion PCs segregated from a much larger notebook PC business. Asustek, or Asus for short, went on to sell millions of these mini-notebooks and soon vaulted to fifth in worldwide PC market share. Just a while ago, rivals had mocked

Jonney Shih and his purse-size laptop computers. Millions of netbooks later, Shih was having the last laugh. In just three short years, Asustek heralded a revolution in the stagnant PC industry and became a US$21-billion-a-year tech conglomerate.

The Eee PC was a hit on the American retail market and it quickly gained popularity worldwide. Asus had no idea the Eee PC's portability would have such broad appeal for kids, first-time computer users, bloggers, students, and mobile workers alike. When the first Eee PC sold over 300,000 units in four months, companies such as Dell and Acer took note and began producing their own inexpensive netbooks. The market was there. Consumers wanted cheap, lightweight laptops. Then all hell broke loose as every electronics manufacturer on the planet jumped the band-wagon, rushing to put out its own no-frills efficiency mobile computer. Netbooks were evolving into super-portable laptops for professionals.

The origins of the netbook could be traced to the "network computer" concept of the mid-1990s. In March 1997, Apple Computer had intro-duced the eMate 300 as a subcompact laptop that was a cross between the Apple's Newton PDA and a conventional laptop computer. The eMate was discontinued along with all other Newton devices in 1998 after Steve Jobs's return to Apple. In retrospect, the success of netbooks could be attributed to the fact that PC technology had now matured enough to allow truly cost-optimized implementations with enough performance to suit the needs of a majority of PC users. The netbook was a classic disruptive innovation in the PC industry. It also proved that the "cloud" is no longer just hype and that it was now inevitable for com-puters that outsourced the difficult work somewhere else.

THE NETBOOK EVOLUTION

Netbooks—not UMPC and MID—became the next big thing at prices so low that some analysts began to see them reshaping the fundamental economics of the PC business. Netbooks were all about shrinking the size and price of mainstream notebooks, the biggest slice of the PC market during 2000s. The netbook market was able to create new price points below traditional notebooks as tech-savvy consumers snatched up US$300 netbooks as secondary machines to their mobile phones. They used the netbook's mobile Internet capability for upgrading Facebook pages, watching YouTube videos, surfing the web, and handling e-mails. The market was caught unaware by the netbook's popularity when consumers took to the products with an enthusiasm that even the leading PC companies failed to foresee. Many ignored, or failed to see, the pent-up demand for cheaper, mobile PCs that could offer longer battery life and instant-on connectivity.

Nearly every company in the PC industry had its game plan uprooted by netbooks, but Intel and Microsoft were especially worried because the idea of inexpensive computing could put a dent in their profit margins. After an initial resistance, however, the twin PC powers embraced the concept and tailored low-cost offerings for the netbook market. Microsoft, for instance, intended to stop selling Windows XP for netbooks in summer 2008 so it could drive customers to its more lucrative Vista operating system. But when Linux roared out of the gate on netbooks, Microsoft quickly backpedaled, extending XP for another two years specifically for netbooks. However, Windows software was still one of the most expensive components in the system, prompting some to dub it "the Redmond tax." Microsoft had initially refused to lower the price of Windows XP for Asustek's Eee PC, the poster child of netbooks, until Asustek started shipping Linux-based systems.

Insiders claim that Redmond charged barely US$15 for XP on a netbook, less than a quarter of what it previously sold for. Likewise, Intel, selling millions of its low-power Atom chips to netbook manufacturers, made only a fraction of the money on an Atom chip as on a more powerful Celeron or Pentium chip in a full-size laptop. The great terror in the PC industry was that it has created a US$300 device so good, most people will simply no longer feel a need to shell out US$1,000 for a notebook computer. In the end, the cannibalization of the regular laptop market that the PC players were so afraid of didn't really happen.

Netbooks enjoyed over two years of unparalleled popularity as people rushed to get their own cheap and portable device for accessing the Internet. During this euphoric period, netbooks continued getting thinner, packing more features and adding larger screens, thus further blurring the line between netbooks and laptops. By mid-year 2008, when the netbooks with Intel Atom processors were introduced, these devices were much more powerful, ran cooler features, and required less power. What actually happened was that when other companies caught on and started producing similar type of devices with more features, it pushed the price of a netbook so up that consumers would be better off buying a full-fledged laptop instead. In the hindsight, as the months went by, more and more users looked at the Eee PC and wanted more functionality. They demanded larger screens and keyboards, longer battery life, more storage, and the ability to run their favorite software programs.

Vendors wanted to meet these needs and grab a share of the new netbook market so they started coming out with 9-, 10-, and even 12-inch models. At the end of 2008, while the Asus Eee PC continued to dominate sales, netbooks were no longer thought of as "subnoteboks" but rather as versatile, understated mini-laptops. At that very point, some industry experts began to reckon that the netbook could sink into the realms of history with the availability of better options. That's when people realized that netbooks had just filled a niche in the market as underpowered PCs and that they performed sluggishly and could not handle many popular software applications. A more skeptical viewpoint called the netbook nothing but a low-cost laptop running big Windows on a remarkable but familiar platform. Most netbooks were based on Windows XP and Intel's Atom—hardly a technological breakthrough.

Some people argued that there were too many trade-offs and that netbooks won't hold up as their own retail class as they would eventually be called what they really are: cheap notebooks. They pointed out that despite a lot of quantity sold the device isn't small enough to be pocketable and not big enough to be a PC. But though netbook sales were slowing, the business was far from dead. As of 2011, netbooks accounted for roughly 10 percent of the total PC market. The bare-bone netbooks continued to enjoy brisk sales in cost-conscious markets, such as China and Brazil, and in schools, which want an affordable way to equip students with computers. Moreover, netbooks could have a play as Internet devices bundled with wireless service through the mobile carriers. By mid-2009, some wireless operators had started offering netbooks to users "free of charge" with an extended service contract purchase. What these deals also signaled was that portable computers were developing the same economics as mobile phones.

Netbook—the miniature notebook PC—had hit the mainstream. Now even if it does disappear, it will surely be remembered as the cheap and cheerful device that shook the whole computer industry into action. Netbooks established a new class of converged products in a few short years and captured the natural middle ground between portability, price, and functionality. As to how the netbook market would evolve into a different form factor, Jonney Shih's instinct told him that the "next netbook" wouldn't come from an engineering specification but from understanding how people use devices to communicate, get work done, and play. So he continued pouring company resources into design innovation. Shih, after creating what had grown into a US$10 billion category in two years, needed to come up with another design breakthrough to claim stakes in the next big thing in computing.

In the early go-go days of the netbook, when virtually every PC manufacturer on the planet, including Dell, Hewlett-Packard, and Toshiba, became part of this earth-shaking force, Apple and Sony remained an exception. Steve Jobs essentially dismissed the idea of an Apple netbook in September 2009; Sony also denied it had plans to market an inexpensive netbook. A Sony executive went so far as to declare netbooks a "race to the bottom" in terms of quality and their tendency to drive down the price of all laptops. That was the time when the rumored Apple tablet was making headlines in the trade press. The netbooks were fine for checking e-mail and surfing the web, but what about editing photos and other cool stuff. Why spend a few hundred dollars for a netbook when people could buy a full-featured tablet for as little as US$500?

For many people, netbooks were merely smaller laptops that fit better on airplane tray tables but didn't capture the power of mobility to transform people's lives the way iPhones did. They said that tablets, not netbooks, would form a better synergy with smartphones in providing the next-generation platform to displace the PC as the center of information universe. The success of tablet computers could drag down sales of notebook as well as netbook computers, shifting some business away from x86 chips and Windows toward ARM-based processors and mobile platforms like Android and iOS. Practically, the netbook market was seen as owned by Intel's Atom and Microsoft's Windows software. The coming chapter delves into the tablet story, where tablets fit into the fight against the PC hegemony, and if they are a friend or a foe to the smartphones.

18 THE TABLET RENAISSANCE

"People are willing to disproportionately spend for these devices because they are becoming so important to their lives."

—Brian Dunn, CEO, Best Buy

More than a half-dozen Japanese business and technology magazines ran cover stories about the iPad's debut, with one phrase, declaring in English, "Here comes the game-changer." Just as the debut of Apple's iPhone in 2007 spawned a host of competing smartphone devices, in April 2010, the iPad seemed set to revitalize the long moribund tablet PC industry. With iPad, Apple was moving closer toward a mobile computing genre that it had helped unleash with the launch of the iPhone a couple of years ago. Spurred by the explosive use of mobile content, the ubiquity of wireless networks, and the increasing affordability of touchscreens and powerful chips, the new portable gizmo would create a brand-new market.

Thirty-three years after Steve Jobs helped to usher in the personal computer age, Apple chief was still making waves. On his watch, the company had transformed digital music with the iPod and iTunes and had shaken up the mobile phone industry with the iPhone. Now Jobs introduced the iPad, a tablet computer that he said could kick off the next computer revolution. The iPad tablet computer was particularly marketed as a platform for audiovisual media such as movies, music, and games, as well as books, periodicals and web content. At about 700 grams, its size and weight were between those of most contemporary smartphones and laptop computers. Some critics called the iPad a giant iPhone on steroids while others predicted the iPad would flop because it didn't do enough. Apple sold 3 million of the devices in eighty days after its launch.

By late 2009, the iPad's release had been rumored for several years. It was mostly referred to as Apple's tablet; the iTablet and the iSlate were among the other speculated names. The iPad was announced on January 27, 2010 by Steve Jobs at an Apple press conference at the Yerba Buena Center for the Arts in San Francisco. Amid excitement and buzz, in the months leading up to the announcement of the iPad, critics and fans alike conjectured and debated the pros and cons of the device that had suddenly resurrected the tablet-PC market. Then, in the months following its release, there had been much consternation over whether or not a tablet resurrection was necessary.

Many critics claimed that the iPad would only be a bigger, less portable version of the iPhone, with less features, and to an extent, this has been true. According to some industry observers, iPad was essentially an iPod Touch with an enhanced display and much increased battery life. Like iPhone and iPod Touch, the iPad was controlled by a multitouch display—a break from most previous tablet computers, which used pressure-triggered stylus.

Initially, the iPad did seem to be a bigger, somewhat less portable version of Apple's iPod Touch and iPhone. However, the similarities lay solely on the surface, for beneath the 9.7-inch touchscreen, the iPad was a wholly different animal. In practice, all these contrasts translated into synergy with Apple's legacy products, which only made the case for the iPad stronger. In retrospect, the popularity of the iPhone helped build excitement about the iPad. Just like the iPhone, the iPad was another marvel of successful integration. Its ten-hour battery life was a result of Apple's ability to have its chip, software, and industrial designers work together to limit un-needed power use.

Apple's iPad was a certifiable hit, having already sold millions of units and spawning tens of thousands of apps tailored for its 9.7-inch screen within months after the launch. The iPad ran the same operating system as the earlier iPod Touch and iPhone, albeit a slightly older version. It could run its own applications as well as ones developed for the iPhone. Without modification, it would only run programs approved by Apple and distributed via its online store. The game-changing focus on aesthetics was one of the biggest sells of the iPhone and iPod Touch. Apple had reentered the portable computing market with its touchscreen-driven iPhone, which shed space-wasting keyboards and buttons and opted for a touch-sensitive interface. This set the stage for the introduction of the iPad three years later. Jobs later admitted that the iPad was developed before the iPhone, but upon realizing that the prototype would work just as well for a mobile phone, he put the development of the iPad on hold and decided to first develop the iPhone instead.

After the successful introduction of portable music player iPod in 2001, Apple revisited the mobile-computing market in 2007 with the iPhone launch. But before that, back in the early 2000s, Jobs had started work on the tablet because he had this idea to get rid of the keyboard. He asked Apple design engineers if they could come up with a multi-touch display that people could type on. About six months later, they showed Jobs a prototype display, which he found amazing. But he put the tablet project on the shelf because the phone was more important: "When we got our wind back, we pulled the tablet off the shelf, took everything we learned from the phone, and went back to work on the tablet."

The iPad's larger screen and more-advanced hardware allowed for greater usage of the multitouch technology, and the tablet design allowed for versatility in what it could be used to do. The iPad, a finished product with a smooth and dependable user experience, had mobility and access to information as its hallmark features. The iPad tablet would eventually change the portable PC market the way the iPhone had altered the mobile phone landscape. The iPad was being hailed as the computing platform of the future when Google, Hewlett-Packard, RIM, and a host of others followed by announcing plans to release their own tablets to compete with Apple's celebrated iPad. Although tablet computers had been around for more than a decade, without a doubt, 2010 was the year of the tablet. In fact, it was the year of the iPad. Interestingly enough, 2010 was arguably the year of the smartphone as well.

Although sales of tablet computers had never been enormous before the iPad arena, they had been used in a number of important, mainly industrial, niche areas where a keyboard was either impractical or unnecessary. But until the iPad came along, none of these devices was remotely cool. Beyond its attractive physical design, where the iPad differed from most existing tablet computers was that it used a variation of the operating system developed for the iPhone rather than one created for desktop or laptop computers. This difference led to a number of analysts creating a new category of devices called "media tablets" to describe the Apple product and its competitors. Tablet competition would soon take off, and subsequently, 2011 would also become known as the "Year of the Tablet."

TABLET 2.0

Apple began taking pre-orders for the device on March 12, 2010 and when the first unit, the Wi-Fi-only iPad was released, thousands stood in lines around the United States to get their hands on the new gadget. Over the course of the next few weeks, a revised model of the iPad with

3G network capability became available amid a renewed enthusiasm. The iPad came in two distinct models, though there were several permutations to be chosen by consumers. But primarily the difference between the two models lay in the wireless configuration. The iPad could also use Wi-Fi network trilateration to provide location information to applications such as Google Maps. The 3G model used assisted GPS technology to allow its position to be calculated with GPS or relative to nearby cell phone towers. For wired connectivity, the iPad had a dock connector; it lacked the Ethernet and USB ports of larger computers.

The iPad used Wi-Fi or 3G mobile data connection to browse the Internet, load and stream media, and install software. The device was managed and synced to iTunes without a USB cable. That made it a potential Internet device that like the netbook could be sold by the wireless service providers. But unlike a netbook, seen as a barely usable version of a laptop computer, the iPad was much more focused and, therefore, much more streamlined machine. By focusing on the quality of the user interface, Apple had made staying in touch with one's social networks via e-mail and the web a truly enjoyable experience. For social media addicts, tablets was an excellent appliance for browsing through Facebook for the latest photos or links that friends had posted, or checking out Twitter to see what the digital hive mind was buzzing about at that moment.

Apple had also made perhaps the first truly exciting, big-enough book interface, mobile game experience, and mobile movie player. The Cupertino, California–based company promised that the iPad would be the ultimate media consumption device of the Internet age. Apple had never made the "everyman computer," and with iPad, it was not trying to replace the laptop for work purposes. The iPad was considered to be Apple's major computing experiment in pushing the boundaries of a connected life. The iPad offered more of an entertainment-oriented consumer experience that felt more like driving a convertible BMW. As an infotainment consumption device, the iPad was heavy on media consumption and light on content creation; the latter was due to the limitations in speed of text input that was a function of the on-screen virtual keyboard.

But combine the iPad's built-in GPS function and its ability to collect electronic signatures with its easy-to-use, always-on capabilities and large screen size, and what one had at hand was a portable device that could also revolutionize the business processes. The iPad came in as a consumer product, and very quickly, the people who actually bought them figured that they could use it for work as well. Again, the iPad's standing as business tool was tied to the availability of apps as Apple could handily provide a thriving community of app developers. Perhaps Apple's central achievement with the iPad has been to tie it into the same ecosystem it had already developed for its range of iPod personal music players and refined for its iPhone handsets. From Apple's own online store it was an easy, intuitive process for users to download music, video and, most notably, apps.

This made the iPad more of a platform for innovation than an innovation in itself. Conceived as a revival of the defunct tablet market, the iPad sold over 8 million units within nine months of its launch. Apple, known for courting consumers with sleek designs and easy-to-use software, was now making inroads with corporations convinced that the iPad could make workers more productive without putting sensitive customer information at risk. This was an iPhone redux except that it happened much faster. Wells Fargo spent two years studying the iPhone before letting its bankers use the device at work. But for the iPad, it took just weeks to be cleared. Apple, having gone through the learning curve, had put in place safeguards against security breaches right from the start.

The pharmaceutical industry, for instance, saw the iPad as a being an appropriate size for a variety of uses. It was small enough to be used unobtrusively by field reps within a physician's office or a hospital setting where notebooks might be obtrusive or prohibited. The large screen made the iPad an ideal platform for showcasing interactive content to physicians. Carmakers like Mercedes-Benz began using the tablet for tasks as varied as accessing work e-mails, approving shipping orders, and calling up on-the-spot auto-finance options. Transport and delivery companies would equip their drivers with iPads to track their routes and thus predict within a one-hour window when they will arrive at a customer's location. The drivers could also show customers photos from the catalog so they could sell accessories during the delivery process.

But the most far-reaching impact of the iPad in business environment related to setting the path to productivity, fulfilling the long-held promise of the paperless office. In the United States, companies spent about US$8 billion on paper in 2007, not counting costs for ink or toner, according to RISI, the organization that tracks the global forest products industry. Moreover, copier giant Xerox estimated that for every dollar spent on printing documents, companies pay an additional US$6 in handling and distribution costs. The iPad could help businesses in going paperless and thus cut down on paper costs. At some restaurants, for instance, diners were handed iPads instead of more traditional menus. Likewise, equipping the delivery staff and salespeople with iPads could dramatically reduce the amount of paper used. In New York City, some beauty salons had started giving clients iPads rather than print magazines to keep them entertained during beauty treatments.

Until the iPad's release, the tablet computer market had been extremely limited, due in part to the constraints of the existing technology. In terms of hardware, the tablets were subject to short battery life, screen problems and damage, and the difficulty of cutting down the size of the machine itself. And tablets have historically had many software issues, from user interface to problems with relaying information via primitive touchscreens. Microsoft's research teams had done a lot of interesting work on the tablet, but it was completely stylus based. Moreover, their tablet was based on a PC, had all the expense of a PC, carried the battery life of a PC, and had the weight of a PC. It used a PC operating system. On the other hand, the tablet that the iPad brought to life had a distinctly different personality than the tablet PCs which had been shipped in low volumes for business users for several years.

Apple's release of the iPad wasn't its first attempt either. The first one was the Newton MessagePad, a very early PDA introduced in 1993, which saw several upgrades until it was discontinued in 1998. The Newton MessagePad was a 1990s-era rival to the Palm Pilot and the fact that it was more ambitious than the Pilot only magnified its shortcomings. Apple also developed a prototype portable computer, called the PenLite, but it wasn't released because the company did not want the introduction of the new device to interfere with the MessagePad's sales. After the MessagePad was discontinued, Apple focused primarily on its desktop- and notebook-based computers, injecting a unique and attractive aesthetic design into their makeup.

THE TABLET CHRONICLE

The tablet PC market was invigorated by Apple through the introduction of the iPad in 2010. While the iPad may not strictly adhere to the personal computer definition because of its restrictions on software installation, its uncompromising attention to the touch interface is considered a milestone in the tablet PC development history. So at this point, it'd be appropriate to trace the genesis of tablet computer. People in the Microsoft camp often point out that Redmond's software giant was first with its tablet PC initiative in the early 2000s. On the other hand, Apple proponents claim that the Newton was the first tablet. But if we want to dig deeper into the history of the tablet computing, we must first separate the concept from the product. It's not uncommon in the world of computing, or any field for that matter, for concepts to be much older than the first actual product, but in the case of the tablet it's all rather obvious.

The origins of the tablet computer design go back decades. The early notion of a tablet computer in the science fiction realm offers a glimpse of things to come. Both *Star Trek* and *2001: A Space Odyssey* showcased a future where computers existed as portable, handheld tablets with full-color displays and touchscreen interfaces. An electronic clipboard with a touchscreen was featured in the original 1966 *Star Trek* television series, and it wasn't long after this utopian version that engineers began working on how to make these fictional devices a reality. They first built a functioning tablet-based input system. Two such tablet interfaces appeared in the 1960s: the Styalator and the RAND tablet. Both allowed for handwriting recognition using an electronic tablet slate and a specialized pen.

In 1968, Alan Kay conceptualized the first detailed concept of a tablet computer in the form of the DynaBook in his article titled "A Personal Computer for Children of All Ages." Kay envisioned the DynaBook as a portable computer with a nearly unlimited power supply that could be used as an educational tool for children. The DynaBook was a tablet form-factor computer perceived long before even the notebook appeared on the scene. Kay also described several other ideas that would become commonplace only decades later. Though the DynaBook never progressed beyond the conceptual stage, roughly thirty-five years later, Kay got actively involved in the One Laptop Per Child (OLPC) initiative that subsequently led to the commercial realization of the Asus netbook.

Enter the 1980s, pen computers came built around handwriting recognition. By then handwriting recognition was already seen as an important future technology. Nobel Prize winner Charles Elbaum started Nestor and developed the NestorWriter handwriting recognizer. Communication Intelligence Corp. created the Handwriter recognition system, and there were many others. For instance, in 1985, Pencept Inc. introduced the Penpad computer. Although not portable by any means, this computer terminal swapped out a keyboard for a touch-based input system that could recognize both handwriting and limited hand gestures. With graphical user interfaces still in their infancy, the Penpad relied on MS-DOS as its operating system. The Penpad was certainly crude by modern standards, but it helped pave the way for a new generation of portable devices.

As the 1980s wore on, more and more technology companies began dipping their toes into a then-promising tablet market. By the close of the decade, early pen computer systems had generated a lot of excitement and there was a time when it was thought they might eventually replace conventional computers with keyboards. In those days, everyone knew how to use a pen and pens were certainly less intimidating than keyboards. GRiD Systems became the first company to offer an actual portable tablet-based computer in 1989 when it introduced the GRiDPad. The GRiDPad weighed just under 1.5 pounds despite offering a large, grayscale, backlit screen; internal floppy drive; fax/modem card; and a PCMCIA slot. Like the Penpad before it, the GRiDPad relied on MS-DOS as its operating system.

Since then the basic concept has altered little although there have been huge improvements in the underlying technologies such as touchscreens, batteries, processors, connectivity, and software. In the start, the early 1990s saw two companies battle it out for tablet OS supremacy. GO Corp. introduced an operating system called PenPoint OS, which allowed for a greater range of gesture recognition. The PenPoint OS debuted on the tablet models in 1992. That same year, Microsoft countered with Windows for Pen Computing. Microsoft, seeing tablets as a potentially serious competition to Windows computers, announced Pen Extensions for Windows 3.1 and eventually called them Windows for Pen Computing.

In 1991, the pen computing hype was at its peak. The pen was seen as a challenge to the mouse, and pen computers as a replacement for desktops. Next year, commercial products arrived. GO released PenPoint; Lexicus introduced the Longhand handwriting recognition system; and

Microsoft launched Windows for Pen Computing. Between 1992 and 1994, a number of companies introduced hardware to run Windows for Pen Computing or PenPoint. Among them were EO, NCR, Samsung, Dauphin, Fujitsu, TelePad, Compaq, Toshiba, and IBM. Few people remember that the original IBM ThinkPad was, as the name implies, a tablet computer.

The trade press, initially very enthusiastic, became critical when pen computers did not sell. Business and tech journalists measured pen computers against desktop PCs with Windows software; most of them also found pen tablets difficult to use. They criticized handwriting recognition and said it did not work. After that, pen computer companies began to fall one by one. Momenta closed in 1992 after sucking up US$40 million of investment in venture capital. Samsung and NCR didn't introduce new products. Pen pioneer GRiD was bought by AST for its manufacturing capacity, and eventually, AST stopped all pen projects. GO was taken over by AT&T, and after the memorable "fax on the beach" TV commercial, Ma Bell closed the company in August 1994. GO had lost almost US$70 million in venture capital. Compaq, IBM, NEC, and Toshiba all stopped making pen products for the consumer market in 1994 and 1995.

PDA AND NETBOOK MUTATIONS

Apple had started working on its own entry into the tablet market during the heady days of pen computing. The Apple Newton had started taking shape as early as 1989. While originally the Newton was intended to be a larger computer along the lines of the GRiDPad, it was eventually shrunk down to a more pocket-friendly size. The Newton ultimately veered away from the tablet arena and became one of the first of a new type of computer called personal digital assistant, or PDA. At that particular junction, the tablet computer faded to the background and Apple's Newton became the inspiration for a new wave of portable computing devices. During the same period, in the late 1990s, Palm proved very successful with its line of Palm Pilot devices. So companies like Dell, Sony, and Compaq began focusing their attention on the PDA market.

All of these devices featured full touch controls, with color screens eventually replacing the monochrome displays of early models. PDAs remained popular until the early 2000s. But then dedicated PDA units started to be replaced by smartphone hybrids that allowed users to connect to the Internet and sync to their PCs without the need for cables. On the other hand, by 1995, pen computing was dead in the consumer market. Microsoft made a half-hearted attempt at "Pen Services" in Windows 95, but slate computers had gone away in consumer markets. However, they lived on in vertical and industrial markets where companies such as Fujitsu, Telxon, Microslate, Intermec, and Symbol Technologies made and sold many pen tablets and pen slates.

Microsoft co-founder Bill Gates had always been a believer in the technology, and one could see slate computers in many of Microsoft's "computing in the future" pitches over the years. According to Microsoft, the primary reason why the earlier attempts were not successful owed to two specific issues. First, the technology required for a pen slate simply wasn't there in the early 1990s. Second, the pen visionaries' idea of replacing keyboard input with handwriting and voice recognition turned out to be far more difficult than anticipated. There were actually some very good recognizers, but they all required training and a good degree of adaptation by the user. It wasn't that a user would just scribble on the screen and the computer would magically understand everything. With the Tablet PC specification, Microsoft downplayed handwriting recognition in favor of "digital ink" as a new data type.

In 2001, Bill Gates took the stage at that year's Comdex trade show and announced the Windows XP Tablet Edition. This was a major shift for the tablet PC. In the past, tablets had relied on proprietary operating systems like PenPoint. These systems were generally simpler and less graphics intensive than those found on standard computers. So once Microsoft reintroduced pen computers as the "Tablet PC" in 2002, slates and notebook convertibles made a partial comeback as upstarts like Motion Computing joined the cluster of vertical and industrial market slate computer specialists. Microsoft cooked up the specification and released several versions of Windows made to work with these computers.

The tablet PC came in a variety of flavors: slates, convertibles, and hybrids. Microsoft created a set of features that computer hardware would need to support to work with its software. This set of guidelines must be met to be called a "Microsoft Tablet PC." Support was added for things like pen-based navigation and digital ink capture. Tablet PCs boasted a wireless adapter for Internet and local network connection. Software applications for tablet PCs included Office suite, web browser, games, and a variety of other applications. Microsoft's initiative had the tech headwinds on its side, and it was evident from the fact that Taiwan's top computer manufacturer Acer had hoped to see its version of the tablet representing 20 percent of its notebook sales. However, by the end of 2003, nearly two years after standing alongside Microsoft's tablet PC camp, Acer had fallen short of its sales targets.

The tablet computer's mass-market promise was still in doldrums and perhaps the biggest barrier was cost and mediocre user experience emanating from heavy weight, slow handwriting recognition and short battery life. These tablets resembled notebook computers in form and function, but they tended to cost much more. The fact that tablet manufacturers remained wishy-washy over the form factor didn't help; most vendors made both a clamshell model with a digital-pen-based touchscreen and a convertible model with a swiveling touchscreen and a laptop-sized keyboard. Another key factor, some experts believed, related to the dearth of applications. Though Microsoft worked with a lot of software houses for applications, the commercial development of the tablet PC concept was taking longer than anticipated also because software developers saw the tablet as a niche area, which it was at that time.

It was at that point that manufacturers began experimenting with low-cost notebooks, first as a means of providing computer access to poorer countries, but later to all consumers. The netbook from Asus was the outcome of that very experimentation. But as time went by, things changed. Hardware got cheaper and processors became faster. Moreover, computers continued getting lighter and more efficient, offering more detailed resolutions and cheaper and better displays. On the other hand, with the clamshell notebook computer, it was not really possible to change the fundamental design of tablet computers. A far more practical innovation came from the development of thinner, lighter, flexible displays. There were plenty of stabs at tablets, but none of which caught the world's imagination because they were all pen-based. In the end, fingertip touch is what made the difference, and that is what Apple has opened the door to.

Bill Gates had declared that he believed tablet computers would outsell traditional laptops within half a decade. It took about ten years for his prognostication to even begin to materialize, and subsequently when it did happen, his old PC nemesis had stolen the show right beneath Microsoft's nose. Microsoft had been pursuing tablet computers for a decade, the very segment Apple dominated in a couple of months. Now Microsoft's chief Steve Ballmer and some other senior company executives regarded tablets as merely a new form of PC. What Microsoft actually did to fulfill Gates' visionary prediction for tablets was a classical Windows-centric recipe—modify the operating system slightly for pen input, then cram a PC into a slate-style case—and it didn't turn out to be good enough. In hindsight, it was just about the replica of what Microsoft had done in the smartphone realm.

THE iPAD ENVY

The early days of personal computing were marked by a spectacular free-for-all as companies big and small fell over one another in a mad rush to become the dominant player in a brand-new business. Now history was repeating itself as one technology firm after another piled into the fledgling market for touchscreen tablet computers, which were apparently set to disrupt the computing industry. Since Apple introduced the iPad, tablets made a big comeback and became the hottest consumer gadget in the market. Tablets were also an important growth opportunity for wireless operators who were looking for new ways to drive data revenue.

The iPad's runaway success, as evidenced by the slew of followers now bringing tablet computers to market, was starting to reach far beyond the consumer market. What began as a coffee table novelty subsequently emerged as a less costly, far slicker, and much more mobile device for corporate executives on the go. It was now seen being used at airports, during presentations, and even in conference rooms. Rarely did people hand a laptop across the table, even though they could, but they did happily share an iPad. Users could conveniently share information with other people because the device itself was culturally acceptable. Apple redefined the tablet market with its iPad device and mobile hardware makers were now rushing to catch up with Apple. The iPad's success, in a way, was in launching a thousand rivals.

Soon consumers were presented a bewildering array of tablet devices, from Samsung's Galaxy Tab to Dell's Streak, with different screen sizes, different operating systems, and various online libraries of software applications or apps. Samsung's tablets were to be made available to all four major U.S. wireless carriers: AT&T, Verizon, Sprint, and T-Mobile. But the launch that really captured trade media's attention after the advent of the iPad was unveiling of RIM's PlayBook tablet. The announcement came at a time when RIM was in the midst of revamping its iconic BlackBerry smartphone—originally made for businesses to handle e-mail—for a market driven more by consumers looking for versatile handsets and cool software apps. The press was abuzz with the crucial question whether the BlackBerry PlayBook was going to be a genuine iPad rival or a mere rookie mistake.

One key factor was the enterprise-level computing as Apple didn't have a strong reputation in business applications or integration with corporate IT infrastructure. The BlackBerry PlayBook announced in September 2010 was an aim at a serious and secure business-oriented device to distinguish it from the more consumer-focused iPad. At that time, when iPad was generally seen as an object of desire for consumers, some industry watchers believed that BlackBerry-powered devices might have more success with the business types. However, the traditional resistance that Apple experienced earlier when trying to break into the corporate world was significantly less this time around given that IT staff members themselves wanted to use the iPad. The corporate IT department geeks loved gadgets, and the iPhone and iPad were proving very attractive devices for them. The iPad was quickly finding favor in corporate settings.

RIM's tablet was likely face stiff competition in an increasingly crowded market. But RIM didn't create the PlayBook to storm the consumer market in the first place. It did so because it had to or faced losing enterprise customers to Apple. Since a tablet was definitely useful in a business environment, people were already buying iPads for enterprise purposes, basically because they didn't have a choice. RIM co-chief executive Mike Lazaridis called the PlayBook a professional tablet and said the original impetus came from RIM's traditional customers in the corporate world. "Corporate technology officers have been asking us to amplify the BlackBerry," Lazaridis told the audience at the launch of the portable computing device.

The premise was that most would prefer a BlackBerry option, so that's what RIM was aiming to provide. RIM was understandably keen to defend its turf against the threat posed by iPhones and iPads, which being stylish and easy to use were starting to make inroads into businesses. RIM was racing to introduce the tablet for its main audience of what co-CEO Jim Balsillie called "busy working people," aiming to prevent the iPad from making further inroads with business users. But, at the same time, RIM was also hoping that the PlayBook's features would help it appeal to a broader audience.

Another legacy challenge for the Canadian company had been convincing software developers to create applications for its BlackBerry phones. The RIM handsets were facing tough competition from Apple's iPhone as well as handsets that ran on Google's Android operating system in the smartphone market. Software developers had been working overtime to churn out apps for the iPad and for the growing number of tablets that would use Android platform. Apple's tablet could also tap into the several hundred thousand apps already available for the popular iPhone. Conversely, the BlackBerry's apps library, with the exception of its business-related offerings, looked anemic alongside those of its main competitors.

Eventually, the apps became a critical part of RIM's strategy to entice consumers to buy the BlackBerry handsets. The company said it would lift many of the fees that it charged developers to sell their apps and offer additional tools to make payment easier. The changes included a new web-based development platform aimed at speeding up the time it took to write an app. RIM also announced a mobile advertising service similar to those offered by Google and Apple. Developers would keep 60 percent of the ad revenue they generated from BlackBerry apps. RIM was now aiming to supplement its applications development framework through the PlayBook launch and use this addition to make App World platform much more appealing to users. It was a bitter pill that RIM was forced to swallow as it rushed to the market to defend a segment that traditionally belonged to the BlackBerry handsets.

In another significant move, RIM's tablet would eschew the recently revamped BlackBerry 6 operating system in favor of a completely new platform built by QNX Software Systems, the embedded software firm that RIM bought in 2010. RIM had launched the BlackBerry 6 OS for smartphones in April 2010, and rolled out the first handset running on it, the BlackBerry Torch, in August that year. Now RIM was adapting to the QNX operating system, which boasted powerful multimedia capabilities. People who had worked with QNX's operating system saw it as a worthy competitor to Apple's and Google's platforms as its architecture was built from the ground-up. They said it could go a long way toward addressing complaints that RIM's devices were slow, unstable, and hard to program for. These developments were harbinger of the fact that the battle over tablet operating systems could be far more intense and complicated than was the fight over smartphone operating systems.

RIM had joined a long line of aspirants including Dell, Samsung, and HP, and the biggest challenge for all these firms touting tablets was to make their devices stand out from the crowd. RIM reckoned it had an edge as it aimed to capitalize on the popularity of its BlackBerry smartphone among corporate road warriors to promote the PlayBook as a must-have business tool. RIM's tablet affair was also testament of the level of how closely next-generation computing devices like tablets were tied to the evolving fortunes of smartphones. The PlayBook tablet had been designed as a companion to the BlackBerry phone. RIM leadership was convinced that PlayBook was compelling because it worked with the BlackBerry, letting those hooked on the BlackBerry to continue using their preferred communications device. In a way, RIM was forced to join the tablet bandwagon to save its position in the smartphone domain.

A TALE OF TWO TABLET SIZES

The introduction of a tablet and a new operating system came at a critical time for RIM, whose BlackBerry phones were facing increasingly tough competition from the iPhone as well as handsets that ran on Android operating system. BlackBerry's share of worldwide smartphone sales had been steadily falling by 2010 while the share of Android and Apple devices rose. Now the Canadian handset maker carried this conviction that its bellwether stature would help it make headway in the embryonic tablet business. RIM's new tablet featured high-definition cameras on the front and back sides, but it differed from its peers in a sense that it wouldn't connect directly to cellular networks; other tablets had these connectivity features built in. The device would support Wi-Fi, Bluetooth, and 3G connectivity, but through tethering the device to a BlackBerry smartphone.

The move implied that either RIM had done work on the run or it wanted to emphasize the "BlackBerry Bridge" as to draw the power from the iconic phone for its new tablet device. That could make the PlayBook less appealing to people who are not already RIM customers. However, the company said future devices would eventually be able to directly connect to 3G and 4G cellular systems. Another feature that made a whole lot of difference was that PlayBook had a 7-inch touchscreen. The 7-inch tablets arguably had a better chance of success than the 9.7-inch tablets looking to go head-to-head with the iPad. They could create a distinct sphere where they compete with each other, rather than with the biggest guy in the room.

This was a classic Apple move: instead of competing in a space where it couldn't win, it created a space where it could do something new. Unlike smartphones, where the iPhone had a high hurdle to overcome in penetrating enterprises dominated by incumbent RIM, the iPad was now the early incumbent in the enterprise, and Android, RIM, and other tablet developers needed to play to their strengths to acquire the number two slot. Also, Apple didn't consider 7-inch tablets as serious competition, at least in the early going. According to Steve Jobs, 7-inch tablets were tweeners: too big to compete with a smartphone and too small to compete with the iPad. But they were at least US$100 cheaper than the entry-level iPad, and the lower cost wasn't the only appeal of going small. Seven-inch tablets were lighter than were the 9.7-inch devices, so they were easier to hold in one hand and they fit into smaller bags.

A 7-inch tablet was also closer to an e-reader, a personal media player, or a handheld gaming device than the iPad was. On the other hand, the iPad looked closer to a mini-notebook than a smaller tablet was. So eventually the iPad competition rallied around the 7-inch tablet, trumpeting it as a clear alternative to Apple's "oversized" iPad. A smaller form factor meant that they could upscale their smartphone software and thus offer a larger canvas to experiment with richer apps. When the dust settled, the iPad and the 7-inch tablet looked more of two totally different classes of devices. Samsung Galaxy Tab was anything but an iPad. At the same time, however, these tablets successfully presented themselves as a legitimate iPad alternative.

But then Hewlett-Packard, with its TouchPad tablet, the first one to be powered by webOS smartphone software it acquired from Palm acquisition, decided to have a 9.7-inch diagonal touchscreen and thus join the iPad camp. HP had kicked off the tablet era with all the usual bells and whistles along with Wi-Fi, front-facing webcam for video calling, and built-in phone service for handling calls on the TouchPad. It was the first step in a long and difficult journey for HP, so to shake up the market, it launched the TouchPad as a bargain offer that came bundled with a printer which could also fax and scan documents. The bar for success was high for the largest PC firm on the planet. It must not only show a compelling product in an already crowded area; it also had to establish a new platform.

In the end, it didn't pan out well and HP had to abandon the TouchPad business altogether. So why did HP go for the 9.7-inch screen and directly challenge iPad's dominance in a new segment that Apple had created in the first place. The logical answer could be that while HP had its own webOS software, it was the only company besides Apple that had its own PCs, smartphones, and operating system. And HP had more buying clout and more retail presence than Apple. Another underlying fact was that the tablets market was developing on a pattern that was similar to the smartphone: Apple defines the market, and Google becomes a come-from-behind market leader. Initially, tablets used versions of the

Android software primarily designed for smartphones and didn't work effectively on larger displays. But Google got its act together through a tablet- and applications-optimized version called Honeycomb, opened up the Android Market, and the Android tablets were poised for a lift-off.

Honeycomb supported various screen sizes. Though Apple had a huge lead, the Android devices were proliferating. A wider range of cheaper devices with Google features like YouTube and Google Maps could probably erode the iPad's market dominance. With OEMs such as HTC, Motorola, and Samsung all adapting the Android OS, the race was on to earn a Wintel-like place in the hearts and minds of consumers. Xoom had emerged as a viable rival to the iPad 2 if being lighter, thinner, and faster were the criteria, and embodied Motorola's dream of making a foray in the tablet market. Samsung had already taken the fast follower slot with its Android-based Galaxy Tab. Samsung was one of few companies to make measurable headway against Apple. The South Korean firm had shipped more than 1 million of Galaxy Tabs within three months of its November 2010 launch in the United States.

CLOSING THE CHASM

Consumer electronics is on the brink of a smart devices revolution. The elements necessary for such a revolution have been maturing over the past decade and would continue to blossom beyond 2011. The key to this smart devices revolution is the ability to adapt to new mass markets for cheap, handheld mobile devices that consume content rather than create it. Scan the mélange of products in the territory between notebook computers and smartphones—e-books, mini-notebooks, netbooks, smart books, tablet PCs, and cloud books—and what can be seen is a consumer choice unprecedented in the history of business. Mainstream notebook PCs and mobile phones are worlds apart when it comes to cost, size and design matrix. But the quest for realization of devices like the netbook shows that the chasm is closing.

There is an increasing drive to bring smartphone-like capabilities into new form factors like tablets and e-books. Devices like media players, netbooks, and tablets will most likely live on until phones can provide those applications without risking their indispensible uses—a.k.a. drain the battery. When one gets down to it, netbooks and tablets are fundamentally the same device; they are designed to meet nearly identical needs, only for consumers with different tastes. Likewise, e-books are a single function fad that tablets could potentially squeeze from the market faster than desktop PCs chased away word-processing machines in the 1980s. All of these mobile computing devices represent shades of gray of the gadget nirvana where creative energy is flowing in all directions.

The iPad, which defined a new computer class, is testament of the fact that convergence of communication and computing is for real and that Apple, progenitor of the iPhone, saw life beyond mobile phones and a role for itself. From a bigger-picture perspective, what people are seeing is the constant assimilation of communication, computing, and consumer electronics segments. The marriage between telecom and computers is the epitome of what the futurists have been talking for decades. Different industries are already converging on this constantly-evolving market: mobile phone makers are launching small laptops, a.k.a. tablets, and computer makers are moving into smartphones. Newcomers like Google first moved into smartphones and then reached out to tablets. Amazon entered the fray with the Kindle and then vowed to offer an iPhone-style "app store" for the Kindle, which would enable it to be more than just an e-reader.

And that was clearly not by design but is how things evolved over the past two decades. The iPhone first got people hooked on the instant-on Internet as the first mass-market computer to deliver immediate access to the Internet. The iPad extended that addiction to a more powerful platform for consuming content. There is a shift from the desktop to the laptop to the netbook on the left side, and on the right side, the smaller handsets are getting bigger and merging with personal media players and personal navigation devices. All these devices are rushing to the center, but each consumer will have different requirements, which opens all kinds of opportunities for device makers.

The devices like the iPad spoke to the societal sea change under way in how consumers use computers and consume content. People's voracious appetite for new media perfectly matched the tablet form factor and the touch interface as it inhabited the space between the notebook and the smartphone. Case in point: when Steve Jobs talked about the development of Apple's upcoming laptop, the MacBook Air, he described it as bringing the best of the iPad to the laptop.

With the iPad transforming media and business landscapes, some people began asking whether the iPhone was really needed. Would the tablet be the smartphone killer? Are we at the dawn of an age in which tablets will become the jewel in the gadget crown, eclipsing the mighty smartphone only a few years into its reign? The ascending of the iPad had taken the information establishment to another turning point where, suddenly, the unthinkable was plausible. The market knew what a tablet was, and the industry can thank the iPad for opening this new space. But the iPad had followed the iPhone triumph, and the two devices shared common heritage. One can't possibly understand a fax machine without first knowing what a telephone is.

Tablets got themselves acknowledged as an important form factor while their anatomy was tied to smartphones via three crucial links: software, touch interface, and relationships with wireless operators. Would they eventually cannibalize smartphones just the way netbooks are known to have partially cannibalized notebooks? In a way, the smartphone was just a smaller tablet—a precursor—because consumers used smartphones primarily for media and data capabilities, not for calling features. The evolution of the smartphone demonstrates its quest for size in a bid to offer quality user experience. Smartphones began with BlackBerry's small 2.25-inch screen and jumped onto the 3.5-inch iPhone screen.

Some Android phone screen sizes have gone as high as 4.3 inches. At the same time, tablet sizes are getting smaller: from iPad's 9.7-inch to 7-inch screens of RIM PlayBook and Samsung Galaxy Tab.

It's worth remembering that the iPad's form factor was almost similar to the iPhone, and so were features; the only real difference was size. Now if two devices mimic each other in features and services and the essential difference is user experience emanating from disparity in screen size, consumers could possibly find it hard to maintain both devices. This is a valid argument but could prove academic in the long run. First, mobility, which is driving this whole revolution, is tied more closely to smartphones than to tablets. Wireless operators are at the vanguard of this mobile nirvana and they are seemingly more comfortable in handling smartphones than they are with assimilating tablets into their networks. Unlike smartphones, tablets aren't built primarily for communicating through wireless networks. Second, most promising applications of the mobile era—mobile payments, m-commerce, and location—inherently better perform on smartphones because of their form factor.

Third and most important aspect relates to the blurring of boundaries between feature phones and smartphones under the premise of cloud computing. While this subject will have a detailed treatment in the next chapter, such symbiosis could ultimately bring down the smartphone prices and make them more affordable for the masses. On the other hand, the PC era is peaking, and if the iPad future really seems to have an impact, it'd most likely be on the PC side of the house. A research report from Goldman Sachs forecasting that tablets could replace one in three PCs in the coming years affirms this premise. Meanwhile, smartphones continue their relentless march toward economy of scale and there is no anecdotal or empirical evidence to suggest that people will ditch their smartphones in droves for tablet computers. The smartphone revolution would most likely go on, and the devices like iPad would find their own place in this fastest-growing giant industry known to the world as mobile business.

19 CRYSTAL BALL IN A CLOUD

"We are not in the PC business; we are in the computing business. We believe this is the way computing is
headed."
>—Alain Mutricy, Motorola's senior vice president for
>mobile devices on the launch of notebook-enabled
>Atrix smartphone

The smartphones of the future won't be able to do everything seen
in science fiction movies, but they have clearly reached a tipping
point. By 2015, the smartphone is likely to change dramatically. How do
we know this for sure? One way to contemplate the future is by looking at the last half-decade as a mirror image of the road ahead for the
adaption of smartphones. Since 2007, the iPhone emerged as an incredible platform, designed with an easy-to-use finger-flicking and made
for loading innovative apps. Accelerometers, touchscreens, GPS-based
location awareness—they all appeared in full force in the last few years
and changed the market entirely. The iPhone has been a phenomenal
success, so it begs the question of what Apple will do to keep the iconic
device in the top spot as king of the smartphones. The answer: lower
the price, add more functionality, and pack future models with software

that has a wow factor. There could be a cloud on horizon to enable all that.

By 2011, smartphones were well on their way to become ubiquitous. The advances in mobile computing profoundly changed what we originally thought of a smartphone. Just as PCs had primarily become terminals for access to Internet-based content and applications, mobile handsets were leaning toward a need of having less and less on-board capability to do the things that made them smart. Many industry watchers considered the iPhone App Store and even iTunes cloud offerings and Apple as one of biggest cloud providers, even though it initially didn't refer to itself as a cloud company. These were the early days of the smartphone in the cloud. Eventually, Apple set to establish a data center in Maiden, North Carolina, which could serve as a hub for Apple's growing family of connected devices and deliver cloud-based media services.

Apple finally made public its cloud ambitions in June 2011, when Steve Jobs took a break from his medical leave of absence to launch the iCloud offering, which will replace iTunes as Apple's digital hub. iCloud took online storage and data synchronization to a new level as it automatically synched users' e-mail, data, songs, and other stuff to the cloud, so that they wouldn't have to do backups, and they would be able to access data from anywhere. iCloud, a free service, will store apps, documents, music, photos, and other content, and will push it to compatible devices such as the iPhone, iPad, and Mac computer. With iCloud, which included 5 Gbytes of free storage for documents, mail, photos, and backup, Apple was clearly driving for broader adaption and greater user loyalty.

A key highlight of Apple's cloud offerings was iTunes Match, an online music service which used "scan and match" technology to analyze a user's hard drive and automatically give access to a copy of any song it can identify, without actually uploading that song. Apple will charge an annual fee of US$25 for online access to thousands of songs in users' personal libraries and split this fee with record companies and music publishers—like apps—keeping 30 percent for itself and giving 70 percent to the music companies. In Wells Fargo's Jason Maynard's view, with iCloud, PC/Mac had been demoted, and the digital hub moved to the cloud. According to BMO Capital Markets' Keith Bachman, the free iCloud service could provide the same stickiness that iTunes for free did for the iPod. Maynard Um of UBS opined: "It's all coming together for Apple. The seamless integration of applications and content across the Apple ecosystem is a major step in the right direction for the company, and one that we believe will set Apple up nicely to continue to drive hardware sales to consumers well into the future."

Once more, Apple's cloud vision was in a stark contrast to that of Google's. Apple chose storage and synching as the centerpiece of its cloud strategy, and regarded the cloud as a hub. For Google, it was cloud-plus-web, as it emphasized streaming and real-time transactions that happen to be within the browser. A testament to this was Google's Chrome OS, which treated the cloud as the computer itself. The cloud, therefore, was a logical extension of these companies' visions of the mobile Internet: while Apple pioneered native apps, Google wanted to move everything to the cloud and drive things through a browser. Apple's move could now allow it to better compete with Google's cloud offerings such as Gmail, Google Docs, photo sharing, and other productivity features. Apple's new iMessage application also emulated BlackBerry's instant messaging capabilities.

CLOUD: THE NEXT INFLECTION POINT

The smartphone was now fueling the mobile industry growth and at the same time was on the path to become an integral part of user's life in connecting him with people, information, and the world. With the sea of hardware, software, and service enablers for the smartphones, it was now also projected to be the unified device equipped with gaming, social networking, photography, health, banking and payments, learning, productivity and other unlimited applications and tools. With all these applications, it became imperative for handsets to provide the best and the most compelling user experience. It also became apparent when consumers woke up to the idea of mobile apps that the lack of processing power and data storage capabilities on smartphones are a stumbling block in making apps more appealing. But to unify all this into a single device is challenging, considering the processing power, form factor, hardware design, software compatibility, and network interoperability limitations.

So is it possible to eliminate some of these limitations using the cloud? Looking at the potential and recent success of cloud computing from technical and business perspective, the answer is yes. For the first time, the mobile phone is a fully functional computer bringing true web browsing, location-based services and thousands of nifty applications to the mobile communications world. Add nearly ubiquitous broadband wireless networks and one has the ability to do real work from anywhere at any time. The ability to have access to all of one's applications and data through communications with remote servers in the computing cloud made computing person-centric, rather than device-centric. In short, people outsource the processing power to the cloud, which in turn, leads to the concept of cloud phones.

The term "mobile cloud computing" refers to an infrastructure where both data processing and data storage happen outside of the mobile device. There are some good examples of mobile cloud computing applications including mobile Gmail, Google Maps, and some navigation apps. However, the majority of applications in early smartphone devices like the iPhone still did most of the data storage and processing on the mobile device itself and not in the cloud. But that's changeing. For a start, smartphones generally house much less on-board memory than do full-fledged computers. Next, no major breakthrough in battery technology is in sight. A key issue in the future designs would be where mobile applications processing power is located: the handset or the Internet cloud? Instead of chasing ever more powerful chips at the expense of battery life, handset makers could count on sufficient mobile bandwidth to access cloud computing for mobile applications heavy lifting.

There are four primary reasons why cloud computing could become a disruptive force in the mobile world. The first reason, as explained above, has to do with how applications are distributed. As of 2011, mobile applications are indirectly tied to a wireless carrier. If someone wants an iPhone app, for example, he or she has to first have a relationship with the mobile operator carrying the iPhone. If that person wants a BlackBerry app, the same rule applies. But with mobile cloud computing applications, as long as a user has access to the web, he or she has access to the mobile application. Once mobile applications begin to store data in the cloud as opposed to on the mobile device, the applications will become far more powerful as their processing power is also offloaded to the cloud. All people will need is a basic mobile Internet connection, and the data could be passed back and forth in near real-time, without having to bog mobile devices with heavy applications and software.

The premise of "behind the cloud" is that the applications reside remotely and are processed remotely over the cloud using the processing abilities of the data center rather than the smartphone. The system would allow the creation of virtual smartphone replicas in the mobile cloud and customize each replica to perform specific tasks. Running applications remotely allows the smartphones to use the power installed in a data center and removes the processing-power, memory, and battery-life limitations of a physical smartphone. This would allow smartphone makers to assign more complex tasks to the cloud while adding other value-added features to the smartphone. So the mobile-cloud model would enable smartphones to continue to evolve and to become more powerful devices as the processing and storage features continue to migrate to the cloud.

Second, the number of mobile users that the wireless industry has the power to reach is far more than the number of smartphone users. The fact that feature phones themselves are becoming more capable with smarter built-in web browsers could have an impact on the mobile cloud's adaption. This could be one way the mobile cloud grows on the cheap. That's because browser is already being seized up as an OS alternative of some kind. So as more apps become truly cloud driven, even feature phones could become more and more cloud compatible. It's this cloud-centric premise that could ultimately provide a logical path for all mobile phones to gradually morph into smartphones. Whether it's a feature phone or a smartphone might not matter much to users as long as it nicely fits in the category of a cloud phone.

The third and the most critical factor is the network effect. There is this never ending strife for increasing mobile processing power with users becoming more and more data hungry and networks transitioning from 3G to 4G to accommodate this tsunami of data. The cloud could be the best possible solution to take the bulk of the network processing power. The AT&T's iPhone data snag is testimony that 3G—even 4G network—won't be able to meet the insatiable demand that smartphones could create. Inevitably, the future of smartphone goes hand in hand with mobile systems that are Internet-native. The best these broadband wireless networks could do is to provide a safe and efficient link to the mobile cloud. The wireless market is getting wild while mobile devices become smarter by the day, but the picture won't be complete without networks that are not bothered by data usage limits.

Finally, there is this notebook effect. Smartphones are now starting to steal the thunder from notebooks as they become capable of offering a user experience similar to a notebook. The transition of the power from a laptop to the palm of one's hand could lead to the demise of an erstwhile companion and bring a disruption of huge scale. However, as smartphones and notebooks converge, one cannot but question as to how a rich array of complex features like analytics, synchronization of documents and contacts, the adaption of augmented reality, and so on could be handled with the limited processing power, gaps in internal memory, and battery life and screen real-estate issues in a handheld device.

A notebook is powerful by order of physics as it has more space to accommodate the essential hardware compared to a palm-size smartphone. The power constraints of smartphones have been behind the much maligned Apple-Adobe feud. Apple's stance that the iPhone cannot support Adobe as decoding Adobe Flash videos consumes more battery life sharply illustrates an inconvenient truth. Yet the smartphone industry continues to grow. From here on, however, the only way it could maintain its relentless march to trump the notebook in terms of features and applications is by harnessing the power of cloud computing. The capacity and scale of the smartphone could, by and large, accelerate through cloud computing.

CLOUD COMPUTING: A BRIEF HISTORY

Cloud computing is a form of Internet-based computing whereby shared resources, software, and information are provided to computers and other devices on demand, like the electricity grid. It is a byproduct and consequence of the easy access to remote computing sites pioneered by the Internet and frequently takes the form of web-based tools or applications that users can access and use through a web browser as if it were a program installed locally on their own computer. So, typical cloud computing providers deliver common business applications online that are accessed from a web service, while the software and data reside on servers. Cloud computing infrastructure mostly comprises of services delivered through data centers built on servers. Clouds often appear as single points of access for consumers' computing needs.

Cloud computing evolved through a number of phases which include grid and utility computing, application service provision (ASP), and

Software as a Service (SaaS). But the overarching concept of delivering computing resources through a global network is rooted in the 1960s when J. C. R. Licklider introduced the idea of an "intergalactic computer network" that would enable everyone use computers and access data anywhere in the world. Licklider was a key figure in the development of Advanced Research Projects Agency Network (ARPANET) in 1969, the precursor of the Internet, and his idea about intergalactic network is known to be inspired from his work on ARPANET.

During the same decade, computer scientist John McCarthy proposed that computation may someday be organized as a public utility. He suggested in a speech given at MIT in 1961 that computing power and even specific applications could be sold through the utility business models like water and electricity. Many of the modern-day characteristics of cloud computing—the comparison to the electricity industry and the use of public, private, government and community forms—were also thoroughly explored in Douglas Parkhill's 1966 book *The Challenge of the Computer Utility*.

The actual term "cloud" borrows from the telecom industry which, during the 1990s, offered dedicated point-to-point data circuits as virtual private network (VPN) setup with comparable quality of service but at a much lower cost. By switching traffic to balance utilization, telecom operators were able to utilize the overall network bandwidth far more effectively. The cloud symbol was used to denote the demarcation point between what was the responsibility of the provider from that of the user. Cloud computing, however, extended this boundary to cover servers as well as the network infrastructure. One of the first milestones for cloud computing was the arrival of Salesforce.com in 1999, which pioneered the concept of delivering enterprise applications via a simple website. The services firm paved the way for both specialist and mainstream software firms to deliver applications over the Internet.

The next major development was Amazon Web Services in 2002, which provided a suite of cloud-based services including storage, computation and even human intelligence through the Amazon Mechanical Turk. Then in 2006, Amazon launched the Elastic Compute Cloud (EC2) as a commercial web service that allowed small companies and individuals to rent computers on which they could run their own IT applications. It was the first widely accessible cloud computing infrastructure service. The e-commerce icon played a key role in the development of cloud computing by modernizing its data centers after the dotcom crash, which, like most large computer systems, were using as little as 10 percent of their capacity at any one time. Amazon later initiated a new product development effort to provide cloud computing to external customers and launched Amazon Web Services (AWS) on a utility computing basis.

The cloud computing era practically took off when large web users such as Google and Amazon built enormous data centers for their own purposes, but then they realized that they could permit others to access these clouds of computing power at relatively attractive prices. Originally the idea was that these clouds of computing would offer processing power and storage; anything else would be added by the customer. But as the concept became more popular, additional functions were added. Some clouds offered systems management while others provided a set of applications as part of the cloud. The IT guys continue to manage their internal computing environments while learning how to secure, manage, and monitor the growing range of external resources residing in the cloud. Commercial offerings are generally expected to meet quality-of-service requirements of customers, and typically include service level agreements (SLAs).

A closer look at the history of computing reveals that IT operations have mostly fluctuated between being centralized one decade and decentralized the next. Now the formation of cloud represents yet one more shift toward centralized systems as cloud computing is moving the power of the commonplace software applications away from the desktop and into the cloud. Software applications were on the desktop because of inadequate connectivity, but the Internet has come a long way from the days of stand-alone PCs. Now scroll back to 1960s and 1970s, when there were mainframes and users accessed the compute power from dumb terminals. Are we moving back to the past, but with a new and improved twist? Will laptops and even cell phones simply become the dumb terminal interface to the applications elsewhere?

OPPORTUNITIES AND CHALLENGES

Cloud is a metaphor for the Internet and cloud computing is a phrase that is being used to describe the act of storing, accessing and sharing data, applications and computing power in cyberspace. In the mobile context, the bottom line is that the phone itself can't do it all; it's the power of the phone's processor plus a combination of network and cloud computing that can make it happen. With a giant server farm, the data get passed from smart devices over to wireless networks to the cluster, where it gets processed and sent back to the device. If the network is relatively slower and is causing latency, the issue could be mitigated with smartphones to a certain extent.

At the same time, however, if wireless network coverage is spotty outside urban areas, leading to broken connection issues and slow speeds, the power of cloud only goes that far. Hence, the lack of speedy mobile Internet access will create potential barriers in the shift toward mobile computing. That's why the access to Wi-Fi is one of the driving forces in the adaption of the smartphone and cloud-based technologies. Then there are new technologies like HTML5 with local caching that could help mobile cloud apps get past network-centric issues.

As evident from the history of cloud computing, the concepts of storing data in remote locations or renting the use of tools only when needed are not new. In practical domain, however, there are positives and negatives of cloud computing in mobile settings that present us with unprecedented opportunities and challenges.

For instance, a boost in productivity, extra convenience, and increased customer responsiveness could represent the early-stage benefits. The mobile apps powered by the cloud on the business front would mostly focus on productivity domain where collaboration, data sharing, multitasking, and scheduling are key features. In the past, field workers had PDAs but not much to do with them because they generally had access to a specialized application. Now they are being supplied with devices like iPhones that act as a virtual personal assistant and are equipped with a number of customized applications useful in a variety of situations. In 2010, corporations were starting to experiment with the combination of the smartphone and cloud computing—an innovation that many see as the most significant since the advent of the Internet.

Take healthcare as a case study: a smartphone tied to the mobile cloud combines access to healthcare information, patient monitoring and mobile communications. Patients and doctors alike could have easy access to healthcare records while away from home and desk, respectively. Then, medical reps could know a lot more about products like medicines and about the uses of these products. If a doctor asks a sales rep a question that he can't answer, to handle that query, he could access a product database or link to an expert who could respond directly to the doctor. Likewise, large healthcare institutes could better manage their hospitals and other facilities scattered across a number of locations by equipping their staff with smartphones so that they could effectively manage tasks like evaluating medical cases and admitting and releasing patients in a coordinated manner.

In businesses, where sales force predominantly relies on telephone calls, face-to-face meetings, and printed marketing brochures, smartphones using the power of cloud could herald a new era of automation-centric efficiency. Sales people, for instance, could access digital marketing material along with customer profiles and preferences, and they could track everything that customer needs to know about the product. Moreover, they could track special incentives, overdue payments and all such information while still on the road. A travel agent, for example, could facilitate booking and ticket arrangements, provide alerts about canceled flight, and offer plans for alternate arrangements via a smartphone tied to a cloud.

For consumers, navigation and mapping applications would likely be the most obvious trendsetters. Then there are personal information, social networking accounts, pictures, videos, articles, blogs, messages, and tweets which could be managed on the cloud and with real-time faster access and almost at no cost. A mobile user could also access data anytime from anywhere even if he loses his device. Apart from the data storage, cloud computing in mobile space holds massive potential in other areas, too. Users could access their bank accounts with real-time transactions, payments, and transfers with single click using mobile phones. The biggest and most popular application could be games with no need to install it on the phone; the processing could be performed by clouds at a higher speed, enabling a breathtaking gaming experience without any glitches or interruptions or processing power limitations.

With services like Google's Voice Actions, a user could hold down the search button for a few seconds and a voice menu automatically waits for a command. He or she could write an e-mail or SMS, pick a song, navigate to a destination, initiate a call, or search the web without having to touch a keypad. The voice recognition interface made available on mobile handsets is a marvel of convergence among telecom, network routing, transcription, and service domains. Such advancements clearly mark a tipping point and make Industry analysts firm in their belief that a new architecture for cloud-based apps would change the way the apps are developed, acquired, and used. The spread of apps and web services into the cloud could create huge opportunities for wireless operators, only if they can transform themselves into the companies that host users' data and apps on their vast servers, accessed across the Internet from phones.

But there is hardly any evidence as of 2011 that mobile phone carriers understand the need to morph from phone companies to data companies. Ultimately, they will most likely have to adapt to cloud computing to survive. Also, while cloud-based services for mobiles remain in their infancy, from an IT perspective, the mobile cloud poses just as many risks as benefits. Data leaks, mobile porn, IP theft, and lost productivity could all result from unfettered mobile cloud access. The notion of cloud computing presents difficult security problems and further exposes private information to governments, corporations, thieves, opportunists, and human and machine errors. Moreover, complex legal issues, including cross-border intellectual property and privacy conflicts, are waiting in the wings.

THE FUTURE OF CLOUD WALKING

Back in the 1980s, the full stakes of the PC market were only dimly perceived, even by its proactive participants. The PC's worldwide adaption eventually cemented its role as the indispensable productivity tool. Now the term "smartphone" is quickly becoming inadequate to describe the future in which individuals, wherever they go, whatever they do, will always have constant, instant access to the resources of the global cloud. Cloud is a wild card as it takes the emphasis off the operating system, launching a genuine computer revolution. There has been even some suggestion that the choice of mobile OS could be almost immaterial in devices such as tablets, designed to be connected wirelessly to the Internet almost permanently. Beyond tablets, there is a host of new mobile devices hoping to break the US$100 retail price barrier; providers of these portable devices ranging from credit card readers to e-books are jockeying to leverage services in the cloud.

The breadth of such devices would be integral to the vision of hyperconnected future. The phenomenon is taking a whole new dimension with remote servers being part of the mobile cloud that are powering services that can even run on simple feature phones. As explained in a previous section, the phones that can connect with the Internet will be able to take advantage of the cloud architecture to deliver an experience that rivals smartphones at a fraction of build and material cost. This coming wave of cloud-based services could translate into a new mobile pedigree in which every mobile phone would be a smartphone. A report from research firm ABI suggests that there will be nearly 1 billion end users accessing the mobile cloud by 2014.

At the same time, however, it's important not to get carried away by the inflated notion of cloud computing as it could give this misplaced impression that cloud computing would simply wipe out anything moving on the smartphone scene. Industry observers don't seem to agree that cloud would make the operating system irrelevant, partly because some users will have specialist needs, such as high security, and partly because it's a more effective way of controlling functions such as built-in cameras. Moreover, portable devices such as tablets would need to be used in an offline mode at certain occasions, so users couldn't always move everything to the cloud. The cloud, as important as it is to always-on capability, is only a part of the "anywhere computing" vision.

Many technology companies are already offering online services to wirelessly manage content over the web. Google, for instance, provides web services for users to automatically sync their e-mails, contacts, and calendars to their phones. Now that the adaption of smartphones is driving the move toward the cloud, industry watchers refer to a future featuring many more different types of networked appliances, all tied into the "Internet of things"—or a world in which everyday objects have their own IP addresses and can be tied together in the same way that people are now tied together by the Internet. The Internet in your hand is a very powerful phenomenon in itself. The services mobile users want to connect with are already residing in the cloud. Once they realize how close they are to access these services, smartphones could break a glass ceiling one service at a time.

EPILOGUE

"This is not the end. It is not even the beginning of the
end. But it is, perhaps, the end of the beginning."
—Winston Churchill, about an entirely different matter

In the early days, smartphones were targeted exclusively at executives
and businessmen for the provision of applications like e-mail, contact
list, and web browsing—probably the embodiment of a long-held tradition that called for access to most of the world's information only to a
handful of influential people. But the advent of the Internet and cheap
mobile phones played a crucial role in an information paradigm shift.
The beginning of twenty-first century marked the smartphone's journey from a niche to mainstream, and now there is no way to look back.
According to Eric Schmidt, smartphone will empower the poor and is
the equivalent to the arrival of television. The smartphone is unmistakably a game-changer, and it's apparent from the fact that the launch
of Apple's iPhone and Google's Android devices is hastening the disap-
pearance of the conventional mobile phone.

The phone is becoming people's everything-computing device. The
era of personal computers is ending, and the future lies with the smartphone. The smartphone as the device of the future has started to upset
traditional business models. The conventional mobile phone is going
the way of black-and-white, low-definition, small-screen TV set adorned
with an antenna as the kind of Swiss-Army knife things that today we
call smartphones and tablets are going to put a lot of pressure on these
single-function boxes. Pagers have demonstrated in the past that anything that performs only a single function is at risk. By designing the
Droid phone, for example, Motorola addressed this subtle shift and
innovated itself out of possible oblivion as a mobile handset provider.

Smartphones are changing into items people put in their pockets before work: keys, money, and credit cards. As a logical consequence of that, many single-use devices—MP3 players, cameras, dash-mounted GPS systems—which revolutionized life in the last few decades are bound for the landfill due to smartphones. This is partly because smartphones are starting to appeal to common consumers who realize that their online lives are now more personal, social, and mobile. And this is partly because the very best engineering efforts now focus on the mobile devices to solve the hardest problems and to accomplish the most adept solutions. The fact that smartphones' always-on network is eradicating the borders between home and office, and is changing the way people work and play isn't a news anymore. The smartphone is not addictive, per se, but its content, namely, the Internet, is.

Smartphones are starting to become a much larger part of people's lives. With wireless data transmission rates reaching blistering speeds and the incorporation of Wi-Fi networks, the sky is the limit regarding what smartphones can do. Users are increasingly attached to smartphones as they constantly send and receive e-mails, check Facebook updates, and play different games or use apps. They occasionally use the GPS maps and even watch TV shows. With a smartphone, the users know who the person is and where he or she is and they don't get that from a desktop application. Possibly the most exciting thing about smartphone technology is that the field is still wide open. It's an idea that probably hasn't found its perfect, real-world implementation yet. Every crop of phones brings new designs and new interface ideas.

No one developer or manufacturer has come up with the perfect shape, size or input method yet. The next "killer app" smartphone could look like a flip phone, a tablet PC, a candy bar, or something no one has conceived yet. Three key trends seem to influence the fundamental change in the role and nature of smartphones over the next few years: broadband everywhere, digitization of all content, and pocket computing power. These trends have played a crucial role in turning cell phones into smartphones and could subsequently drive the evolution of smart-phones into devices that are always connected, that have access to all of a person's digital content as well as to the web, and that provide a wider array of personal services.

According to industry figures from Informa, by 2008, 3 billion people owned a mobile device, a number that surpassed television sets and the fixed Internet connections. The whole idea of combining the convenience of the fastest-growing consumer device in history—the cellular phone—with the reach and richness of the Internet provides wireless industry a technological field of dreams. Nearly all the leading firms in the communications, computing and consumer electronics industries are jostling to form alliances they thought they needed. And if the flow of deals and mushrooming of startups are anything to go by, there are few more exciting industries than the smartphone market.

VANTAGE POINT

This book has asserted time and again that the iPhone was a real turning point in the smartphone's journey from an expensive luxury to a common tool. Apple transformed the wireless industry by taking power away from mobile phone operators and by enabling software distribution, without going through operator stores, and thus created a brand new app business. Mobile phone companies didn't sell many apps, because they were generally expensive and mediocre. But while Apple has been doing so much to take power away from mobile phone operators, Android gave some of it back through its Android model. Nevertheless, it was Apple and Google who changed the wireless business from being handset-focused to software-focused. According to 148Apps, an iPhone apps review blog, as of January 2011, more than 10 billion apps had been downloaded from Apple's App Store, which featured around 500,000 apps. The closest contender, Android Market, boasted 294,000 apps and 3 billion app downloads.

Apple's iOS 5—featuring over-the-air updates and notifications, among other things—further raised the software bar by creating greater synergy with the cloud and the larger ecosystem, and welcomed the iPhone 4S with speech-enabled Siri user interface. Google, on its part, was streamlining issues that had haunted Android; for instance, providing a better navigability of apps. Also, Android was finally over its fixation of being seen as an open platform. Google's next version of Android—Ice Cream Sandwich—would unify the mobile platform across devices with smaller displays and larger-screen tablets. But then Google's bid for Motorola's handset business brought a new set of challenges to the Android scheme of things.

Meanwhile, the so-called third horse, Microsoft, burned midnight oil to get its tablet act together. Its new operating system, Windows 8, would support both touchscreen applications on tablets and legacy Windows applications with a traditional PC mouse and keyboard. Windows 8 wasn't going to be just about tablets or smartphones; instead, it would open the door for a greater variety of form factors, some of which didn't even exist yet.

In the midst of all these exciting developments, Eric Schmidt handed over the reins of Google to Larry Page and Steve Jobs ended his memorable stint as Apple CEO for health reasons to make way for successor Tim Cook. A little more than a month after passing the baton to Cook, Steve Jobs, the father of the modern smartphone, passed away on October 5, 2011 at age 56. He had been battling pancreatic cancer for some years. If there was a single biggest contributor to the smartphone story, it was Jobs. He not only built Apple into one of the world's most valuable companies, Jobs also transformed the communications, computing and consumer electronics industries in profound ways. He was a visionary and a creative genius, and the pioneering aura that he left behind impacted not only technology but media and retail as well.

In the final analysis, mobile phones that can function as a PC, and tablets emulating the PC with the mobility and connectivity of a mobile phone, are rewriting IT business rules. The smartphone is convergence at its best. Consider the case of HTC's Evo 3D handset, which can shoot and display 3D images and videos. Then there is this integration of social networking sites like Facebook and Twitter with smartphones, which could spawn a new generation of services and applications. There won't be a Facebook phone, according to the company sources, but the mobile has become one of Facebook's top priorities. Facebook has also started collaborating with wireless companies for launching mobile devices focused on social networking features.

We can also look forward to some other exciting smartphone innovations in the coming years. For instance, voice recognition, artificial intelligence, and other improvements could make smartphones a lot smarter than they are, as of 2011. In this book, so much has been said about the smartphone and the ecosystem of new-era devices and applications that they are driving to the next level. Still, it's probably the beginning, not the end. The advent of the smartphone has created a myriad of losers and winners, and this will most likely continue in the coming years. The technology surrounding smartphones is constantly changing. The companies that will survive and thrive in the smartphone era will be the ones that demonstrate innovation both in technological and business model domains.

ACKNOWLEDGMENTS

I would like to thank book editor Wendy Jo as well as design and production teams at CreateSpace, who, in a true spirit of collaboration, brought this book to fruition. A big thanks to Jenny Legun, senior publishing consultant at CreateSpace, for providing valuable support throughout the span of this book project. I also want to thank Alex Bishop and Alice Jackson at Racepoint Group, who on behalf of ARM, provided artwork for the cover of this book.

I am deeply grateful to the former president of eMedia Asia and former publisher of *EE Times Asia*, Mark Saunderson, for teaching many invaluable insights about the publishing business, the worth of quality reading being one of them. I would also like to extend warm thanks to Paul Miller, CEO, UBM Electronics and UBM Cannon (Publishing), who took time to write the prologue of this book despite his busy schedule. A special thanks to Jim Shelton, a good friend and former colleague, for providing crucial help in the final editing stage.

Finally and especially, my dear mother, Tahira Nazeer Malik, has my deepest gratitude and appreciation. I am indebted to her for many reasons that go well beyond this book, but specifically in this instance, for her unconditional support and steadfast encouragement. During her last illness, while she was in hospital, she insisted that I should stay home and carry on with my work on this book. She passed away shortly before the book was complete. She is not in this world now, but her kind memories will remain with me. Blessed be her soul.

NOTES

Chapter 1

Andrea Butter and David Pogue, *Piloting Palm*, (New York: John Wiley & Sons, Inc., 2002).

"In search of smart phones," *The Economist*, October 7, 1999.

Robert Keenan, "The Mighty Shall Fall," *Communication Systems Design*.

Chapter 2

Bolaji Ojo, "Motorola loves the iPhone," *EE Times*, September 6, 2007.

Darcy Travlos, "The Four Horsemen Of The Mobile Internet," *Forbes*, December 3, 2009.

Diane Brady and Arik Hesseldahl, "RIM Needs a Lift with BlackBerry 6," *Bloomberg Businessweek*, July 7, 2010.

Don Robers, "The History of The Smartphone," *Sooper Articles*.

Elena Malykhina and Chris Murphy, "Why The iPhone Won't Make Apple A Player In Business IT," *InformationWeek*, January 13, 2007.

"History of the iPhone, Android and BlackBerry," *Wikipedia*, January 12, 2011.

Holman W. Jenkins, Jr., "Google and the Serach for the Future," *Wall Street Journal*, August 14, 2010.

James Collins, "High Stakes Winners," *Time*, February 26, 1996.

Jessie Scanlon and Helen Walters, "The Real Genius of Apple's iPhone," *Bloomberg Businessweek*, January 12, 2007.

Junko Yoshida, "Analysis: Is Nokia under pressure?" *EE Times*, May 23, 2008.

Nick Wingfield, "Microsoft's Bach to Leave in Shake-Up of Gadget Unit," *Wall Street Journal*, May 26, 2010.

Sara Silver, "Motorola Bets Big on Google, Verizon," *Wall Street Journal*, May 28, 2010.

Tish Williams, "Handheld Makers See a Wireless Future," *TheStreet.com*, February 4, 2002.

Chapter 3

Adam Creed, "Wireless Net Access Will Be 'Mostly Free'—Hawkins," *Washington Technology*, November 14, 2001.

Amy Doan, "Palm flop," *Forbes*, November 9, 1999.

Arik Hesseldahl, "Wireless PDAs Circa 2007," *Forbes*, April 25, 2002.

Carmen Noble, "Palm's Yankowski Calls it Quits," *Interactive Week*, November 8, 2001.

Charles Haddad, "Why Apple Shouldn't Renew Newton," *Bloomberg Businessweek*, June 6, 2001.

Cliff Edwards, "No Cartwheels for Handspring," *Bloomberg Businessweek*, April 2, 2001.

Jeff Kirvin, "PDA Evolution," *Writingonyourpalm.net*, December 23, 2002.

John Simons, "Has Palm Lost Its Grip?" *Fortune*, May 28, 2001.

"PDA, RIP," *The Economist*, October 16, 2003.

"PDA to Smartphone Evolution," *Techsplosive*, February 11, 2009.

Penelope Patsuris, "Palm Gives Pepsi Bottler A Pop," *Forbes*, April 22, 2002.

Pui-Wing Tam, "Tech Tables Turn on Palm CEO," *Wall Street Journal*.

Pui-Wing Tam, "The Palm at the End of the Mind," *Wall Street Journal*.

Rick Merritt, "Ubiquitous computing: slow going," *EE Times*, March 31, 2003.

Stephen Baker, "Smart Phones," *Bloomberg Businessweek*, October 18, 1999.

Tish Williams, "Wintel and Nokia Squaring Off on Smart Phones, *TheStreet.com*, February 26, 2002.

"Unbound," *The Economist*, February 13, 1999.

Walter S. Mossberg, "Garmin Devises PDA That Can Show Way To Your Destination," *Wall Street Journal*.

Chapter 4

"A mobile future," *The Economist*, October 11, 2001.

Andy Reinhardt, "Who Needs 3G Anyway," *Bloomberg Businessweek*, March 26, 2001.

Barnaby Page, "After Party, Europe's Telecoms Face Hangover," *TechWeb*, December 27, 2000.

Irene M. Kunii, "I-Way Bumps," *Bloomberg Businessweek*, May 15, 2000.

Irene M. Kunii, "Telecom Tremors," *Bloomberg Businessweek*, October 16, 2000.

Loring Wirbel, "Don't expect a false start," *EE Times*, December 19, 2000.

Loring Wirbel, "The road ahead," *EE Times*, September 27, 2000.

"Peering round the corner," *The Economist*, October 11, 2001.

Peter Elstrom, "Does Galvin Have The Right Stuff," *Bloomberg Businessweek*, March 17, 1997.

Rana Faroohar, "The Other Bubble," *Newsweek*, May 28, 2001.

Richard Comerford, "Handhelds duke it out for the Internet," *IEEE Spectrum*, August 2000.

Rick Merritt, "Software technologist sees cellular shakup coming," *EE Times*, September 26, 2002.

"Snap happy," *The Economist*, April 25, 2002.

"Son of Netscape," *The Economist*, August 10, 2000.

Stephen Backer and Kerry Capell, "Europe's Wireless Auctions: Give The Money Back," *Bloomberg Businessweek*, February 12, 2001.

"The Internet, untethered," *The Economist*, October 11, 2001.

"The wireless gamble," *The Economist*, October 14, 2000.

William Sweet, "Cell phones answer Internet's call," *IEEE Spectrum*, August 2000.

Yoshiko Hara, "Tomihisa Kamada—Tying home electronics to the Internet," *EE Times*, September 26, 2000.

Chapter 5

"A Finnish fable," *The Economist*, October 14, 2000.

Austin Carr, "Adobe CTO on MacBook Air, HTML5," *Fast Company*, November 8, 2010.

Bolaji Ojo, "Consumer electronics at Apple's core," *EE Times*, July 9, 2007.

Brian X. Chen, "Will the Mobile Web Kill Off the App Store," *Wired*, December 18, 2009.

Chris Anderson and Michael Wolff, "The Web Is Dead. Long Live the Internet," *Wired*, August 17, 2010.

Claudine Beaumont, "RIM CEO tells Apple: 'You don't need an app for the web'" *The Telegraph*, November 17, 2010.

Dan Butcher, "Smartphone apps: the future of mobile advertising," *Mobile Marketer*, July 14, 2009.

David Rowan, "Mobile Makes Million—But It's Not as Simple as It Seems," *Wired*, April 10, 2011.

David Sarno, "BlackBerry vs. iPhone: What's in your pocket?" *Los Angeles Times*, December 5, 2010.

"Flash in the pan," *The Economist*, April 16, 2010.

Fred Vogelstein, "Behold, the Next Media Titans: Apple, Google, Facebook, Amazon," *Wired*, October 25, 2010.

Jane Black, "Will Inverstors Spring for Handspring?" *Bloomberg Businessweek*, March 14, 2002.

Jon Brodkin, "Google CEO Eric Schmidt: Smartphones will outsell PCs in two years," *Network World*, September 28, 2010.

JP Mangalindan, "HTML5: not ready for primetime, but getting very close," *Fortune*, December 3, 2010.

JP Mangalindan, "Why rivals Google and Apple agree on HTML5?" *Fortune*, December 6, 2010.

Michael V. Copeland and Seth Weintraub, "Google: The search party is over," *Fortune*, July 29, 2010.

Richard Waters and Joseph Menn, "Space invader," *Financial Times*, June 6, 2010.

Rick Merritt, "Beyond the iPhone," *EE Times*, August 11, 2008.

Rick Merritt, "Opinion: Mobile Industry will eat Nokia's lunch," *EE Times*, September 1, 2009.

Seth Weintraub, "Larry Page: Jobs is rewriting history," *Fortune*, July 9, 2010.

Seth Weintraub, "Android less about money, more about iPhone disruption," *Fortune*, August 17, 2010.

"Seven mass media," *Wikipedia,* January 20, 2011.

"Understanding Smartphone Market Share? Battle not for phones, is for platform!" Tomi Ahonen's blog, July 30, 2010.

Chapter 6

Daniel Lyons, "Android Invasion," *Newsweek*, October 3, 2010.

Duncan Geere, "Google Android: Its history and uncertain future," *Wired*, April 15, 2010.

Harry McCracken, "Apple's iPhone OS 4.0: Afterthoughts," *PC World*, April 10, 2010.

Jeff Bercovici, "Steve Jobs on Apple Earnings Call: iPhone Rulez, Android Droolz," *Forbes*, October 18, 2010.

Jenna Wortham, "Google's Coup Shifts Mobile Alliances," *New York Times*, August 15, 2011.

Michael Dorf, "Is Android Fragmentation a Myth?" *Ezine Articles*.

Natasha Lomas, "A Talk with Symbian CEO Nigel Clifford," *Bloomberg Businessweek*, October 8, 2008.

Philip Elmer-DeWitt, "Smartphone wars: The big picture," *Fortune*, August 3, 2010.

Prince McLean, "Inside Google's Android and Apple's iPhone OS as software markets," *Apple Insider*, November 21, 2009.

Richard C. Morais, "Bloody but unbowed," *Forbes*, May 14, 2001.

Robert Lane Greene, "Apple V Google," *More Intelligent Life*, Winter 2010.

Scott Gilbertson, "Is Android Open?" *Wired*, October 22, 2010.

Seth Weintraub, "Sub-$100 Android phone hits U.S. shores," *Fortune*, January 24, 2011.

Stephen H. Wildstrom, "Palm's Slipping Grip," *Bloomberg Businessweek*, November 12, 2001.

Walter S. Mossberg, "Nokia Steps Into Race For 'Communicator' With a Weak Start," *Wall Street Journal*.

Chapter 7

Brian X. Chen, "How Microsoft Hit CTRL+ALT+DEL on Windows Phone," *Wired*, November 8, 2010.

Claudine Beaumont, "Windows Phone 7 handset range revealed," *The Thelegraph*, October 23, 2010.

Jessi Hempel, "Microsoft Windows Phone 7: Simply (Not That) Different," *Fortune*, October 11, 2010.

Junko Yoshida, "Opinion: Nokia vs. Nokia," *EE Times*, May 1, 2009.

Junko Yoshida, "Chasing Apple, Nokia calls up all developers," *EE Times*, April 29, 2009.

Kevin J. O'Brien, "Nokia's New Chief Faces Culture of Complacency," *New York Times*, September 26, 2010.

Kevin J. O'Brien, "Together, Nokia and Microsoft Renew a Push in Smartphones," *New York Times*, February 11, 2011.

Nick Wingfield and Christopher Lawton, "Nokia's Flirtations Put the Fear of Google Into Microsoft," *Wall Street Journal*, February 18, 2011.

Priya Ganapati, "4 Reasons Why Microsoft's Kin Phones Failed," *Wired*, June 30, 2010.

Priya Ganapati, "Symbian OS Is Broken. Can It Be Fixed?" *Wired*, October 25, 2010.

Rich Jaroslovsky, "New Crop of Windows Smartphones," *Bloomberg Businessweek*, October 28, 2010.

Seth Weintraub, "Nokia-Microsoft deal is good news for Android," *Fortune*, February 11, 2011.

Seth Weintraub, "Steve Jobs confirms: Android outselling iPhone," *Fortune*, July 16, 2010.

"The fight for digital dominance," *The Economist*, November 21, 2002.

Tish Williams, "Microsoft's Wireless Deal Could Send in the Clones," *TheStreet.com*, March 15, 2002.

Chapter 8

Alan Cane, "Understanding 3G—The Next Generation," *Financial Times*, May 31, 2001.

Anne-Francoise Pele, "ST-Ericsson likes its odds in race to smarter smartphones," *EE Times*, February 9, 2010.

Chris Meadows, "PDA vs smartphone: History repeats…in reverse?" *TeleRead*, September 11, 2009.

Cliff Edwards, "Time to Rename the Cell Phone?" *Bloomberg Businessweek*, November 27, 2006.

Jack Ganssle, "Convergence in Your Pocket," *Embedded Systems*, January 2, 2003.

Jane Black, "Cell Phones at the Crossroads," *Bloomberg Businessweek*, February 15, 2002.

Dylan McGrath, "Broadcom sees feature phones getting smarter," *EE Times*, February 11, 2010.

Jeff Bier, "Camera of Frankenstein," *EE Times*, August 26, 2002.

Junko Yoshida, "It's anybody's game in cellular," *EE Times*, September 9, 2002.

"N-Cage and Smartphone History," *Wikipedia*, February 16, 2011.

R. Collin Johnson, "Touch mania swipes across markets," *EE Times*, June 7, 2011.

Rick Merritt, "Internet calling," *EE Times*, August 13, 2001.

Stewart Alsop, "Apple Of Sun's Eye," *Time*, February 5, 1995.

Chapter 9

Amber Bouman, "The 20 Most Important Moments in Mobile Phone History," *Maximum PC*, May 26, 2010.

Anupreeta Das and Nick Wingfield, "Microsoft Near Deal to Acquire Skype," *Wall Street Journal*, May 10, 2011.

Benj Edwards, "Evolution of the Cell Phone," *PC World*, October 4, 2009.

"Camera Phone and Videophone History," *Wikipedia*, February 24, 2011.

"Convergence Contest? For the 5 Trillion Trophy," Tomi Ahonen's blog, August 30, 2010.

"FaceTime: Is Apple the new revolution?" *New Cell Phones*, July 9, 2010.

John Tilak and Harro ten Wolde, "Garmin mulls future of smartphone unit," *Reuters*, September 9, 2010.

Peter Burrows, "Video Phones Are Coming. And This Time It's for Real," *Bloomberg Businessweek*, April 29, 2010.

Peter Landers, "Video Phones: The Hype of Future's Past," *Wall Street Journal*.

Stewart Alsop, "Not Only Have I Seen the Future, I've Got It Ringing in My Pocket," *Fortune*, February 18, 2003.

Tim Bajarin, "How Steve Jobs Redefined the Smartphone...Again," *PC Magazine*, June 14, 2010.

Yoshiko Hara, "Video's mobile in Japan," *EE Times*, July 29, 2002.

Chapter 10

Adam Lashinsky, "Palm fights back (against Apple)," *Fortune*, May 26, 2009.

"A different way of working," *The Economist*, October 11, 2001.

Al Sacco, "Palm Pre, WebOS: Super Software, Sad Device," *PC World*, June 23, 2009.

Brent Schlender, "The App-Phone Revolution is Well Underway. So Where's Microsoft," *BNET*, December 9, 2009.

Cliff Edwards, "Palm's Market Starts to Melt in its Hands," *Bloomberg Businessweek*, June 4, 2001.

Harry McCracken, "Can Microsoft Get Its Mojo Back?" *Time*, November 2, 2010.

Jay R. Galbraith, "Will Microsoft become the General Motors of software?" *Fortune*, November 6, 2009.

JP Mangalindan, "What's next for the post-Ozzie Microsoft? No one knows," *Fortune*, October 26, 2010.

Laura M. Holson, "Palm, Once a Leader, Seeks Path in Smartphone Jungle," *New York Times*, August 20, 2008.

Matthew Lynn, "How Nokia Fell From Grace," *Bloomberg Businessweek*, September 15, 2010.

Michael Meyer, "Culture Club," *Newsweek*, July 11, 1994.

Rachael King, "Companies Shed Initial Resistance to iPhone," *Bloomberg Businessweek*, May 4, 2009.

Rich Jaroslovsky, "The Torch 9800: BlackBerry's Big Jump," *Bloomberg Businessweek*, August 12, 2010.

Sascha Segan, "Motorola: We're Now an Android Smartphone House," *PC Magazine*, June 10, 2009.

Shin Kiju, "How LG lost the smartphone race," *Fortune*, February 15, 2011.

Yukari Iwatani Kane and Ben Worthen, "New Fronts Open Up in Smartphone Turf War," *Wall Street Journal*, June 23, 2010.

Chapter 11

"3G Necessary for M-Commerce Success," *Wireless Week*, October 26, 2001.

Arik Hesseldahl, "Apple's iDecade," *Bloomberg Businessweek*, April 26, 2010.

Dan Whipple, "Leap of Faith," *Interactive Week*, March 5, 2001.

Jesse Berst, "Don't Be Seduced By the M-Commerce Siren Song," *ZDNet*, April 17, 2000.

Junko Yoshida, "Nokia, not Google, sees itself reshaping the Internet," *EE Times*, February 11, 2008.

Junko Yoshida, "Nokia's naked ambition: Moving beyond cellphones," *EE Times*, October 23, 2008.

"Looking for the pot of gold," *The Economist*, October 11, 2001.

"Net-enabled cell phones create m-commerce," *Tribune*, October 19, 2000.

"No Quick Cure for Ailing M-commerce Market," *Wireless Week*, October 26, 2001.

"Why mobile is different," *The Economist*, October 11, 2001.

Chapter 12

Anne Marie Squeo, "U.S. Hopes GPS Leads Bombs to Specific Targets," *Wall Street Journal*.

Arik Hesseldahl, "Whence it came," *Electronics Supply & Manufacturing*, December 2004.

Bill Siwicki, "160 Characters or Less," *Internet Retailer*, December 29, 2008.

Bill Siwicki, "E-commerce on the move," *Internet Retailer*, February 28, 2008.

Bill Siwicki, "The smart set," *Internet Retailer*, February 26, 2009.

"GPS, Mobile Commerce, and SMS History," *Wikipedia*, March 10, 2011.

"Helping Out With a Text," *Newsweek*, October 10, 2010.

James Gleick, "Inescapably Connected: Life in the Wireless Age," *New York Times*, April 22, 2001.

Junko Yoshida, "Urgent call to carriers: Deploy enhanced 911," *EE Times*, September 17, 2001.

Katherine Rosman, "Y U Luv Texts, H8 Calls," *Wall Street Journal*, October 14, 2010.

Matt Lake, "Global Positioning: Getting There With Help From Above," *New York Times*, March 29, 2001.

Meghan Casserly, "Geoloqi, Foursquare's Biggest Threat?" *Forbes*, December 9, 2010.

Mike France and Dennis K. Berman, "Big Brother Calling," *Bloomberg Businessweek*, September 25, 2000.

Sharon Gaudin, "Future smartphone will be assistants, companions," *Computerworld*, September 15, 2010.

Stephen Labaton and Riva D. Atlas, "Against All Odds, a Telecom Rebirth," *New York Times*, July 15, 2001.

"Texting the television," *The Economist*, October 17, 2002.

"The joy of text," *The Economist*, September 15, 2001.

Tom Kaneshige, "Haiti Donations: A Turning Point in Mobile Commerce?" *CIO*, February 10, 2010.

Will Wade, "Sirf to buy Conexant's global-positioning technology," *EE Times*, July 9, 2001.

Chapter 13

Amir Efrati and Robin Sidel, "Google Sets Role in Mobile Payment," *Wall Street Journal*, March 28, 2011.

Brad Stone and Olga Kharif, "Pay as You Go with Smartphones," *Bloomberg Businessweek*, July 14, 2011.

Brian X. Chen, "Will the iPhone Become Your iWallet?" *Wired*, August 16, 2010.

Dirk Smillie, "The App Is The Ad," *Forbes*, March 18, 2010.

Douglas MacMillan, "The iPhone as a Cash Register," *Bloomberg Businessweek*, February 10, 2011.

Jason Ankeny, "Ads in apps signal Apple's next mobile revolution," *Fierce Mobile Content*, April 8, 2010.

Jeff Thul, "Needs-Based Innovation Reigns," *Bloomberg Businessweek*, April 9, 2010.

Kathleen Richards, "iTunes App Store: Mobile Revolution?" *Application Development Trends*, August 26, 2008.

Marin Perez, "Mobile Internet To Grow Rapidly By 2010," *InformationWeek*, January 16, 2009.

Olga Kharif, "M-Commerce's Big Moment," *Bloomberg Businessweek*, October 11, 2009.

Olga Kharif, "Smartphones Make Mobile Radio Sing," *Bloomberg Businessweek*, July 12, 2010.

Paul Jones, "Apps That Help You Manage Your Life," *Forbes*, October 26, 2010.

Phred Dvorak and Stuart Weinberg, "RIM, Carriers Fight Over Digital Wallet," *Wall Street Journal*, March 18, 2011.

Rachael King, "Mobile Business Apps Flourish at IBM, Google," *Bloomberg Businessweek*, November 1, 2010.

R. Colin Johnson, "MEMS-packed smartphones are world's biggest wireless sensor net," *EE Times*, August 16, 2010.

Rohin Dharmakumar, "The Midas Touch," *Forbes*, May 24, 2010.

Sam Gustin, "Near Field Communication's Big (Money) Moment," *Wired*, May 25, 2011.

NOTES | 453

Spencer E. Ante, "On the Menu—Future of Mobile Apps," *Wall Street Journal*, February 9, 2011.

"The iPhone Software Revolution," Jeff Atwood's blog, June 24, 2009.

Tom Kaneshige, "iPhone in 2010: Mobile Commerce, Social Networking Star," *CIO*, January 26, 2010.

Yukari Iwatani Kane and Emily Steel, "Apple Fights Rival Google on New Turf," *Wall Street Journal*, April 8, 2010.

Chapter 14

Apostolis K. Salkintzis and Christodoulos Chamzas, "An Insight into MOBITEX Architecture," *IEEE Persoanl Communications*, February 1997.

Brendan Greeley, "Overstating Smartphone Data Hogs?" *Bloomberg Businessweek*, February 9, 2011.

David Ayala, "Dreaming Up the Smartphone of the Future," *GeekTech*, February 18, 2010.

David Pringle, "Don't Expect Cure For All Wireless Ills From GPRS Phone," *Wall Street Journal*, June 15, 2001.

Dulan McGrath, "AT&T adopts tiered pricing for smartphone data access," *EE Times*, June 2, 2010.

"GPRS as the gateway to 3G," Northstream AB press release, February 20, 2002.

Joilie O'Dell, "What Makes a Smartphone a Superphone?" *Mashable*, July 12, 2010.

JP Mangalindan, "Unlimited data, R.I.P.," *Fortune*, June 4, 2010.

Kendra Wall, "Splitting the spectrum," *Upside*, August 31, 2001.

Kevin C. Tofel, "Flat-Rate Data Plans Are Dead. Is That a Good Thing?" *GigaOM*, November 4, 2010.

"Pass the painkillers," *The Economist*, May 3, 2001.

Richard Ernsberger Jr., "Electronic Mail on the Airwaves," *Newsweek*, April 5, 1993.

Stephen H. Wildstrom, "Wireless Data: Call Back Later," *Bloomberg Businessweek*, March 4, 2002.

Stuart Corner, "Wireless networks struggling to cope with smartphone & data boom," *iTWire*, August 3, 2009.

Thomas Claburn, "Is a closed iPhone doomed to fail?" *EE Times*, January 11, 2007.

Chapter 15

"3G and 4G History," *Wikipedia*, March 30, 2011.

Ben Charny, "Wi-Fi phones make a splash," *CNET News*, August 5, 2004.

Christopher Ryan, "The Next iPhone: Are We Ready for 4G?" *GigaOM*, March 22, 2010.

Craig Mathias, "Getting early handle on 4G," *EE Times*, November 12, 2001.

Craig Matsumoto, "Panel decries lethargy in 3G," *EE Times*, June 11, 2001.

Dan Briody, "3G is no laughing matter," *Red Herring*, February 27, 2001.

Dan Devine, "Wireless data network dominance behind AT&T, Verizon smartphone wars," *TechTarget*, October 22, 2009.

"Generational change," *The Economist*, December 3, 2010.

"LTE, WiMAX and the end of history," Matt Hatton's blog, June 15, 2009.

Lucas van Grinsven, "Motorola sees no European 3G adoption before 2004," *Reuters*, October 25, 2001.

Priya Ganapati, "Everything You Need to Know About 4G Wireless," *Wired*, June 4, 2010.

Stacey Higginbotham, "Will the Real 4G Please Stand Up?" *GigaOM*, November 4, 2010.

Tony Daltorio, "Long Term Evolution vs. WiMax: The 4G Technology Showdown," *Investment U Research*, August 23, 2010.

"Verizon: Long Term Evolution of 4G and the Internet," *C114*, July 27, 2010.

Chapter 16

"ARM, MID and UMPC History," *Wikipedia*, April 14, 2011.

Darcy Travlos, "Will The iPad Kill The Laptop?" *Forbes*, November 2, 2010.

Jonah Probell, "MIPS in handsets—why not?" *EE Times*, October 12, 2009.

Junko Yoshida, "Dissecting MID, netbook, smartbook and 'cloudbook'" *EE Times*, October 12, 2009.

Mark LaPedus, "McAfee deal brings Intel into mobile security," *EE Times*, August 19, 2010.

Matt Hamblen, "Mobile Internet Devices: Just getting started or dead in the water?" *Computerworld*, September 25, 2009.

Nick Wingfield and Don Clark, "Microsoft Alliance With Intel Shows Age," *Wall Street Journal*, January 4, 2011.

Paul O Shea, "Here we go again—battery life blues," *EE Times*, March 10, 2006.

Peter Clarke, "Intel's wireless move no guarantee of success," *EE Times*, August 29, 2010.

Peter Clarke, "Ten years ago: the point when ARM broke through," *EE Times*, January 9, 2009.

Peter Glaskowsky, "The mobile Internet device: in search of itself," *CNET News*, January 22, 2009.

Richard Martin, "Intel bets big on the ultramobile PC," *InformationWeek*, June 16, 2007.

Rich Jaroslovsky, "Google Aims Twin Daggers at Microsoft's Heart," *Bloomberg Businessweek*, December 16, 2010.

Rick Merritt, "Nvidia: ARM smartphones will bury x86 PCs," *EE Times*, September 23, 2010.

Steve Hamm and Cliff Edwards, "Is It A Smartphone Or A Laptop?" *Bloomberg Businessweek*, June 11, 2007.

Steve Paine, "A Short History of Ultra Mobile Computing," *Knol*, December 15, 2008.

Steve Tobak, "A new era beckons," *Electronics Supply & Manufacturing*, July 1, 2004.

"The end of Wintel," *The Economist*, July 29, 2010.

"The UMPC is dead; long live the netbook!" *DeviceGuru*, August 29, 2008.

Chapter 17

Andy Reinhardt, "Eavesdropping at Europe's Wireless Bash," *Bloomberg Businessweek*, March 6, 2002.

Brian Proffitt, "Hey! You've Got Your Moblin in My Maemo!" *IT World*, February 17, 2010.

Brian X. Chan, "Why Google Should Cool It With Chrome OS," *Wired*, November 20, 2009.

Clive Thompson, "The Netbook Effect: How Cheap Little Laptops Hit the Big Time," *Wired*, February 23, 2009.

Dirk Meyer, "Why The PC Is Not Dead," *Forbes*, October 4, 2010.

Harry McCracken, "Apple's New Goal: The Computer as Appliance," *Time*, October 26, 2010.

Ian Fried, "Handhelds gaining the upper hand," *CNET News*, April 17, 2001.

Michael V. Copeland, "The man behind the netbook craze," *Fortune*, November 20, 2009.

Mike Feibus, "Mobile gadgets: Three's a crowd," *EE Times*, February 22, 2010.

Mike Feibus, "Worried about a little netbook? Don't be," *EE Times*, September 19, 2008.

Nick Wingfield, "Ballmer Jabs Jobs Over PCs," *Wall Street Journal*, June 4, 2010.

Priya Ganapati, "Intel's MeeGo OS Runs Into Rough Weather," *Wired*, October 8, 2010.

Seth Weintraub, "HP bought Palm to compete with ChromeOS?" *Fortune*, June 2, 2010.

Steve Lohr, "Netbooks Lose Status as Tablets Like the iPad Rise," *New York Times*, February 13, 2011.

Yukari Iwatani Kane and Ian Sherr, "Apple Updates MacBook Air, Macintosh Software," *Wall Street Journal*, October 21, 2010.

Chapter 18

"A Brief History of Apple's iPad," *Inforgraphic*, May 19, 2010.

Conrad Blickenstorfer, "A Brief History of Tablet PCs," *WebProNews*, December 20, 2005.

Daisuke Wakabayashi, "Japan's iPad Frenzy Signals a Sea Change," *Wall Street Journal*, May 27, 2010.

Darrell Etherington, "BlackBerry PlayBook: iPad Rival or Rookie Mistake?" *GigaOM*, September 28, 2010.

Gary Kovacs, "Why Tablet Computers Will Succeed Now," *Forbes*, September 29, 2010.

Jesse Schedeen, "The History of the Tablet PC," *IGN*, April 1, 2010.

John C Abell, "Are Tablets the Smartphone Killer?" *Wired*, November 9, 2010.

Phred Dvorak and Ian Sherr, "BlackBerry Maker RIM Enters Tablet Scrum," *Wall Street Journal*, September 26, 2010.

Rachael King, "Apple's iPad Sets Path to Productivity, Paperless Office," *Bloomberg Businessweek*, July 6, 2010.

Rachael King, "Apple's iPad Wins Corporate Converts at Wells Fargo, SAP," *Bloomberg Businessweek*, July 6, 2010.

R. Colin Johnson, "How to make next-gen media tablet an iPad killer," *EE Times*, June 28, 2010.

Scott Stropkay, "The Apple iPad: Your future third device?" *EE Times*, January 29, 2010.

Thom Howerda, "A Short History of the Tablet Computer," *OSNews*, January 15, 2010.

Tim Carmody, "How 7-Inch Android Tablets Can Succeed," *Wired*, October 19, 2010.

Walter S. Mossberg, "The iPad: Past, Present, Future," *Wall Street Journal*, June 7, 2010.

Chapter 19

Arif Mohamed, "A history of cloud computing," *Computer Weekly*, March 27, 2009.

"Cloud Computing History," *Wikipedia*, April 28, 2011.

Duane Craig, "Cloud Computing History 101," *Construction Cloud Computing*, August 14, 2010.

Jeff Vance, "Mobile Cloud Computing: Is You Phone Drifting to the Cloud?" *Datamation*, October 2, 2009.

Mark Jacobstein, "Every Phone a Smart Phone," *Fortune*, September 14, 2009.

Michael V. Copeland, "The Motorola phone that becomes a laptop," *Fortune*, January 5, 2011.

Philip Elmer-DeWitt, "Steve Jobs' iCloud: The view from 10,000 feet," *Fortune*, June 10, 2011.

Philip Elmer-DeWitt, "What's Apple's iCloud cut?" *Fortune*, June 3, 2011.

Sarah Perez, "Why Cloud Computing is the Future of Mobile," *ReadWriteWeb*, August 4, 2009.

"Smartphones expected to drive most work from the desktop to 'the cloud'" *Pew Research Center*, June 11, 2010.

Epilogue

Joel Stonington, "Old Tech Never Fades Away; It Just Dies," *Bloomberg Businessweek*, October 21, 2010.

Richard Martin, "Cell phones face extinction as smartphones take over," *EE Times*, April 1, 2008.

INDEX

www.ingramcontent.com/pod-product-compliance
Lightning Source LLC
Chambersburg PA
CBHW071355050326
40689CB00010B/1657